Developments in Photoelasticity

A renaissance

IOP Series in Advances in Optics, Photonics and Optoelectronics

SERIES EDITOR

 Professor Rajpal S Sirohi Consultant Scientist

About the Editor

Rajpal S Sirohi is currently working as a faculty member in the Department of Physics, Alabama A&M University, Huntsville, Alabama (USA). Prior to this, he was a consultant scientist at the Indian Institute of Science Bangalore, and before that he was chair professor in the Department of Physics, Tezpur University, Assam. During 2000–11, he was academic administrator, being vice chancellor to a couple of universities and the director of the Indian Institute of Technology Delhi. He is the recipient of many international and national awards and the author of more than 400 papers. Dr Sirohi is involved with research concerning optical metrology, optical instrumentation, holography, and speckle phenomenon.

About the series

Optics, photonics and optoelectronics are enabling technologies in many branches of science, engineering, medicine and agriculture. These technologies have reshaped our outlook, our way of interaction with each other and brought people closer. They help us to understand many phenomena better and provide a deeper insight in the functioning of nature. Further, these technologies themselves are evolving at a rapid rate. Their applications encompass very large spatial scales from nanometers to astronomical and a very large temporal range from picoseconds to billions of years. The series on the advances on optics, photonics and optoelectronics aims at covering topics that are of interest to both academia and industry. Some of the topics that the books in the series will cover include bio-photonics and medical imaging, devices, electromagnetics, fiber optics, information storage, instrumentation, light sources, CCD and CMOS imagers, metamaterials, optical metrology, optical networks, photovoltaics, freeform optics and its evaluation, singular optics, cryptography and sensors.

About IOP ebooks

The authors are encouraged to take advantage of the features made possible by electronic publication to enhance the reader experience through the use of colour, animation and video, and incorporating supplementary files in their work.

Do you have an idea of a book you'd like to explore?

For further information and details of submitting book proposals see iopscience.org/books or contact Ashley Gasque on Ashley.gasque@iop.org.

Developments in Photoelasticity

A renaissance

K Ramesh

Department of Applied Mechanics, IIT Madras, Chennai, India

IOP Publishing, Bristol, UK

ISBN   978-0-7503-2472-4 (ebook)
ISBN   978-0-7503-2470-0 (print)
ISBN   978-0-7503-2473-1 (myPrint)
ISBN   978-0-7503-2471-7 (mobi)

DOI   10.1088/978-0-7503-2472-4

Version: 20211001

IOP ebooks

British Library Cataloguing-in-Publication Data: A catalogue record for this book is available from the British Library.

Published by IOP Publishing, wholly owned by The Institute of Physics, London

IOP Publishing, Temple Circus, Temple Way, Bristol, BS1 6HG, UK

US Office: IOP Publishing, Inc., 190 North Independence Mall West, Suite 601, Philadelphia, PA 19106, USA

This book is dedicated to the memory of my parents

Sri M Krishnamurthi, a great teacher and a Scholar of Tamil

and

Smt Jamuna Krishnamurthi, a Vainika coming in the lineage of

Umayalpuram Sri Krishna Bhagavatar—a direct disciple of Saint Thyagaraja

and

*Immensely blessed by their Holinesses Sri Sankaracharyas of
Kanchi Kamakoti Peetum*

Contents

Preface

Photoelasticity has witnessed rapid strides in the last three decades, ever since the emergence of digital computers for image acquisition and subsequent processing by digital means. In the last decade, there have been several applications of photoelastic principles to solve a wide variety of problems ranging from biology to multi-physics domain, from aerospace components to silicon wafers/chips using infrared photoelasticity. Although several classics exist in the field of photoelasticity detailing the basic principles, since the onset of the digital era, only one book exists that excellently summarises the early developments titled *Digital Photoelasticity— Advanced Techniques and Applications*, Springer Nature publications, that has been written by me in 2000. The digital techniques have matured over the last two decades and a need is felt to succinctly summarise the key and time-tested developments in the digital domain in the interest of providing a platform for scientists from various disciplines to adopt and draw the benefit of photoelasticity to solve their research problems. This book bridges this gap and will surely inspire the reader by briefing on the possible diverse applications of photoelasticity and encouraging them to further apply photoelasticity innovatively in their domains. The book is divided into five chapters and the details are briefly discussed next.

The basics of photoelasticity and photoplasticity are briefly discussed in chapter 1 and a new specimen for photoplastic calibration is highlighted. The essence of variants of transmission photoelasticity such as three-dimensional photoelasticity with stress freezing and slicing as well as reflection photoelasticity for analysing opaque prototypes are discussed. A new way to calculate the loads necessary for stress-freezing so that the model does not break or excessively deform, or buckle is discussed. A new scheme of post processing the results of numerical studies to plot photoelastic fringe contours, thereby verifying its validity, is emphasised as the way forward for utilising the results of photoelasticity comfortably.

The use of fringe multiplication and fringe thinning is still relevant to solve a class of problems in photoelasticity. Chapter 2 begins with the discussion on the basics of digital fringe multiplication and digital fringe thinning. Application of these methods for evaluating the stress field in the neighbourhood of crack-tips in general and in particular for interacting crack-tips is discussed in detail with a brief overview of fracture mechanics. Attention is paid to finer issues such as data extraction from the field and minimization of errors due to origin selection. Empirical equations are provided for a designer to assess the severity of interacting crack-tips. A very systematic discussion on the nature of residual stresses in a typical glass plate is discussed from the perspective of mechanics of solids. The elegance of carrier fringes method to find the residual stress in polycarbonate sheets is presented. The use of carrier fringes in glass stress analysis to evaluate the thickness stress, edge stress and for calibration of glass is subsequently discussed. Finally, how the flow induced residual stresses can alter the stress field parameters in the neighbourhood of a crack is presented.

Phase shifting techniques (PSTs) for extracting the photoelastic data are discussed in chapter 3. The importance of calibrating a polariscope for digital analysis is highlighted. The difficulties associated with the extraction of two different phase data relating to isochromatics and isoclinics and their associated mathematical intricacies along with a way to address these issues is well brought out in this chapter. The presence of *inconsistent* zones in isoclinic phasemaps and *ambiguous* zones in isochromatic phasemaps and their interdependence are discussed in detail. The development of the ten-step PST in both monochromatic and colour domain are discussed with example problems that are quite complex in terms of fringe density, nature of their variations, and model geometries to establish the ease with which the modern developments could tackle such issues. The details of quality guided phase unwrapping of isoclinics and isochromatics, and the need for selection of proper schemes for smoothing the data forms part of this chapter. The extension of the basic unwrapping to simultaneous multiple seeded unwrapping has been demonstrated for a complex aerospace component having eight independent segments.

Traditionally, colour information of the fringes has been mostly used for identifying the fringe gradient direction in photoelasticity. With advancements in processing colour data coupled with sophisticated scanning schemes, it is now possible to extract the isochromatic data using a single image of isochromatics recorded in colour. The details of the methodology, the inadequacy of simple scanning schemes, and a need for accommodating the fringe density in directing the process of scanning are discussed in chapter 4. The results obtained for the problem of a plate with a hole subjected to a high level of stress with complex fringe features compare well with the results of PST to extract fringes up to 12 fringe orders. The applicability of the methodology to solve industrial problems is considerably enhanced with the development of colour adaptation techniques. The results obtained for a slice from a stereolithographic model and a problem dealing with microbubble have demonstrated its utility for handling problem situations with multiple random noise and different length scales. It is also demonstrated by solving a problem in orthodontic dentistry that a set of just four polarisation stepped images recorded in colour in a plane polariscope is sufficient to get both isochromatics and isoclinic data over the domain.

The diverse applications of photoelasticity to various domains ranging from engineering to physics, neurobiology to plant root stress studies have been summarised in chapter 5. Most complex problems in engineering deal with the evaluation of residual stresses and the role of photoelasticity in solving such problems with examples from precision glass moulding, manufacturing of plastic bottles, and swelling of composites are discussed. Assembly stresses are again quite difficult to model numerically and the innovative use of photoelasticity to diagnose the failure of chains and the use of stress freezing to solve the problem of seals are discussed. As problems are becoming more complex, combined use of experimental and numerical techniques is essential to solve and, in this chapter, the elegance and importance of photoelasticity to improve numerical modelling is brought out through various examples. Granular materials are quite common and the force transfer in such materials has been elegantly solved using an exceptionally large number of photoelastic grains—the essential details of these studies and where to look for additional information are presented in this chapter. How photoelasticity has effectively contributed to solving problems in neurobiology and how it can be used to enhance food security by evaluating the root stresses is also discussed. Defects in silicon and residual stress field during manufacturing are possible to be evaluated by infrared (IR) photoelasticity. Ten-step PST is used to its advantage in IR photoelasticity, which is discussed. Finally, the elegance of photoelasticity to teach subtle principles of solid mechanics and its indispensable role in engineering education is brought out in this chapter.

Photoelasticity has always embraced new technologies starting from the use of digital computers for analysis and later digital cameras for image acquisition and processing. Now, it is embracing deep learning tools to solve complex problems of evaluating the force transfer in each grain of the granular media. This decade would see many applications of deep learning in solving problems using photoelasticity.

Writing this book would not have been possible but for the gentle and persistent persuasions by Professor R S Sirohi, who is also the editor of this series, and my first thanks go to him. Professor Venkitanarayanan of IIT Kanpur has patiently gone through my manuscript and his comments and views have enriched the same. My former students Dr Hariprasad, Mr Jeby Philip, Dr Subramanyam Reddy, Dr Tarkes Dora, Dr Vivek Ramakrishnan and Dr Vivekanandan have read the manuscript consisting of their research work and have helped me to report the relevant details with clarity. My current students Mr Anand, Mr Ganesh Ramaswamy, Mr Naman Verma, Mr Sachin Sasikumar and Mr Shins were my constant companions at every stage of the manuscript development and their useful inputs have helped me to make the manuscript intelligible for an uninitiated reader. Their involvement has helped them to learn the subject for themselves and their originality and freedom with which they corrected has rendered the manuscript error free in all aspects. My thanks are also due to Ms Ashley Gasque, Mr Chris Benson and Mr Robert Trevelyan for helping me bring this manuscript to a book form. The various studies of my research reported in this book result from my work at IIT Kanpur and IIT Madras and my various project sponsors such as Aeronautical Research and Development Board, Department of Science and Technology, Government of India, European Union (FP7 project), Indian Space Research Organization, Indian National Academy of Engineering and thanks to all of them and the former students and project associates who have contributed to my research. My thanks are also due to Mr L Kannan for making suitable models at IIT Madras and Mr Radhey Shyam at IIT Kanpur.

My doctoral thesis advisor was Professor L S Srinath, a great educator in all aspects of the term, who was a student of Professor M M Frocht considered as the father of photoelasticity who was in turn the student of the illustrious Professor Stephen Timoshenko. I also had the privilege of showing my PhD thesis on photoelasticity to His Holiness Sri Kanchi Sankaracharya known as Kanchi Maha Swamigal in 1987 and when I explained that my work deals with polymer, the Great Acharyal simplified 'is it on transparent plastic?'—succinctly striking at the root of my work with a mere glance. So, from an Indian standpoint, I am blessed with a great lineage and the blessings of a Great Sage, and I hope this book brings out their spark and motivates innovation in the minds of its readers.

Many welcome changes have happened whilst writing this book in my family with my daughter Ms Karthika Ramesh getting married, and my son Mr Sankara Prasad who was a tiny tot while I was writing my earlier book on *Digital Photoelasticity*, has now grown up to the extent that he could enhance the impact of certain portions of the book through his corrections in English! My son-in-law Mr Santhosh and particularly his mother Ms Subbulakshmi Sriram evinced great interest in the progress of the book from time to time. The support provided by my wife Ms Anuradha Ramesh during this testing times of the pandemic and an excellent working environment provided at home has enabled me to complete the book in time.

K Ramesh
14 May 2021
Chennai

Author biography

K Ramesh

 **K Ramesh** is currently the K Mahesh Chair Professor at the Department of Applied Mechanics, IIT Madras and formerly a Professor at the Department of Mechanical Engineering, IIT Kanpur. He has made significant contributions to the advancement of *Digital Photoelasticity*. This has resulted in a Monograph entitled *Digital Photoelasticity—Advanced Techniques and Applications* (2000), Springer, a chapter in *Photoelasticity* in the *Springer Handbook of Experimental Solid Mechanics* (2009) and a chapter called *Digital Photoelasticity* in the book *Digital Optical Measurement Techniques and Applications* (2015), Artech House, London. He has over 190 publications to date of which two have been reproduced in the *Milestone Series* of SPIE. His research has been funded by organizations such as ARDB, ISRO, DST, EU (FP7), and NSF. He received the Zandman award for the year 2012, the first Indian to receive it since its inception in 1989, instituted by the Society for Experimental Mechanics, USA for his outstanding research contributions in applications utilizing photoelastic coatings.

He has pioneered a new paradigm in Engineering Education by writing innovative e-Books on *Engineering Fracture Mechanics* and *Experimental Stress Analysis* published by IIT Madras. He has also given video lectures of 40 hrs each on *Experimental Stress Analysis*, *Engineering Fracture Mechanics* and *Engineering Mechanics* as part of the National Program for Technology Enhanced Learning (NPTEL), India. These lectures are available free on YouTube. Professor Ramesh has also developed several educational pieces of software such as P_Scope®, *Digi*TFP®, *Digi*Photo and PSIF for photoelastic analysis.

He has been a Fellow of the Indian National Academy of Engineering since 2006. He received several awards such as Distinguished Alumnus Award of NIT, Trichy (2008), President of India Cash Prize (1984). He has been a member of the Editorial Boards of the International Journals: *Strain* (since 2001), *Journal of Strain Analysis for Engineering Design* (2009–10), *Optics and Lasers in Engineering*, and a steering committee member of Asian Society for Experimental Mechanics since its inception in 2000.

For details see: https://home.iitm.ac.in/kramesh/index.html

IOP Publishing

Developments in Photoelasticity
A renaissance
K Ramesh

Chapter 1

Basics of photoelasticity and photoplasticity

The basics of photoelasticity relating to transmission photoelasticity, reflection photoelasticity and three-dimensional photoelasticity employing stress freezing and slicing as well as introduction to photoplasticity are presented. The phenomenon of temporary birefringence, which is key to all variants of photoelasticity is presented lucidly. Similitude relations, interpretation of experimental results done on plastics to metallic prototypes are presented. Experimental studies have always been essential to validate the numerical models and hitherto the focus has been to extract stress components from photoelasticity for comparison. This is simplified by a new approach of post processing the numerical results to plot photoelastic contours, which are discussed. The current understanding on calculating the necessary loads for stress freezing and the use of a new specimen for photoplastic calibration are highlighted.

1.1 Introduction

Photoelasticity, in view of its remarkably simple optical arrangement and no stringent requirements for vibration isolation has gained wider acceptance and has established itself as an excellent tool in visualising/quantifying stress/strain fields and as a teaching aid for stress analysis. The formation of fringes in all variants of photoelasticity [1] is due to the phenomenon of temporary birefringence exhibited by the models/coatings. The discovery of the phenomenon of temporary birefringence, which has been normally credited to Brewster, has now been credited to Seebeck also in a recent paper by Aben [2]. Frocht is considered as the father of photoelasticity through his seminal contributions and for his classics on photoelasticity [3, 4].

Since 1990, the new developments in data acquisition have fundamentally changed the way data can be acquired and processed in photoelasticity. Early developments of many of these techniques were nicely summarised as *digital photoelasticity* [5] to emphasise the role of digital computers and digital image processing methods for processing the data. One of the aspects that was focussed on in the 1990s was evolving newer optical arrangements; but now, after two decades of research, it has stabilised with the view that even conventional optical arrangements are quite sufficient for complete data reduction using modern approaches. Several books [5–7] and chapters [1, 8] are already available on the basics of photoelasticity. This chapter provides a summary of photoelasticity and photoplasticity highlighting some of the key concepts, as well as some subtle ones that have not previously been emphasised in books. The reader is advised to look at the references for a detailed treatment.

1.2 Birefringence and its use in photoelasticity

Birefringence is the double refraction of light in a transparent, molecularly ordered material. Crystals display permanent birefringence, whereas photoelastically sensitive polymers exhibit temporary birefringence that is precipitated by stress or strain induced on them. Light is an electromagnetic disturbance consisting of electric and magnetic vectors that are mutually perpendicular. It can be expressed either as a magnetic or an electric vector for mathematical treatment. Expressing it as an electric vector is commonly used. A natural light source emits light such that the magnitude and direction of the light vector changes randomly. If the tip of the light vector is constrained to lie in a plane, then the light is said to be plane polarised. Figure 1.1(a) shows the word 'Beam' and if it is viewed through a calcite crystal, one would see, in general, two images as in figure 1.1(b) indicating that there are two refracted light beams instead of one! The two refracted beams are not unpolarised light beams but are linearly polarised with their directions of polarisations mutually perpendicular. This is brought out by placing a linear polariser called analyser, which selectively cuts off one of the light beams indicating that they are linearly polarised, and one perceives the word 'Beam' clearly (figure 1.1(c and d)). Of the two beams refracted, one of the beams that faithfully follows Snell's law is labelled as ordinary ray ('o') and the other light ray that does not obey Snell's law is labelled as extraordinary ray ('e').

In general, whether one would see two light beams depends on the angle of incidence of light with respect to the optic axis of the crystal. Figure 1.2 shows three cases and in figure 1.2(a), when the unpolarised light beam is incident along the optic

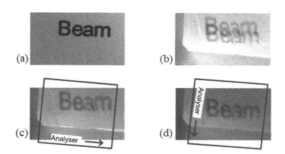

Figure 1.1. Phenomenon of birefringence. (a) The word 'Beam' seen directly, (b) the word 'Beam' viewed through a calcite crystal, (c) one of the refracted images is fully cut by orienting the analyser suitably, (d) the other refracted image is seen by keeping the analyser $\perp^r$ to the case in 'c'. Courtesy: [7, 9].

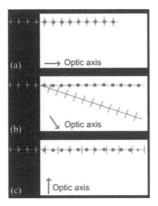

Figure 1.2. Unpolarised light incident on a crystal at various angles of incidence: (a) along the optic axis; (b) at an angle to the optic axis; (c) perpendicular to the optic axis. Courtesy: [7, 9].

axis, there is no difference between the refractive indices of the 'o' and the 'e' rays, and they travel in the same direction unaltered. In figure 1.2(b), when the angle of incidence is arbitrary, one observes two refracted beams as in figure 1.1(b), and they travel with different velocities as the different refractive indices of the two beams can also be linked to their velocities. The third case shown in figure 1.2(c) is interesting in that, when the incident beam is perpendicular to the optic axis, the two refracted rays travel along the same direction but with different velocities. This aspect is well exploited in photoelasticity. The retardation acquired within the crystal is linked to one physical parameter and the inclination of the optic axis of the crystal is linked to another physical parameter in photoelasticity.

It is to be noted that a crystal always supports and transmits only a polarised beam within it. An unpolarised beam incident on it gets polarised and this property is effectively used in crystal polarisers to generate polarised light of remarkably high quality but with a small field of view. Commercial polariscopes need large fields of view, depending on the applications, and sheets of 300 mm diameter are quite common. These are made from a sheet of polyvinyl alcohol in which the long chain molecules are oriented by stretching to achieve birefringence. These long chains absorb molecules of iodine dopant by selective attachments. The sheet thus prepared, selectively absorbs one of the refracted rays, thereby achieving linear polarisation close to 95%. With the emergence of photoelastic studies of granular media by the physics community, much larger sized polaroid sheets are easily available now commercially.

1.3 Retardation plates

Consider a crystalline plate of thickness h. Let a plane-polarised light of a single wavelength (λ) be incident normally as shown in figure 1.3(a). In all our subsequent discussions, unless stated explicitly, the light source will be a monochromatic source. Let the polarising axes of the plate be oriented at angles θ and $\theta + \pi/2$ with the horizontal. Since the incident ray is perpendicular to the optic axes (polarising axes),

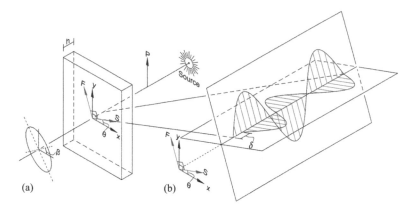

Figure 1.3. (a) Linearly polarised light incident on a retarder. (b) Inside the retarder two plane polarised lights, whose planes of polarisation coinciding with the fast and slow axes of the retarder, travel with different velocities.

the ordinary and the extraordinary rays travel in the same direction but with different velocities of v_1 and v_2. These two rays will have a net phase difference of δ upon emergence from the crystal plate. Since the velocities of propagation within the crystal are different for these two rays, they will take h/v_1 and h/v_2 seconds, respectively, to traverse the plate. This time difference contributes to the phase difference. Let the frequency of the light be f, then,

$$\delta = 2\pi f \left(\frac{h}{v_1} - \frac{h}{v_2} \right) = 2\pi h \frac{c}{\lambda} \left(\frac{1}{v_1} - \frac{1}{v_2} \right)$$
$$= \frac{2\pi h}{\lambda}(n_1 - n_2)$$

(1.1)

where, c is the velocity of light in vacuum, n_1 and n_2 are the refractive indices.

As the planes of polarisation of the two beams are mutually perpendicular, when they emerge out of the crystal, it is the interaction of two simple harmonic motions that are mutually perpendicular with an acquired phase difference. It can be shown mathematically that this would result in an ellipse, and the emergent light is in general elliptically polarised [5]. If the thickness of the plate is such as to produce a phase difference of $\pi/2$ radians, then, it is a quarter-wave plate ($\lambda/4$) and if it is π radians, then it is a half-wave plate. If the retardation is 2π, then, one gets a full-wave plate, and the incident light is unaltered. This condition is the most useful one for interpreting the fringes seen in a polariscope.

Wave-plates or retarders have two polarising axes and one of these axes is labelled as the fast axis (F) and the other as the slow axis (S). As the quarter-wave plate introduces a retardation of $\pi/2$ radians, the axes of the light ellipse coincide with the axes of the quarter-wave plate. If the amplitude of vibrations along the axes of the quarter-wave plate are equal, then one gets a circularly polarised light. Using wave plates, a linearly polarised light can be changed to a circularly or elliptically polarised light of any ellipticity (ratio of minor to major diameter of an ellipse) and azimuth (orientation of the major axis of the ellipse with respect to x-axis). This is left as an exercise to the reader.

1.4 Stress-optic law

The stress or strain imposed on a photoelastically sensitive material would dictate the birefringence at the point of interest, which is temporarily induced, thereby altering the phase difference and hence the elliptical polarisation of the light passing through it. Unlike in a crystal plate where the birefringence is the same at all points, in a 2-D photoelastic model, the phase difference introduced will be governed by the induced stress/strain field at the point of interest.

Consider a transparent model made of a high polymer (a polymer composed of many monomers) subjected to a state of plane stress. Let the state of stress at a point be characterised by the principal stresses σ_1, σ_2 and their orientation be θ with reference to a set of axes. Let n_1 and n_2 be the refractive indices for vibrations corresponding to these two directions. Following Maxwell's formulation, the relative retardation in terms of principal stress difference can be written as [5, 6]

$$\delta = \frac{2\pi h}{\lambda}(n_1 - n_2) = \frac{2\pi h}{\lambda}C(\sigma_1 - \sigma_2) \tag{1.2}$$

where, C is the relative stress-optic coefficient, usually assumed to be a constant for a material. However, various studies have shown that this coefficient depends on the wavelength and should be used with care. Equation (1.2) can be rewritten in terms of fringe order N as

$$N = \frac{\delta}{2\pi} = h\frac{C}{\lambda}(\sigma_1 - \sigma_2) \tag{1.3}$$

which can be recast as

$$\sigma_1 - \sigma_2 = \frac{NF_\sigma}{h} \tag{1.4}$$

where,

$$F_\sigma = \frac{\lambda}{C} \tag{1.5}$$

which is known as the material stress fringe value with units N/mm/fringe. Equation (1.4) is known as stress-optic law as it relates the stress information to optical measurement. Equation (1.5) shows the dependence of F_σ on the wavelength. Equation (1.4) implicitly gives the indication that F_σ and ($\sigma_1-\sigma_2$) are linearly related. However, at higher stress levels, this relationship is non-linear, and equation (1.4) should be used with care.

The principal stresses are labelled such that σ_1 is always algebraically greater than σ_2. In view of this, the L.H.S of equation (1.4) is always positive. Thus, in photoelasticity fringe order N is always positive. If the fringe order and the material stress fringe values are known, then, one can obtain the principal stress difference.

If the principal stress direction at the point of interest is known, then one can find the in-plane shear stress as well as normal stress difference using the Mohr's circle [7].

1.5 Optical arrangements and fringe fields in conventional photoelasticity

The focus of an optical arrangement is to reveal the stress state in the model visually. Usually, either a plane or a circularly polarised light is used to investigate a stressed photoelastic model. One of the simplest optical arrangements possible is a plane polariscope (figure 1.4(a)). The polariser and analyser are indicated by their polarising axes in the diagram. A ring under diametral compression is kept in the field of view. As the plane polarised light passes through the model, the state of polarization changes from point to point depending on the magnitude of the principal stress difference and the principal stress direction at the point of interest. The analyser is materially the same as the polariser and is kept with its axis perpendicular to that of the polariser and will cut off the light completely when the incident plane polarised light is unaltered. The region outside the model will be dark as the incident plane

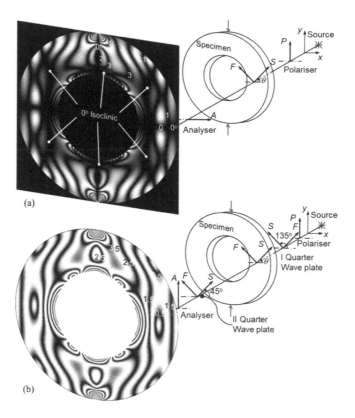

Figure 1.4. Ring under diametral compression illuminated by a monochromatic light source: (a) dark field plane polariscope showing both isochromatics ($N = 0, 1, ...$) and zero degree isoclinic. (b) Bright field circular polariscope showing only isochromatics ($N = 0.5, 1.5, ...$).

polarised light is completely cut off by the analyser (figure 1.4(a)). The corresponding optical arrangement is known as *dark field* plane polariscope.

One has to see what can cause the incident plane polarised light impinging on the stressed model to pass without any alterations. This can happen for two cases: one in which the point of interest acts like a full-wave plate for the incident wavelength of light and another when one of the principal stress directions at the point of interest coincides with the incident plane polarised light. As the stress field is continuous, when such conditions are satisfied, one observes instead of random dark points, contours of dark points as fringes. In a plane polariscope one observes two sets of fringe contours viz., *isoclinics* (points where principal stress direction coincides with the incident plane polarised light) and *isochromatics* (points at which the retardation is in multiples of 2π radians that are related to the principal stress or strain difference). The deficiency of the recording medium makes the fringes appear as bands and the width is dictated by the gradient of the variable.

If a circularly polarised light is used to investigate the model by keeping a quarter-wave plate after the polariser at 135°/45°, the elements after the model need to be complementary (figure 1.4(b)) to form the fringe patterns. In a circular polariscope, one observes only one set of contours that correspond to *isochromatics*. A simplistic investigation of the light as done for a plane polariscope is not possible here. By sending a circularly polarised light, the directional dependence of the incident light is eliminated and hence, the *isoclinic* fringes get eliminated. One can appreciate this better after the derivation of the mathematical expressions for the light transmitted (sections 1.7 and 1.8).

The basic optical arrangement shown (figure 1.4(b)) is for a *bright field* circular polariscope in which the quarter-wave plates are kept crossed and polariser and analyser are kept parallel. If the analyser is kept at 0°, the polariser and analyser will be in a crossed position and the background would become dark. Such an arrangement is termed as *dark field* circular polariscope. Although bright and dark fields could be achieved through different relative orientations of the optical elements [5, 7], only those arrangements wherein the quarter-wave plates are kept crossed are preferred, as it reduces the error due to mismatch of quarter-wave plates. Mismatch of a quarter-wave plate implies that it is not introducing a retardation of $\pi/2$ radians for the given light source as its thickness is matched for a particular wavelength (equation (1.1)), which may be different from the light source, and there will be a small deviation of retardation from $\pi/2$ radians. In large-sized quarter-wave plates, this error may also change from point to point due to manufacturing difficulties. While constructing a polariscope, the manufacturer usually tries to select a quarter-wave plate that matches with the monochromatic light source as closely as possible. Only while using a white light source can the deviations be significant, and more details regarding the influence of quarter-wave plate error are discussed in chapter 4.

1.6 Jones calculus

Certain observations on fringe patterns are possible through logical explanations without recourse to any mathematics like what is discussed in the previous section for plane polariscope. However, for quantitative appreciation, a systematic mathematical analysis is desirable. Conventionally, one can take recourse to trigonometry to trace the polarisation change of the light as it passes through various optical elements [6]. Such an approach is straightforward for a plane polariscope but becomes cumbersome even to analyse a simple circular polariscope and will be too complicated for a generic polariscope that would be discussed in chapter 3. R C Jones published a series of papers in the 1940s to trace the polarisation change of light comfortably by expressing the changes introduced by the optical elements as matrix operators [10]. In his honour, such an approach is aptly termed as Jones calculus.

In general, an optical element in a polariscope introduces rotation and retardation. As the orientations of the fast (F) and slow (S) axis of the model, in general, are different with respect to the incident polarised light, when the polarised beam impinges upon the model, rotation is introduced. A rotation matrix is useful to find the components of a vector if the reference axes are rotated by an arbitrary angle θ. It is given by,

$$\begin{bmatrix} \cos\theta & \sin\theta \\ -\sin\theta & \cos\theta \end{bmatrix} \tag{1.6}$$

The usual convention of measuring angles is valid for constructing a rotation matrix and the angle is positive if it is measured anticlockwise.

The retardation is acquired when light traverses within the model. Let the phase difference introduced at the point of interest be δ after passing through the stressed model. Using complex number notations, the emerging light can be obtained as a matrix operation of the incident light as [5]

$$\begin{Bmatrix} u' \\ v' \end{Bmatrix} = \mathfrak{R} \begin{bmatrix} e^{-i\delta/2} & 0 \\ 0 & e^{i\delta/2} \end{bmatrix} \begin{Bmatrix} a_1 e^{i\alpha_1} \\ a_2 e^{i\alpha_2} \end{Bmatrix} e^{i\omega t} \tag{1.7}$$

The first matrix on the right side of equation (1.7) is called the retardation matrix. The vector after that and the exponential term express the unpolarised incident light. On the explicit understanding that one deals with real parts only, the notation $\mathfrak{R}$ is usually left out. It is to be noted that in the above representation, the u' axis is considered as the slow axis.

As elements of a generic polariscope can be considered as retarders, it is desirable to obtain a matrix representation of the retarder. Let the slow axis of the retarder be oriented at an angle θ with the x-axis and let the total relative retardation introduced by the retarder be δ. To trace the transformation of light through the retarder, the incident light must be rotated by an angle θ when it enters the retarder, then the relative retardation of δ is to be introduced and finally, the emerging light is to be represented with respect to the original reference axes. These operations are represented in matrix form as,

$$\begin{bmatrix} \cos\theta & -\sin\theta \\ \sin\theta & \cos\theta \end{bmatrix} \begin{bmatrix} e^{-i\delta/2} & 0 \\ 0 & e^{i\delta/2} \end{bmatrix} \begin{bmatrix} \cos\theta & \sin\theta \\ -\sin\theta & \cos\theta \end{bmatrix} \tag{1.8}$$

The product of the matrices in equation (1.8) gives the Jones matrix for a retarder as follows:

$$
\begin{bmatrix}
\cos\dfrac{\delta}{2} - i\sin\dfrac{\delta}{2}\cos 2\theta & -i\sin\dfrac{\delta}{2}\sin 2\theta \\[2mm]
-i\sin\dfrac{\delta}{2}\sin 2\theta & \cos\dfrac{\delta}{2} + i\sin\dfrac{\delta}{2}\cos 2\theta
\end{bmatrix}
\tag{1.9}
$$

If the values of the retardation and the orientation of the slow axis with the x-axis are known, then one gets a remarkably simple matrix for computations. Equation (1.9) is particularly useful to quickly write the Jones matrix for retardation plates.

1.7 Analysis of plane polariscope by Jones calculus

Using Jones calculus, the components of light vector along the analyser axis and perpendicular to the analyser axis for the plane polariscope arrangement are obtained as

$$
\begin{Bmatrix} E_x \\ E_y \end{Bmatrix} =
\begin{bmatrix}
\cos\dfrac{\delta}{2} - i\sin\dfrac{\delta}{2}\cos 2\theta & -i\sin\dfrac{\delta}{2}\sin 2\theta \\[2mm]
-i\sin\dfrac{\delta}{2}\sin 2\theta & \cos\dfrac{\delta}{2} + i\sin\dfrac{\delta}{2}\cos 2\theta
\end{bmatrix}
\begin{Bmatrix} 0 \\ 1 \end{Bmatrix} ke^{i\omega t}
\tag{1.10}
$$

where $ke^{i\omega t}$ is the incident light vector, E_x and E_y are the components of light vector along the analyser axis and perpendicular to the analyser axis, respectively. In equation (1.10), the polariser is represented as a vector and the model is represented as a retarder. The intensity of light transmitted (I_p) is the product $E_x E_x^*$ where E_x^* denotes the complex conjugate of E_x and the intensity of light transmitted is,

$$
I_p = I_a \sin^2\frac{\delta}{2}\sin^2 2\theta
\tag{1.11}
$$

where I_a accounts for the amplitude of the incident light vector.

1.8 Analysis of circular polariscope by Jones calculus

To get the components of light along the analyser axis and perpendicular to it, the individual matrices of the various optical elements must be multiplied in the order they are placed in the polariscope (see figure 1.4(b)). A quarter-wave plate provides a retardation of $\pi/2$ and its slow axis is oriented either at 135° or 45°. Their relevant matrices can be obtained by substituting these values in equation (1.9) and the final expression is as follows:

$$
\begin{Bmatrix} E_x \\ E_y \end{Bmatrix} = \frac{1}{2}\begin{bmatrix} 1 & -i \\ -i & 1 \end{bmatrix}
\begin{bmatrix}
\cos\dfrac{\delta}{2} - i\sin\dfrac{\delta}{2}\cos 2\theta & -i\sin\dfrac{\delta}{2}\sin 2\theta \\[2mm]
-i\sin\dfrac{\delta}{2}\sin 2\theta & \cos\dfrac{\delta}{2} + i\sin\dfrac{\delta}{2}\cos 2\theta
\end{bmatrix}
\tag{1.12}
$$

$$
\times \begin{bmatrix} 1 & i \\ i & 1 \end{bmatrix} \begin{Bmatrix} 0 \\ 1 \end{Bmatrix} ke^{i\omega t}
$$

Upon simplification one gets,

$$\begin{Bmatrix} E_x \\ E_y \end{Bmatrix} = \begin{Bmatrix} \sin \dfrac{\delta}{2} e^{-i2\theta} \\ \cos \dfrac{\delta}{2} \end{Bmatrix} k e^{i\omega t} \tag{1.13}$$

Let the intensity of light transmitted in bright field be denoted as I_ℓ, and obtained as the product of $E_y E_y^*$, which is simplified as

$$I_\ell = I_a \cos^2 \frac{\delta}{2} \tag{1.14}$$

By rotating the analyser and keeping it along the horizontal, one gets a dark background. The fringes in the model also shift when the analyser is rotated. In this case, the analyser will transmit the E_x component of equation (1.13) and denoting the intensity of light as I_d, one gets

$$I_d = I_a \sin^2 \frac{\delta}{2} \tag{1.15}$$

1.9 Fringe contours and their numbering in photoelasticity

The intensity equations obtained in the previous sections support the qualitative discussions in section 1.5. For the plane polariscope, the intensity is a function of both the magnitude of the principal stress difference (δ) and its orientation (θ) (equation (1.11)). In a circular polariscope, for both bright and dark fields (equations (1.14) and (1.15)), intensity of light transmitted is only a function of δ. This mathematically explains why both *isochromatics* and *isoclinics* are seen in a plane polariscope (figure 1.4(a)) whereas, only *isochromatics* are seen in a circular polariscope (figure 1.4(b)).

In optical techniques, the numbering of fringes is not trivial. Based on the intensity equations, one can arrive at the nature of numbering in dark and bright field arrangements. In dark field arrangements of both plane and circular polariscopes, intensity is zero when $\delta = 2m\pi$ ($m = 0, 1, 2, \ldots$) and fringe orders correspond to 0, 1, 2, ... (figure 1.4(a)). In bright field arrangement, intensity is zero when $\delta = (2m + 1)\pi$, i.e., when the retardation is an odd multiple of half-wave lengths. The fringe orders correspond to 0.5, 1.5, 2.5, ... (figure 1.4(b)). To translate this understanding to the actual labelling of fringes in the model, one should know what the gradient direction is and the fringe order for at least one of the fringes.

Instead of using a monochromatic source, if one uses white light, then the isochromatics will appear as distinct contours in colour (figure 1.5(a)). It has been pointed out earlier that the wave plates are wavelength dependent. Hence, when white light is incident on the model, at any point, only a single wavelength is cut off. In view of this, one observes white light minus the extinct colour over the field. *Iso* means constant and *chroma* means colour. Thus, the term *isochromatics* for contours of ($\sigma_1 - \sigma_2$) is more appropriate when white light is used as a source. In conventional

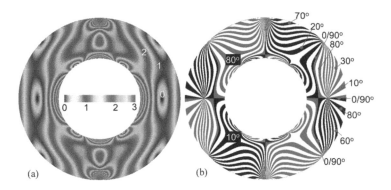

Figure 1.5. Ring under diametral compression: (a) isochromatics observed in dark field circular polariscope in white light with the black background removed for clarity. (b) Assembled whole field isoclinics in steps of 10° simulated with major principal stress direction in black and minor principal stress direction in blue simulated using P_Scope®. Courtesy: [11].

photoelasticity, by using a colour code from a standard problem like a beam under pure bending, the colours can be interpreted as integer order fringes approximately (see the inset in figure 1.5(a)). Zeroth fringe order will appear as black even with a white light source, fringe order 1 is a dull-red-blue transition, fringe order 2 is pink-red-green transition, and fringe order 3 is rose-red-green transition and so on. This knowledge can facilitate fringe ordering at least approximately in a model domain. For accurate determination, the reader is advised to refer to the methods given in chapter 4.

Isoclinics are usually numbered with the angles they denote, such as 0°, 10°, 15° etc and they appear as dark fringes even while using a white light, as the light extinction condition is independent of the wavelength. These are determined using a plane polariscope. The isoclinic value at a point of interest can be determined by monitoring the intensity of light when the crossed polariser and analyser combination is rotated until the intensity goes to zero. The corresponding angular positions of the polariser and analyser represent the principal stress directions at the point of interest. To determine whether the polariser represents the major or minor principal stress direction, the polariscope needs to be calibrated [5, 7]. For each crossed arrangement of plane polariscope, one gets only one isoclinic and if one wants to get it for the whole field, one must repeat the rotation of crossed polariser–analyser combination several times. If all that data is collated in steps of 10°, the isoclinics field over the model domain will appear as shown in figure 1.5(b). As it is simulated theoretically, the orientations of major and minor principal stress directions are represented as black and blue, respectively. This is a unique way of representing the isoclinics and will help the reader to appreciate the nuances of digital photoelasticity in the later chapters. In a model domain, one may come across points at which all isoclinics merge—such points are called *isotropic* points—meaning that at these points every direction is a principal stress direction, and the principal stress magnitudes are equal. In view of this, the isochromatic fringe order is zero at an *isotropic* point. In figure 1.5(a), the fringe order marked as zero is an *isotropic* point, which can be verified by comparing it with figure 1.5(b). All isoclinics merge at load application points also, but they are not *isotropic points*. The ring under diametral compression has ten *isotropic points* and it is left as an exercise to the reader to identify them. At an *isotropic point*, the result that isochromatic fringe order is zero is useful information for fringe ordering.

Although in general, isochromatics and isoclinics can be easily identified using white light, a tie always exists between zeroth fringe order and the isoclinics as both appear as black. A simple trick to differentiate the two is to view the model in a plane polariscope and rotate the crossed polariser–analyser combination, the fringes that move with it are isoclinics.

Isoclinics means contours of constant inclination and the principal stress direction on all points lying on an *isoclinic* is constant. However, the magnitude of $(\sigma_1-\sigma_2)$ may change from point to point on an isoclinic. On an *isochromatic* fringe, the value of principal stress difference $(\sigma_1-\sigma_2)$ remains the same but the orientation of the principal stress direction may change. Several nuances of conventional photoelasticity can be well understood by using the simulation software virtual polariscope named as P_Scope® [11]. This is developed based on Jones calculus and one can even see the change of polarisation state of the light as it passes through each optical element. The basic optical arrangements in conventional and digital photoelasticity can be instantly arranged and its effect on the appearance of retardation pattern can be recorded. It is the best tool currently available to learn both conventional and digital photoelasticity.

Assigning *isochromatic* fringe orders is a challenging task in conventional photoelasticity. A detailed discussion on how to use the fringe features of isoclinics, isochromatics and concepts of mechanics of solids to order the fringes is given in reference [12]. Such an exercise may help in labelling the fringes, but for labelling in non-fringe areas, one must resort to point-by-point methods known as compensation techniques (section 1.11). This will make the data acquisition process quite cumbersome for the entire model.

Thus, if there is a demand to get photoelastic data all over the model, conventional photoelasticity is inadequate, cumbersome and time consuming. This issue is fully addressed now with modern developments in data acquisition, which is the subject matter discussed in chapters 2–4.

Determination of the principal stress/strain direction at any point of interest is quite simple in conventional photoelasticity, but it remained a challenge in digital photoelasticity.

1.10 Calibration of model materials

Determination of the material stress fringe value (equation (1.4)) is known as calibration of a photoelastic material. The stress-fringe values of model materials vary with time and from batch to batch. Hence, it is necessary to calibrate each sheet or casting at the time of the experiment. As the focus is on finding out the material stress fringe value in equation (1.4), the value of principal stress difference needs to be known by a theoretical solution and the model loading should be such as to simulate the theoretical conditions closely in the calibration test. Also, the fringe order needs to be conveniently measured from the experiment accurately. Among the various specimens such as a beam under pure bending or a simple tension specimen or a circular disc under diametral compression, the use of a circular disc under diametral compression is preferred for calibration. This is because the specimen is compact, easy to machine, the loading conditions are easy to simulate accurately as per theory and it is easy to find the fringe order for a quick calibration. In modelling granular materials, circular discs are widely used, and it is worthwhile to see the procedure for the theoretical evaluation of the stresses. This is beneficial from the point of view of developing a greater understanding of the principle of superposition in solid mechanics as well.

The focus of the subsequent discussion will be only on how to utilise the principle of superposition effectively rather than getting into the complete set of mathematical steps that can be worked out easily by the reader as an exercise.

The solution is constructed based on Boussinesq's solution for a semi-infinite plate subjected to a concentrated vertical load and Lame's problem judiciously (figure 1.6). From the semi-infinite plate, one must cull out the circular disc and apply a load from the top and bottom to simulate diametral compression. In addition, as the boundary of the disc is traction-free, one should apply suitable loading to cancel out the stress field generated by the diametral loads on the boundary of the disc. While evaluating the load on the boundary of the disc, the subtle understanding of solid mechanics with respect to principal stress directions is well utilised.

It is well known that Boussinesq's problem produces a simple radial distribution of stress field as,

$$\sigma_r = \frac{-2P}{\pi h} \frac{\cos \theta_i}{r_i}; \ \sigma_\theta = 0; \ \tau_{r\theta} = 0 \tag{1.16}$$

For a circle of diameter 'D' as shown in figure 1.6(a), the stress at any point on the boundary of the circle (figure 1.6(b)) turns out to be,

$$\sigma_r = \frac{-2P}{\pi h D} \text{ since } D = \frac{r_i}{\cos \theta_i} \tag{1.17}$$

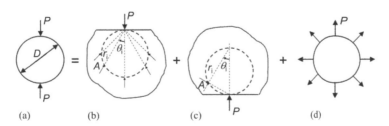

(a)   (b)   (c)   (d)

Figure 1.6. Solution to the problem of a disc under diametral compression as superposition of three simpler problems: (a) disc under diametral compression; (b) Boussinesq's problem 1; (c) Boussinesq's problem 2; (d) Lame's problem.

For any point on the boundary of the circle, the magnitude of the stress is the same, and is directed towards the load application point as it is taken as the origin of the coordinate system (figure 1.6(b)). Now consider the inverted semi-infinite plate to introduce the concentrated load. For the typical point A (figure 1.6(c)), this will also give the stress magnitude towards the load application point as

$$\sigma_r = \frac{-2P}{\pi h D} \tag{1.18}$$

Due to the property of the circle, the stress components from these two loadings are mutually perpendicular. These magnitudes are the same and from the knowledge of stress tensor, every direction at this point is the principal stress direction. This effect can be construed as a Lame's problem of radial compression which can be nullified by adding a radial tension to keep the boundary of the disc traction-free (figure 1.6(d)). Thus, the principle of superposition is cleverly used in this problem.

From equation (1.17), a semi-infinite plate subjected to a concentrated vertical load, will have the isochromatic fringes representing the contours of constant radial stress, as circles touching the top surface, as shown in figure 1.7(a). This is left as an exercise for the reader to work out. Nature loves to have nice geometric patterns [13].

The stress tensor at a generic point Q (figure 1.7(b)) is the superposition of the three problems (figure 1.6) after expressing them in Cartesian coordinates. Keeping the centre of the disc as the origin, and denoting the radius of the disc as R, the stress components are obtained as,

$$\begin{Bmatrix} \sigma_x \\ \sigma_y \\ \tau_{xy} \end{Bmatrix} = -\frac{2P}{\pi h} \begin{Bmatrix} \dfrac{(R-y)x^2}{r_1^4} + \dfrac{(R+y)x^2}{r_2^4} - \dfrac{1}{D} \\[3mm] \dfrac{(R-y)^3}{r_1^4} + \dfrac{(R+y)^3}{r_2^4} - \dfrac{1}{D} \\[3mm] \dfrac{(R+y)^2 x}{r_2^4} - \dfrac{(R-y)^2 x}{r_1^4} \end{Bmatrix} \tag{1.19}$$

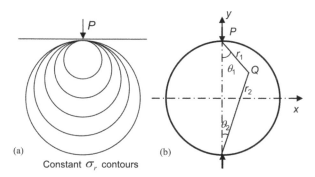

(a) Constant σ_r contours

(b)

Figure 1.7. (a) Contours of constant radial stress appear as a family of circles touching the top of the semi-infinite plate at the load application point, (b) disc under diametral compression representing an arbitrary point Q to superimpose the stress solutions with disc centre as the origin.

where,

$$r_1 = \sqrt{x^2 + (R - y)^2} \text{ and } r_2 = \sqrt{x^2 + (R + y)^2}$$

The principal stress difference $(\sigma_1-\sigma_2)$ at any point in the disc can be expressed as

$$(\sigma_1 - \sigma_2) = \frac{4PR}{\pi h} \frac{R^2 - (x^2 + y^2)}{(x^2 + y^2 + R^2)^2 - 4y^2 R^2} \tag{1.20}$$

At the centre of the disc, using equations (1.20) and (1.4), the material stress fringe value can be expressed as,

$$F_\sigma = \frac{8P}{\pi D N} \tag{1.21}$$

For a quick calibration, it is enough that one finds the fringe order at the centre of the disc and uses equation (1.21). It will be discussed later that determining the fringe order accurately at the centre of the disc using Tardy's method of compensation is quite simple. Thus, the use of a circular disc under diametral compression has gained prominence in conventional photoelasticity. The value of material stress fringe value is further improved by obtaining the fringe orders at the centre for both increasing and decreasing loads and evaluating the ratio P/N by a best fit straight line mimicking a least squares fit graphically.

In digital photoelasticity, as the fringe data can be easily obtained over the model (chapters 3 and 4), the use of many data points from the field is the preferred approach for calibration. It is desirable that the results are nearly independent of the choice of data points from the field. To achieve this, the least squares technique is usually combined with a random sampling process. It is worthwhile to account for residual birefringence also in the analysis. Let the residual birefringence expressed in fringe orders be assumed as a linear function in x and y as [14]

$$N_r(x, y) = Ax + By + C \tag{1.22}$$

The material stress fringe value is evaluated in a least squares sense by solving the following equation.

$$[b]^T[b] \{u\} = [b]^T\{N\} \tag{1.23}$$

where,

$$[b] = \begin{bmatrix} S_1 & x_1 & y_1 & 1 \\ S_2 & x_2 & y_2 & 1 \\ \vdots & \vdots & \vdots & \vdots \\ S_M & x_M & y_M & 1 \end{bmatrix}, \{u\} = \begin{Bmatrix} 1/F_\sigma \\ A \\ B \\ C \end{Bmatrix}, \{N\} = \begin{Bmatrix} N_1 \\ N_2 \\ \vdots \\ N_M \end{Bmatrix} \tag{1.24}$$

$$\text{and } S(x, y) = \frac{4PR}{\pi} \frac{R^2 - (x^2 + y^2)}{(x^2 + y^2 + R^2)^2 - 4y^2 R^2} \tag{1.25}$$

The unknown coefficient vector $\{u\}$ can be easily evaluated using the standard Gaussian elimination procedure. The construction of equation (1.23) is quite simple from experimental data and hence this method has found wide acceptance.

Least squares technique by itself does not guarantee that the result is physically admissible. Evaluation of the parameter values is complete only if the theoretically reconstructed fringe patterns agree well with the experimentally obtained ones [7].

1.11 Tardy's method of compensation

As noted earlier, in non-fringe areas, one must use compensation techniques to determine the value of fringe order. Amongst the several compensation techniques available, Tardy's method of compensation uses the analyser itself as a compensator. Hence, an understanding of it paves the way for appreciating phase shifting techniques in photoelasticity (chapter 3).

In both circular and plane polariscopes, the light extinction dictates the formation of fringes. So, at a point in the non-fringe areas, the light intensity would be a shade of grey, which should become dark by a suitable, measurable compensation of retardation. It becomes easier to quantify if the retardation is applied along the principal stress directions at the point. This makes the compensation techniques a point-by-point approach. Initially, the principal stress direction at the point of interest is determined in a plane polariscope arrangement and a circular polariscope is then formed such that the polariser is kept at the isoclinic angle and all the other optical arrangements are appropriately positioned. In other words, the loaded model is rotated by the isoclinic angle, which greatly simplifies the Jones retarder matrix with $\theta = 0$ (equation (1.9)). The analyser is rotated until the intensity at the point of interest becomes zero. Let the angle of rotation be β (figure 1.8). The intensity of light transmitted by the analyser can be evaluated as,

$$\begin{Bmatrix} E_{-\beta} \\ E_{-\beta+\pi/2} \end{Bmatrix} = \frac{1}{2} \begin{bmatrix} \cos\beta & -\sin\beta \\ \sin\beta & \cos\beta \end{bmatrix} \begin{bmatrix} 1 & -i \\ -i & 1 \end{bmatrix} \begin{bmatrix} \cos\dfrac{\delta}{2} - i\sin\dfrac{\delta}{2} & 0 \\ 0 & \cos\dfrac{\delta}{2} + i\sin\dfrac{\delta}{2} \end{bmatrix}$$

$$\times \begin{bmatrix} 1 & i \\ i & 1 \end{bmatrix} \begin{Bmatrix} 0 \\ 1 \end{Bmatrix} ke^{i\omega t} \qquad (1.26)$$

Upon simplification, one gets,

$$\begin{Bmatrix} E_{-\beta} \\ E_{-\beta+\pi/2} \end{Bmatrix} = \begin{bmatrix} \cos\beta & -\sin\beta \\ \sin\beta & \cos\beta \end{bmatrix} \begin{Bmatrix} \sin\dfrac{\delta}{2} \\ \cos\dfrac{\delta}{2} \end{Bmatrix} ke^{i\omega t} \qquad (1.27)$$

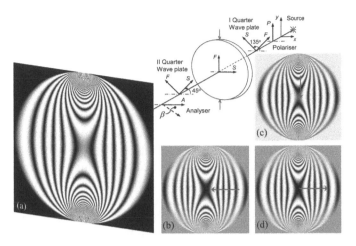

Figure 1.8. Tardy's method of compensation applied to the centre of a disc (horizontal and vertical directions are principal stress directions) under diametral compression. (a) Dark field circular polariscope. The centre fringe order lies between 5 and 6 but closer to 5. (b) By rotating the analyser clockwise by 36°, the lower fringe order has moved to the centre, forming the figure of '8'. The movement of the fringes is shown by an arrow in this and subsequent figures. (c) By rotating the analyser anti-clockwise by 120°, the higher fringe order is moving down. (d) At 144°, the higher fringe order forms a figure of '8'. Even the change of background light intensity is well captured by the software P_Scope® [11]. Though images (b) and (d) look identical, the data interpretation is restricted to the centre from the way the ordered fringes have moved in the process.

By invoking the requirement of light extinction, a relationship can be established between the angle β and δ as follows:

$$\cos \beta \sin \frac{\delta}{2} = \sin \beta \cos \frac{\delta}{2}$$

$$\tan \frac{\delta}{2} = \tan \beta$$

(1.28)

If β is measured in degrees, the fractional fringe order δ_N is obtained as

$$\delta_N = \pm \frac{|\beta|}{180°}$$

(1.29)

The fractional fringe order δ_N is taken as positive if a lower fringe order moves to the point of interest and vice versa. In conventional photoelasticity, determination of fringe order at the centre of the circular disc becomes quite simple as the polariser and analyser are pre-aligned to the principal stress directions at the centre of the disc and hence a simple rotation of the analyser is enough to estimate the fringe order quite accurately (figure 1.8) and hence the material stress fringe value using equation (1.21). In the current example, the fractional fringe order is +0.20 for clockwise rotation of the analyser and −0.80 for the anti-clockwise rotation of the analyser. Both operations finally yield the final fringe order at the centre to be 5.2. Thus, while employing Tardy's method of compensation, the movement of fringes need to be noted carefully by the experimenter. This aspect is missing in automated processing of data and that remained a challenge in the initial days of development of *digital photoelasticity*. This has been completely resolved and the details can be seen in chapter 3.

1.12 Three-dimensional photoelasticity

Among the experimental techniques, photoelasticity is unique to offer determination of stresses interior to the model. A mathematically intensive approach is possible through integrated photoelasticity—that accounts for the variation of principal stress directions along the light path [15]. With suitable simplifications, the technique is suitable for glass stress analysis [16]. A more generic approach, albeit, experimentally intensive is possible through the unique property of certain epoxies in which the stresses can be locked-in by subjecting the loaded model through a controlled thermal cycling process. The loaded model is heated gradually and soaked at the stress freezing temperature for a few hours depending on the size of the model and then furnace cooled. After the thermal cycling process, the loads are removed, and the model is carefully sliced to view and interpret the fringe patterns as a 2-D model for analysis. Provided no machining stresses are introduced while slicing, the slice retains the stress field as in the original 3-D model for a 2-D analysis.

At a macroscopic scale, freezing a loaded spring surrounded by water is easily understood—even after the removal of the loads, the spring retains its loaded position. In photoelasticity, it is achieved at a microscale during stress freezing by a thermal cycling process [5, 6, 17]. At stress freezing temperatures, the Young's modulus decreases rapidly, and the material stress fringe value also reduces by an order of magnitude. So even for small loads, the model can deform excessively affecting the similitude conditions. The biggest challenge is estimating the load at stress freezing temperatures so that the loaded model is not unduly deformed, and yet the slices cut have enough fringes for analysis. Figure 1.9(a) shows a 3-D aerospace model of a wing and fuselage, which is reversed while loading in the stress freezing furnace to simulate the lift loads (figure 1.9(b)). If loads are excessive, the model can break during the thermal cycling process (figure 1.9(c)), experience excessive deformation (figure 1.9(d)) or buckle at thin cross sections (figure 1.9 (e and f)). This has been a common occurrence until a systematic study by Swain *et al* [17] in 2016 established the importance of thermal strains in estimating total permissible strains at stress freezing temperatures. The permissible strain ε_{per} at the stress freezing temperature is to be calculated based on

$$\varepsilon_{per} = \alpha \Delta T + \sigma_{per}/E_f \qquad (1.30)$$

where, α is the coefficient of thermal expansion, ΔT is the difference between stress freezing and room temperatures, σ_{per} is the permissible stress on the stress frozen model and E_f is the Young's modulus at the stress freezing temperature. With modern developments in data acquisition and data processing, even low retardation data can be efficiently processed (chapters 3 and 4).

The loading at stress freezing temperatures must be applied through dead weights only and researchers have used a mercury column to simulate pressures in studying

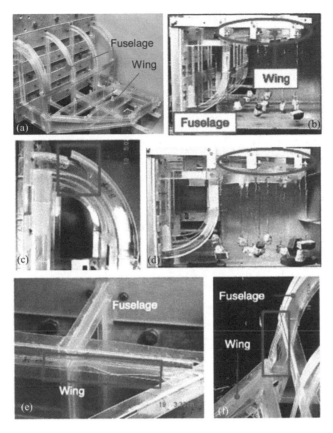

Figure 1.9. (a) 3-D model of an aerospace fuselage and wing, (b) the model is reversed to apply lift loads in the wing inside the stress freezing oven, (c) breakage of fuselage, (d) excessive deformation of the wing, (e) buckling in one zone, (f) buckling in another zone. Courtesy: [17]. Reprinted by permission from Springer Nature.

nuclear reactors [6]. Innovative loading arrangements in the analysis of seals have been developed (section 5.7.2). The main challenge has always been how to fabricate complicated 3-D models as one would require a new model for each type of loading. This has been greatly simplified now with the development of rapid tooling and rapid prototyping technologies (section 5.7.3).

Stress components that lie in a plane perpendicular to the incident light path alone contribute to the photoelastic effect. With this in view, the fringe patterns of the slices/sub-slices cut from a 3-D model need to be interpreted [1]. As principal stress difference contributes to the formation of isochromatics, the stress components that lie in the plane perpendicular to the incident light are used to calculate the pseudo principal stresses, termed as secondary principal stresses. The thickness of the slice considered is sufficiently small so that there is no variation of the secondary principal stresses over the thickness.

1.13 Interpretation of results obtained from plastics to metallic prototypes

Photoelasticity employs transparent polymers for model studies and the question often asked is how to extrapolate the results to metallic prototypes whose stiffness is quite high. This question can be answered only with the knowledge of governing equations from theory of elasticity (TOE). Stress formulation is one approach in which, the stress field is obtained by satisfying the boundary conditions, equilibrium, and compatibility conditions. The compatibility conditions, expressed in terms of stresses, provide an insight into how the elastic constants influence the results. For a 3-D stress field, the compatibility conditions known as Beltrami–Mitchell equations are as follows:

$$\nabla^2 \sigma_{xx} + \frac{1}{(1+\nu)} \frac{\partial^2}{\partial x^2} I_1 = \frac{-\nu}{1-\nu} \left(\frac{\partial}{\partial x} F_x + \frac{\partial}{\partial y} F_y + \frac{\partial}{\partial z} F_z \right) - 2\frac{\partial}{\partial x} F_x$$

$$\nabla^2 \sigma_{yy} + \frac{1}{(1+\nu)} \frac{\partial^2}{\partial y^2} I_1 = \frac{-\nu}{1-\nu} \left(\frac{\partial}{\partial x} F_x + \frac{\partial}{\partial y} F_y + \frac{\partial}{\partial z} F_z \right) - 2\frac{\partial}{\partial y} F_y$$

$$\nabla^2 \sigma_{zz} + \frac{1}{(1+\nu)} \frac{\partial^2}{\partial z^2} I_1 = \frac{-\nu}{1-\nu} \left(\frac{\partial}{\partial x} F_x + \frac{\partial}{\partial y} F_y + \frac{\partial}{\partial z} F_z \right) - 2\frac{\partial}{\partial z} F_z$$

$$\nabla^2 \sigma_{xy} + \frac{1}{(1+\nu)} \frac{\partial^2}{\partial x \partial y} I_1 = -\left(\frac{\partial}{\partial y} F_x + \frac{\partial}{\partial x} F_y \right) \qquad (1.31)$$

$$\nabla^2 \sigma_{yz} + \frac{1}{(1+\nu)} \frac{\partial^2}{\partial y \partial z} I_1 = -\left(\frac{\partial}{\partial z} F_y + \frac{\partial}{\partial y} F_z \right)$$

$$\nabla^2 \sigma_{xz} + \frac{1}{(1+\nu)} \frac{\partial^2}{\partial x \partial z} I_1 = -\left(\frac{\partial}{\partial z} F_x + \frac{\partial}{\partial x} F_z \right)$$

where,

$I_1 = \sigma_x + \sigma_y + \sigma_z$ and $\nabla^2 = \frac{\partial^2}{\partial x^2} + \frac{\partial^2}{\partial y^2} + \frac{\partial^2}{\partial z^2}$; F_x, F_y, and F_z are the body force components per unit volume. In this set of equations, one can observe that the only elastic constant appearing is the Poisson's ratio of the material. The stress magnitude and its distribution are influenced by the value of Poisson's ratio. A live loaded, 3-D model will have a Poisson's ratio of the parent plastic of around 0.35, whereas for a stress frozen 3-D model made of conventional photoelastic materials, the Poisson's ratio will approach 0.5. As most metals have Poisson's ratio around 0.24, in principle, the results of the stress frozen model would deviate somewhat. Several studies have shown that this deviation affects only the major principal stress and not the minor principal stress and the results are taken as a close approximation of reality in complex problem situations.

As long as the load applied on the model and prototype produces a similar level of strains, the only disturbing factor in the interpretation of 3-D stress analysis is the mismatch of Poisson's ratio. With modern improvements in rapid prototyping, the resins used are reported to exhibit an interesting property that their Poisson's ratio reduces from 0.38 to 0.295 at stress freezing temperature [18], which is desirable, as it becomes closer to metallic prototypes. Hence, the results of plastics can be directly interpreted for metallic prototypes using similitude relations.

Instead of a 3-D stress field, if one considers the problem of either plane stress or plane strain, the compatibility conditions reduce to,

$$\left(\frac{\partial^2}{\partial x^2} + \frac{\partial^2}{\partial y^2}\right)(\sigma_x + \sigma_y) = -(1 + \nu)\left(\frac{\partial F_x}{\partial x} + \frac{\partial F_y}{\partial y}\right)$$

$$\left(\frac{\partial^2}{\partial x^2} + \frac{\partial^2}{\partial y^2}\right)(\sigma_x + \sigma_y) = -\frac{1}{(1 - \nu)}\left(\frac{\partial F_x}{\partial x} + \frac{\partial F_y}{\partial y}\right) \qquad (1.32)$$

For the special case, when the body forces are zero or constant, the right-hand side of the equations goes to zero and the governing equations become independent of the elastic constants. In other words, if one takes a circular disc of epoxy, aluminium and steel and applies the same load meant for the epoxy disc, the stresses produced in all three discs will be identical. However, the displacements in each of these will be different dictated by their Young's moduli. As most engineers would like to simplify a problem preferably to either 2-D or 1-D, the experimental results of photoelasticity on polymers can be directly used for the metallic prototypes.

1.14 Similitude relations

One cannot obviously apply the same load as in a prototype to the plastic models used in photoelasticity. Model studies in general involve smaller models than prototypes. In some special situations, one may use larger models, and, in some cases, it can be of the same size as the prototype. If, the experiment is conducted on prototypes directly, such as glass articles, polycarbonate structures, etc as they are photoelastically sensitive, the actual loads can be applied for conducting the studies.

Therefore, one of the first issues is what should be the geometric scale factor? This decides the load ratio and in addition, the other key requirement is that the strain levels introduced in the model and the prototype be similar. This is quite important. If the model experiences large deformation, the governing equations will be non-linear, and the results cannot be used for prototypes. This becomes crucial for stress frozen studies. By assuming that there is no mismatch in Poisson's ratio, the force scale is obtained as,

$$F_m = F_p \left(\frac{L_m}{L_p}\right)^2 \frac{E_m}{E_p} \qquad (1.33)$$

where the subscript 'm' stands for the model and 'p' for prototypes, L denotes the characteristic length and E denotes the Young's modulus. Mismatch of Poisson's ratio appears in different forms in most of the experimental techniques, which is nearly unavoidable, and one should use appropriate correction factors when it can significantly influence the results.

1.15 Photoelastic results and methods for comparison

Photoelasticity directly provides only the principal stress difference and the principal stress direction at a point of interest. If one combines these results with Mohr's circle, then one can get the normal stress difference and the in-plane shear stress at a point. With modern data acquisition methods, one can get this information for the complete model domain.

In fact, the basic information obtained from a photoelastic test can be profitably used for comparative designs. This has been extensively used to decide what kind of fillet is more suitable for a particular application as the fringes themselves provide enough data to evaluate the comparisons [19]. If a material follows Tresca yield criteria, as $(\sigma_1-\sigma_2) = \sigma_{ys}$ is the yield surface, then the photoelastic contours represent the yield surface as well.

Evaluation of stress concentration factors (SCF) has also been one of the crucial data that designers are looking for and these can be determined very easily by photoelasticity based on the relevant fringe orders (SCF $= N_{max}/N_{far\text{-}field}$). With the birth of fracture mechanics, the focus has shifted to the evaluation of stress intensity factors (SIFs). These are also readily obtainable by suitably processing the positional co-ordinates and the corresponding fringe orders, which will be discussed in detail in chapter 2. In contact-stress problems, one would be interested in contact length and friction developed at the contacts. These are obtainable from basic photoelastic data for Hertzian contact, Hertzian contact with friction as well as for the more general case of a flat punch in contact [20].

As one is not getting individual stress components in photoelasticity, in the early development of photoelasticity, extensive research has been focussed upon stress separation techniques. Such techniques involved integration of the governing equations as in shear-difference technique or use of results from another experimental technique or numerical solutions. Individual stress components thus obtained are used to compare the results of photoelasticity with analytical or numerical methods. In fact, to validate any numerical technique, experimental comparisons are essential. It is true that the separation of stresses is cumbersome, but if one poses the problem differently, comparisons can be carried out through much simpler ways.

If results of numerical techniques could be post-processed to get photoelastic contours, validation of the numerical modelling is possible through photoelasticity. In fact, this approach is used in some of the examples studied later in this book. Many commercial packages of finite elements (FE) do not have a facility to provide fringe contours. Ramesh *et al* in 1995 [21] published a software code in turbo-pascal to post-process the results of FE where the domain is discretised by 8-noded isoparametric element. In this approach, the shape functions themselves are used as interpolation functions to plot fringe contours in black and white using a scanning scheme. This has helped to avoid solving any non-linear equations to plot the contours. The variation in the fringe thickness as observed in the experiments is mimicked very easily and even closed contours within an element are plottable through the scanning scheme. The method is generic and has been applied to plot photoelastic fringe contours, isopachics from holography, and u and v displacement contours from FE results.

Validation of FE modelling [22] and discretization schemes [23] have been shown by comparing the results from photoelastic contours using the methodology developed. The methodology has been extended to plot the fringe patterns in grey scale and colour using a twenty-noded brick element for a slice cut from a 3-D model as well as the integrated retardation of the 3-D model [24]. The challenge here is to properly collate the data concerning a slice from the FE results that should consider the elements involved in a light path, consider contributions from all the connected elements and selection of the scanning increment to get a smooth contour. The procedures have been extended to both normal and oblique incidences as encountered in photoelasticity for a slice as well as a sub-slice [25]. The scheme has also been extended for reflection photoelasticity [26]. All these approaches are restricted to process data from particular element types.

A much simpler approach to post-process results from a standard FE software ABAQUS has also been reported [27]. In this, the colour spectrum of the plotting module is changed to mimic the photoelastic contours in grey scale and in colour. The colour spectrum is represented in the format '#$rrggbb$' where rr, gg and bb represent the hexadecimal values of the intensities of the red, green, and blue image planes, respectively. The decimal R, G and B values range from 0–255 and their hexadecimal equivalents range from 00 to FF. Table 1.1 gives this for grey scale as well as for plotting in colour. The colour spectrum is given up to three fringe orders and beyond this, the scheme would plot the fringes in red colour. Figure 1.10 gives the example of a spanner tightening a nut solved by FE and postprocessed as fringe patterns in colour as well as grey scale using table 1.1.

The boundary conditions to be used in an FE analysis can be tricky even for seemingly simple problems. The boundary condition for one of the edges of a plate coming into contact with a hot plate is ably solved by post-processing the FE results and comparing it with experimental results (section 5.4.2). Post-processing to plot integrated retardation patterns is more involved. This has been done to plot fringe patterns in precision glass lens moulding using four-noded axisymmetric thermally coupled quadrilateral, bilinear displacement, and temperature element for simple

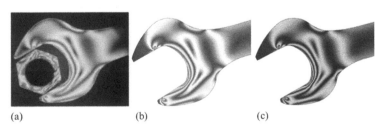

(a)   (b)   (c)

Figure 1.10. (a) Experimental isochromatics of a spanner tightening a nut. Post-processed result of the FE analysis of the spanner showing isochromatics: (b) in colour (c) in grey scale. Courtesy: [7, 28].

Table 1.1. Spectrum to plot fringes

(N)	Three fringe colour spectrum Decimal (R, G, B)	Hexadecimal	Grey scale Hexadecimal
0	53, 58, 59	'#353A3B'	'#3A3A3A'
0.1	73, 83, 81	'#495351'	'#535353'
0.2	116, 130, 125	'#74827D'	'#828282'
0.3	182, 191, 179	'#B6BFB3'	'#BFBFBF'
0.4	227, 224, 190	'#E3E0BE'	'#E0E0E0'
0.5	251, 230, 166	'#FBE6A6'	'#E6E6E6'
0.6	252, 205, 118	'#FCCD76'	'#CDCDCD'
0.7	232, 159, 70	'#E89F46'	'#9F9F9F'
0.8	175, 94, 49	'#AF5E31'	'#5E5E5E'
0.9	121, 59, 72	'#793B48'	'#3B3B3B'
1	74, 51, 118	'#4A3376'	'#333333'
1.1	46, 84, 172	'#2E54AC'	'#545454'
1.2	55, 133, 184	'#3785B8'	'#858585'
1.3	101, 192, 155	'#65C09B'	'#C0C0C0'
1.4	153, 222, 113	'#99DE71'	'#DEDEDE'
1.5	206, 226, 75	'#CEE24B'	'#E2E2E2'
1.6	241, 203, 61	'#F1CB3D'	'#CBCBCB'
1.7	252, 159, 76	'#FC9F4C'	'#9F9F9F'
1.8	240, 99, 121	'#F06379'	'#636363'
1.9	203, 66, 154	'#CB429A'	'#424242'
2	153, 58, 166	'#993AA6'	'#3A3A3A'
2.1	95, 86, 151	'#5F5697'	'#565656'
2.2	58, 142, 115	'#3A8E73'	'#8E8E8E'
2.3	56, 183, 90	'#38B75A'	'#B7B7B7'
2.4	78, 213, 82	'#4ED552'	'#D5D5D5'
2.5	133, 218, 99	'#85DA63'	'#DADADA'
2.6	184, 194, 122	'#B8C27A'	'#C2C2C2'
2.7	226, 154, 141	'#E29A8D'	'#9A9A9A'
2.8	246, 111, 147	'#F66F93'	'#6F6F6F'
2.9	244, 74, 136	'#F44A88'	'#4A4A4A'
3	219, 70, 121	'#DB4679'	'#464646'
$N > 3$	255, 0, 0	#FF0000	Problem specific

glass discs with rectangular elements [29] and for plano-convex lenses with arbitrarily shaped elements [30]. In fact, the use of photoelasticity to measure the residual birefringence in glass lenses has greatly improved the FE process simulation of the precision moulding process. The key contribution has been in the scientific approach to evaluate the thermal boundary conditions that have been assumed arbitrarily hitherto (section 5.4.1).

1.16 Reflection photoelasticity

Working directly on a prototype has many advantages as the boundary and loading conditions are as per the application envisaged and the experimental results reflect the response of the structure truthfully. For materials like glass, polycarbonate, and silicon, transmission photoelasticity in the visible or infrared regime itself can be used for prototype analysis. If, however, components like aircraft landing gear, aircraft tyres, masonry structures, assembly of tall lamp posts need to be studied, reflection photoelasticity offers hope for performing the test on these prototypes using bondable and contourable birefringent coatings with a reflective backing. In fact, photoelasticity is the preferred method for design optimisation of landing gear to reduce its weight leading to an enormous amount of fuel saving. Failure of tall lamp posts has been due to assembly stresses, which are rather difficult and at times, impossible to model analytically, and photoelastic coatings have helped to improve the design of their attachments to the ground.

In view of the use of homogeneous coatings, the data interpretation is simple. It has been used for various materials ranging from a human skull (to study crash response) to titanium alloys used in aerospace structures to fibre reinforced composites used in a variety of applications. Suitable coating materials applicable for various prototype materials have been developed and the user must carefully select the coating and its thickness for a given application [7].

Reflection photoelasticity is truly an engineering approach and approximations start right from the data acquisition stage. Normally, in photoelasticity, one demands normal light incidence on the model, and this is violated in the design of commercial polariscopes. To minimise the angle of oblique (to within 4°), the model to equipment distance is kept large (figure 1.11). The use of telephoto lenses is recommended for recording the fringes.

Since a coating is pasted on the model, the bonding should be perfect to transfer the strains generated in the model to the coating faithfully. The light traverses twice within the model (neglecting the small angle of oblique) and since strains cause the formation of fringes, in reflection photoelasticity the strain-optic law is obtained as,

$$\varepsilon_1^c - \varepsilon_2^c = \frac{N F_\varepsilon}{2 h_c} \qquad (1.34)$$

where F_ε is material strain fringe value related to the wavelength as λ/K, where K is the strain coefficient supplied by the manufacturer of the coating or needs to be calibrated if the coating is made in-house. The letter 'c' denotes the quantities related to the coating and in the subsequent discussions letter 's' would be used to denote the quantities related to the specimen either as a subscript or a superscript.

From stress-strain relations, the principal stress difference can be related to principal strain difference as:

$$\sigma_1^c - \sigma_2^c = \frac{E_c}{1 + \nu_c}(\varepsilon_1^c - \varepsilon_2^c) \qquad (1.35)$$

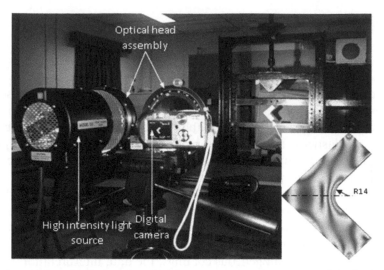

Figure 1.11. Overall view of reflection photoelasticity arrangement using a commercial reflection polariscope. The optical head assembly has a combination of a polariser and a quarter-wave plate each. Here a polycarbonate model with a reflective backing is being analysed and the arrangement gives an idea of the distance between the model and the polariscope to minimise the angle of oblique. The inset shows the reflection photoelasticity fringes. A representative schematic forms part of figure 1.12.

The ultimate objective is to find the stresses in the prototype specimen from a coating analysis. The principal stress difference of specimen stresses is denoted by,

$$\sigma_1^s - \sigma_2^s = \frac{E_s}{E_c}\frac{1 + \nu_c}{1 + \nu_s}(\sigma_1^c - \sigma_2^c) \qquad (1.36)$$

The usual thickness of the coating in many applications would be of the order of 3 mm, which is considerable in comparison to other coating techniques like Moiré, Digital Image Correlation (DIC) or even strain gauges where the thickness would be a fraction of a millimetre. In view of thick coatings, Zandman and his co-workers have systematically studied the influence such as reinforcement, variation of strain over the thickness, etc and have come up with suitable correction factors for data interpretation [1]. By combining equations (1.34) and (1.35) and replacing F_e as λ/K, the revised form of evaluating principal stress difference of the specimen is:

$$\sigma_1^s - \sigma_2^s = R_f \frac{E_s}{1 + \nu_s}\frac{N}{2h_c}\frac{\lambda}{K} \qquad (1.37)$$

where R_f is the correction factor for a given loading condition. The correction factors are influenced by the coating and model materials, and the type of loading. Although the correction factors have the Poisson's ratios of the specimen and the coating, the mismatch of Poisson's ratio is not handled in their development. Mismatch of Poisson's ratio becomes significant in evaluating SCFs. Photoelasticity makes it very efficient and simple to evaluate the SCF and in transmission photoelasticity, it is just the ratio of $N_{\max}$ to $N_{\text{far-field}}$. The corrected equation for SCF evaluation in reflection photoelasticity is as follows:

$$\text{SCF} = \frac{N_{\max}}{N_{\text{far-field}}} \frac{(1 + \nu_s)}{(1 + \nu_c)} \qquad (1.38)$$

The number of fringes seen in a reflection photoelastic coating test is considerably less and one of the issues is to find the maximum fringe order obtainable in a photoelastic coating test. If one considers that the specimen obeys Tresca yield criteria, then the maximum specimen principal stress difference will be the yield strength σ_{ys}. The maximum fringe order $N_{\max}$ obtainable is:

$$N_{\max} = \frac{1 + \nu_s}{E_s} \frac{2h_c K}{\lambda} \sigma_{ys} \qquad (1.39)$$

Usually, one uses manufacturer supplied sheets or contourable plastics that have only a few days of shelf-life. They provide the strain coefficient K. Only rarely is there a need for calibration. If sheets are made in-house, then calibration is a must. Unlike circular discs for calibration in transmission photoelasticity, for all coating techniques, the preferred model is a cantilever beam with an end load, effected through displacement control. An aluminium specimen is normally used and is pasted with the coating material and deflected to a known extent; from the fringe orders observed, the strain coefficient K is obtained as,

$$K = \frac{R_f^b}{1 + \nu_s} \frac{L_s^3 \lambda}{3Lh_s h_c} \frac{N}{y_o} \qquad (1.40)$$

It is important to note that even for calibration, one has to use the correction factor applicable for bending. The symbol y_o denotes the end deflection, L and L_s are distances measured from the point of load application to the point of interest and to the fixed end of the beam, respectively. The correction for bending is given as,

$$R_f^b = \frac{(1 + eg)}{(1 + g)} \left[4(1 + eg^3) - \frac{3(1 - eg^2)^2}{(1 + eg)} \right] \left[\frac{1 + \nu_s}{1 + \nu_c} \right] \qquad (1.41)$$

where,

$$e = \frac{E_c}{E_s}, \quad \text{and} \quad g = \frac{h_c}{h_s} \qquad (1.42)$$

1.17 Photoplasticity

The use of strain induced birefringence for data reduction beyond the yielding of the model material comes under the purview of photoplasticity. In many instances, one would be interested in local yielding surrounded by elastic material, where the problem is elasto-plastic rather than fully plastic. The problems of yielding near the crack tip or such stress raisers are of interest in engineering design. In these problems, the strain encountered will be within 6%. The existence of Dugdale plastic zone ahead of a crack-tip as observed in metals is reported for a thin polycarbonate specimen by Brinson [31]. Excessive plastic deformations take place in manufacturing processes such as forming operations. In these, strains will be well beyond 20%.

In transmission photoelasticity, one is accustomed to linking the optical information to the principal stress difference. Though transmission analysis is applicable in photoplasticity, here, one would normally relate the optical information to the principal strain difference. Several materials such as polyethylene, polystyrene, Lucite, Plexiglass, nylon, and celluloid have been tried for photoplastic studies of which celluloid has been found to be more suitable in the initial developments of photoplasticity [32]. Celluloid is optically inhomogeneous as received from the suppliers and needs to be suitably heat treated to eliminate the initial fringe patterns. Use of a mixture of 'rigid' and 'flexible' Laminac polyester resins 4116 and 4134 in the ratios 60/40 or 70/30 by weight have also been suggested for photoplastic studies in the study of forming operations [32].

The use of polycarbonate as the material for photoplasticity was first suggested by Ito in 1962 [33]. Three-dimensional photoplasticity was first demonstrated by Dally and Mulc [34] by slicing the unloaded strain-frozen models of polycarbonate. The resulting birefringence due to plastic flow of polycarbonate is permanent and it is locked at the molecular scale in the model much like stress-freezing as employed in conventional 3-D photoelasticity [1]. Shimamoto and Takahashi in 1990 [35] compared celluloid, cellulose acetate and polycarbonate and reported that polycarbonate is the most preferred material as it offers linear optical and stress or strain difference relationship in both elastic and plastic regions. In the elastic region, the fringe order is linearly related to principal stress difference and in the plastic region, fringe order is linearly related to the principal strain difference. In their studies, the strain in the plastic region is limited to 6%. Evaluation of the material strain fringe value is important, and it requires the evaluation of the principal strain difference for calibration. As one encounters large values of strain in the plastic region, Shimamoto and Takahashi [35] used the technique of Moiré for the evaluation of the strains.

A novel approach to determine the strain-optic coefficient of polycarbonate for photoplastic studies is done by combining reflection photoelasticity for fringe data and DIC for extracting the principal strain difference [36] (figure 1.12). The polycarbonate specimen is transparent and one of its sides is coated with aluminium paint to make it reflective and thus suitable for reflection photoelastic studies. Once the paint is dried, speckles are sprayed on it to be used for DIC measurements. The reflection photoelasticity used a DSLR camera (Canon EOS 450D) with a monochromator of wavelength 575 nm and the 2-D DIC system (VIC 2-D, Correlated Solutions, USA) used a Point Grey camera (GRAS-20S4M-C) of resolution 1624 × 1224 pixels with a 35 mm lens. The scale factor for both reflection and DIC is made equal by suitable choice of lenses and extenders and it is 0.082 mm pixel^{-1}.

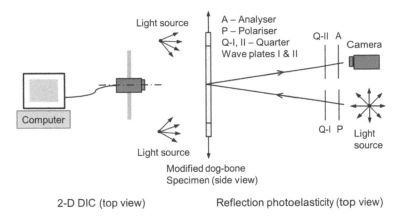

Figure 1.12. Novel experimental setup to simultaneously measure the fringe data by reflection photoelasticity and principal strain difference by DIC. Courtesy: [36]. Reprinted by permission from Springer Nature.

To determine the photoplastic constant, both reflection photoelasticity and DIC experiments are simultaneously performed on the specimen.

For determining the strain-optic coefficient till about 6% of strain, a new tensile specimen that facilitates easy ordering of fringes has been developed. A conventional tensile specimen provides just one data point for the tensile portion, whereas the dog-bone specimen, in view of its cross-sectional variations, provides more data points but fringe ordering is quite difficult. It is well known that the fringe order is zero at a free outward corner (section 5.13) and the dog-bone specimen is modified to have a free outward corner (figure 1.13(a)). The gauge length of the modified dog-bone specimen has a curvature of radius 125 mm with the dimensions of the minimum cross section being 13 mm × 1.2 mm. Due to the variation of width in the gauge length region; the applied load will result in the failure of the specimen at the minimum cross section without any onset of localized yielding.

The modified dog-bone specimen is loaded at a rate of 1 mm min^{-1} using a 5 kN UTM. No localized yielding is observed in the specimen before its failure. The results of the experiments at displacement loads of 1, 1.5, 2, 3 and 3.3 mm are shown in figures 1.13(b–f). The fringe orders are easy to label starting from the free outward corner, where it is zero, and these are marked in the figures. The DIC technique basically provides the displacements by correlating the pixel grey scale intensity pattern of a square subset in the reference image with its corresponding subset in the deformed image using the normalized sum of squared difference criterion (χ_{NSSD}) given as,

$$\chi_{\mathrm{NSSD}}^2 = \sum \left(\frac{\sum F_i G_i}{\sum G_i^2} G_i - F_i \right)^2 \qquad (1.43)$$

where F and G are grey scale intensity values in the reference and deformed configurations, respectively. After a parametric study, a subset size of 25 and a step

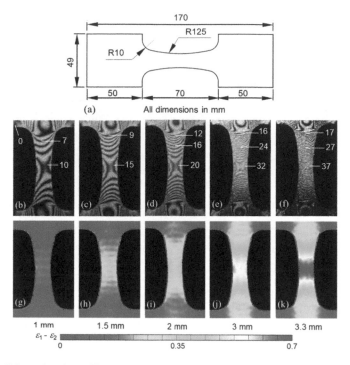

Figure 1.13. (a) Schematic of modified dog-bone specimen. Isochromatic fringes obtained from reflection photoelasticity for displacement loads, (b) 1 mm, (c) 1.5 mm, (d) 2 mm, (e) 3 mm, (f) 3.3 mm. Contour plots of $(\varepsilon_1-\varepsilon_2)$ from DIC for displacement loads (g) 1 mm, (h) 1.5 mm, (i) 2 mm, (j) 3 mm, (k) 3.3 mm. Courtesy: [39].

size of 3 are selected to evaluate the displacement fields. From the displacement fields, Hencky or true strain is evaluated using,

$$\varepsilon = \ln\left(\frac{l}{l_o}\right) \tag{1.44}$$

where l is the current length and l_o is the initial length. Although when deformations are small, both Lagrangian (engineering strain) and Hencky strains (true strain) would give the same strain values, when deformations are large, Hencky strain is the preferred form of measurement [37, 38]. The strain values determined at each of the data points are smoothed using a 90% center weighted Gaussian filter of size 9. The strain data points are separated by the step size, which is 3, and hence the total smoothing diameter for the strain data is 27 pixels. Figures 1.13(g–k) show the corresponding $(\varepsilon_1-\varepsilon_2)$ contours obtained from DIC experiments.

Data points corresponding to integral fringe orders are extracted from reflection photoelasticity interactively. For these co-ordinates, DIC results of principal strain difference are then extracted. Since reflection photoelasticity is used for determining the principal strain difference, the strain-optic law is,

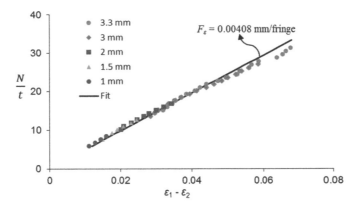

Figure 1.14. Variation of N/t with $(\varepsilon_1-\varepsilon_2)$ obtained from the images captured at 1, 1.5, 2, 3 and 3.3 mm displacement loading of the modified dog-bone specimen. Courtesy: [36]. Reprinted by permission from Springer Nature.

$$\varepsilon_1 - \varepsilon_2 = \frac{NF_\varepsilon}{2t} \qquad (1.45)$$

where $\varepsilon_1-\varepsilon_2$ is the principal strain difference, N is the fringe order, F_ε is the photoplastic constant of the material and t is the thickness. Figure 1.14 shows the variation of N/t with averaged $\varepsilon_1-\varepsilon_2$ for displacement loadings of 1, 1.5, 2, 3 and 3.3 mm. The variation of averaged $\varepsilon_1-\varepsilon_2$ values with N/t is fit to a line and F_ε (slope of the line) is evaluated to be 0.004 08 mm/fringe. A maximum principal strain difference of 6.8% is reached during the experiment. The change in specimen thickness during the experiment is small (1.5% of the original thickness) and does not significantly affect the F_ε value.

The similitude requirements for photoplastic studies are tricky as the model and prototype material should have the same yield behavior and flow rules. Metals display equal yield strengths for tension and compression but photoplastic materials generally have higher yield strength for compression than tension. Raghava *et al* [40] have demonstrated that polycarbonate obeys the modified von Mises criterion that accommodates the differences in tensile and compressive behavior and also accounts for the hydrostatic component of the applied stress state on the yielding which is given by,

$$(\sigma_1 - \sigma_2)^2 + (\sigma_2 - \sigma_3)^2 + (\sigma_3 - \sigma_1)^2 + 2(\sigma_{ys}^c - \sigma_{ys}^t)(\sigma_1 + \sigma_2 + \sigma_3) = 2\sigma_{ys}^c\sigma_{ys}^t \quad (1.46)$$

where σ_{ys}^c and σ_{ys}^t are the absolute yield strength values in compression and tension, respectively. If the tensile and compressive yield strengths are equal, then the equation reduces to the usual form of von Mises criterion.

Understanding the stress–strain behaviour of polycarbonate has received much attention [40, 41]. Polycarbonate exhibits a pronounced yield point and reaches a shear instability at the maximum load, and if the crosshead extension is continued, the load falls below the maximum and additional displacements produce progressively greater strain causing the shear bands to stabilize into a necked profile. Upon further crosshead extension, the neck stabilises to a constant profile and travels along the specimen, which is called cold drawing. After the neck has propagated completely through the length of the specimen, the stress required to produce further strain increases. Subramanyam Reddy and Ramesh used DIC [36] to capture the large evolution of strain in a standard tensile specimen and carefully analysed the unloaded specimen using transmission photoplastic analysis. The high-density fringes at the transition zone are revealed by immersing the specimen in a matching liquid of the same refractive index and established a bilinear fit (figure 1.15(a)) and a power law type non-linear fit (figure 1.15(b)) for the material strain fringe value. The power law equation captures the non-linear variation of N/t with $(\varepsilon_1-\varepsilon_2)$ and is given as,

$$\varepsilon_1 - \varepsilon_2 = (\varepsilon_1 - \varepsilon_2)_0 + F_\varepsilon^{tz}\left(\frac{N}{t}\right) + \left(\frac{1}{C}\frac{N}{t}\right)^m \tag{1.47}$$

where $(\varepsilon_1-\varepsilon_2)_0$ is the principal strain difference at the beginning of the transition zone, F_ε^{tz} is the initial slope of the curve in figure 1.15(b) depicting the transition

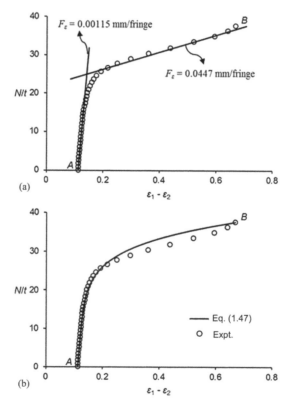

Figure 1.15. Variation of N/t with $(\varepsilon_1-\varepsilon_2)$ in the transition and plastic zones fit to: (a) bi-linear equation, (b) power law equation. Courtesy: [36]. Reprinted by permission from Springer Nature.

zone, C and m are to be determined from the regression analysis. The transition strain value $(\varepsilon_1-\varepsilon_2)_o$ is shown to be a property of the material and is determined as 0.11 for the polycarbonate material used. The values of F_ε^{tz}, C and m are 0.001 15 mm/fringe, 42 and 6, respectively [36]. This is the first study that has attempted to determine the variation of material strain fringe value of polycarbonate up to 60% strain, which is made possible by combining DIC with transmission photoplastic study. The results can hopefully be used to extend the use of photoplastic studies for metal forming.

1.18 Closure

In this chapter, the basics of photoelasticity, both transmission and reflection, three-dimensional photoelasticity and photoplasticity are discussed. The concept of birefringence, which is the fundamental physics underlying all the variants of photoelasticity has been explained succinctly. A new insight in calculating the requisite load for stress-freezing is provided to facilitate the safety of the models and the validity of similitude relations. The fundamental question as to how the results from plastics can be attributed to metallic prototypes has been addressed and justification is brought out with the help of governing equations of theory of elasticity. The purpose of experiments is to validate a design in comparison to other verification methods like numerical modelling. Hitherto, photoelasticity literature paid excessive emphasis on stress/strain separation for comparisons with other methods as photoelasticity basically provided only the principal stress or strain difference. A novel and more pragmatic approach for comparisons through post-processing the numerical results to get photoelastic contours is emphasised in this chapter and liberally used in the case studies discussed later in chapter 5.

Exercises

1. Buy a calcite crystal and a small sheet of polariser and recreate the experiment corresponding to figure 1.1 and convince yourself of the presence of double refraction and learn that the crystal sustains only polarised light.
2. A photoelastic model behaves like a crystal when stressed. Within a crystal, the incident unpolarised light is polarised. If so, why does one have to use polarised light in photoelasticity?
3. A light ellipse of ellipticity 0.3 and an azimuth of 30° needs to be produced. If you have access to a polariser and a quarter-wave plate, how do you use them to get such a light ellipse?
4. With a monochromatic light source how would you distinguish isoclinics and isochromatics in a plane polariscope? Describe at least two methods.
5. A circular polariscope is constructed using various optical arrangements of the elements. Establish using Jones Calculus, which of the given arrangements correspond to bright and dark fields. In the table, α is the orientation

of the polariser, ξ is the orientation of the slow axis of the quarter-wave plate-I, η is the orientation of the slow axis of the quarter-wave plate-II, and β is the orientation of the analyser.

α	ξ	η	β
$\pi/2$	$3\pi/4$	0	$\pi/4$
$\pi/2$	$3\pi/4$	$\pi/4$	$\pi/2$

6. Two discs one made of epoxy ($F_\sigma = 12$ N/mm/fringe) and another made of polycarbonate ($F_\sigma = 8$ N/mm/fringe) are subjected to diametral compression in a photoelasticity experiment. The diameter and thickness of both the discs are 60 mm and 6 mm, respectively. The load applied in both the cases is 100 N. If the fringe order (N) at a particular point of the epoxy disc is 1.5, what would be the fringe order at the same point in the polycarbonate disc?

7. One of the key steps in photoelastic analysis is the calibration of the model material. An ideal model to conduct calibration is a disc under diametral compression. A disc made of epoxy (diameter 60 mm) was loaded in compression and the fringe order at the centre was accurately estimated by Tardy's method of compensation. The observations are tabulated as follows:

Load (P) applied (N)	Rotation of analyser (in degrees)	Fringe order moved
246	145	0
328	12	1
410	60	1
554	144.5	1

It is found that the lower fringe order has moved to the centre in all the measurements. The fringe orders at the centre point are determined and a graph is plotted between the load P and fringe order N. Find the material stress fringe value in N/mm/fringe.

8. At a point P in a photoelastic material, a fringe order of 5 was observed. The isoclinic parameter at the same point was 30°. The material stress fringe value of the model material is 8 N/mm/fringe. The model material's Young's modulus is 2.6 GPa and Poisson's ratio is 0.28. Before the application of the loads, the thickness of the model was 6 mm; and after the application of the loads, the thickness at P measured by a lateral extensometer was 6.01 mm. Find out the following for point P: Principal

stress difference, normal stress difference, in-plane shear stress, the individual stress components σ_{xx} and σ_{yy}.

9. The figure shows the fringe patterns observed for two chain links of slightly different designs for an axial pull of 160 N. Comment on the stress fields in each of the links. If you have to choose a link for heavy duty purposes which of them, would you choose and why?

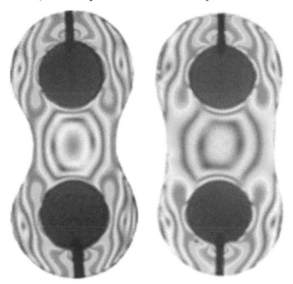

10. One of the simplest ways to minimise the SCF is to introduce stress relief holes. The figure shows the fringe patterns for a finite plate, subjected to an axial tensile load of 288 N, having dimensions (Length = 230 mm, width = 36 mm, thickness = 6 mm) with two smaller holes of 8 mm diameter introduced on either side of the 12 mm hole at a distance of 15 mm. The far-field fringe order for this load is found to be 0.56. In the smaller hole, fringe order 1 has moved to its boundary when the analyser is rotated by 70 degrees. For the bigger hole, the analyser needed to be rotated by an angle of 147 degrees. With this data find the SCF at both the holes. If there had been only a single large hole, the maximum fringe order at the boundary is found to be 2.2 fringe orders. Calculate the percentage reduction in SCF due to the relief hole for both the holes. To estimate the boundary stresses what additional data do you require?

11. Figures (a) to (d) show the isoclinics and isochromatics of a beam subjected to three-point bending.
   a. Identify and mark the isotropic point(s) from isoclinic pattern in figures a and b.
   b. Mark the fringe orders along the section *AB* in figure (d). Specify the criteria used to order fringes.
   c. What is the variation of shear stresses along a typical section *AB* predicted by simple beam theory? State and discuss the result qualitatively.
   d. Determine the values of in-plane shear stress using the zoomed isoclinic and isochromatic patterns in figures (b) and (d) and plot its variation along lines *AB* and *CD*. Take material stress fringe value as 13 N/mm/fringe and model thickness as 6 mm.
   e. Does the result contradict the prediction by simple beam theory? Can you give an explanation for this discrepancy?

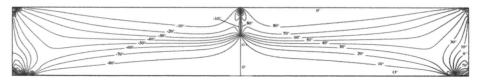

(a) Isoclinic pattern

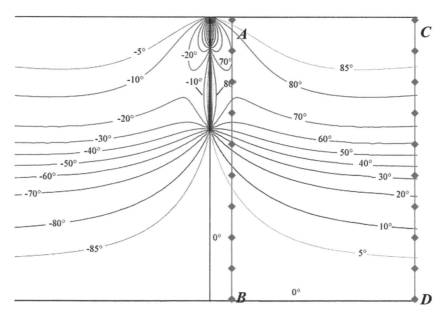

(b) Zoomed isoclinic pattern

(c) Isochromatic fringe pattern

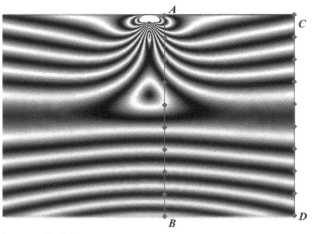

(d) Zoomed Isochromatic fringe pattern

12. A steel turbine blade is subjected to a load of 1500 N. For geometric similarity, the characteristic length can be taken as 20 cm. An epoxy model of the blade having a 6 cm characteristic length is used to analyse the whole field stress using photoelasticity. If Young's modulus of steel and epoxy are 200 and 3 GPa, respectively, what load (in N) must be applied to the model to get a geometrically similar deformation. (Neglect the Poisson's ratio mismatch)

13. The maximum fringe order observed in a plate with a hole subjected to tension is 2.6. The average fringe order observed is 0.9. The coating material used is PS1 sheet of 3-mm thickness. Determine the SCF if the specimen material is (a) HT1040 steel (b) 2024 aluminium (c) Glass fibre reinforced plastic (d) If the loading is directly done on the PS1 sheet with a hole (Note that the sheet has a reflective backing). The following are the material properties:

Material	E (GPa)	v	K
PS1	2.5	0.38	0.15
HT1040	207	0.292	
2024 Al	71	0.334	
Glass fiber reinforced plastic	100	0.15	

14. An industrial component made of aluminium 1100 H16 (E^s = 71 GPa; v^s = 0.33; S_Y = 140 MPa) is studied using a photoelastic coating (E^c = 2.5 GPa, v^c = 0.38, K = 0.15 and of thickness h^c = 2.0 mm) using white light. The specimen is tested until yielding occurs in the region of interest. From a numerical study of the component, it is learnt that the specimen principal stresses are of opposite signs in the area where the photoelastic coating is pasted. The dark field isochromatics observed are as shown.

Can you find the region of yielding in the test specimen from the interpretation of the fringe patterns? If so, mark the line demarcating the elastic and plastic region approximately using the colour code shown as an inset to figure 1.5(a).

15. It is proposed to conduct a photoelastic coating test of an aircraft landing gear. One of the first steps in a photoelastic coating test is to determine the strain coefficient K of the coating. Displacement controlled cantilever beam is an ideal model for calibration. The cantilever beam (thickness 5 mm, width 25 mm and length 250 mm) is of the same aluminium alloy (E_s = 71 GPa; v_s = 0.33) as that of the landing gear.

A small strip of photoelastic coating material (E_c = 2600 MPa; v_c = 0.38) of thickness 3 mm is bonded onto the aluminium alloy cantilever specimen. The model is illuminated with white light and the fringe orders obtained at a section 50 mm from the fixed end is as follows:

Displacement, mm	1	2	3	4	5	6
Fringe order, N	0.3	0.5	0.8	1	1.3	1.6

Determine the strain-optic coefficient K.

Instead of white light (577 nm) if the experiment is conducted with a sodium vapour lamp (589.3 nm) what modifications need to be done for data interpretation?

16. It is proposed to demonstrate the phenomenon of stress freezing to the students by conducting a suitable experiment. A circular disc of dia. = 60 mm, thickness = 6 mm and $F_\sigma = 12$ N/mm/fringe develops a fringe order of 2.83 at the centre under a diametral load of 800 N. If the F_σ at the stress freezing temperature is expected to be 0.35 N/mm/fringe, what should be the load applied to get fringe order of 3 at the center of the disc?

17. A detailed experiment for analysis is planned by loading an aerospace component consisting of several members, as shown in figure 1.9. The prototype is made of high strength aluminum having yield strength of 350 MPa and the strain at yield is 5000 µε. The component needs to be proof tested which is usually done at 110% of the service load and one can load it only up to the yield strain for testing purposes.

It is proposed to conduct a model study using epoxy through stress freezing. Assume that in order to maintain geometric similarity, the epoxy model also can be loaded up to 5000 µε. The chief engineer with a wealth of experience has simplified the loading calculations by basing them on a rectangular cross-section of one of the members. Unlike the previous problem, where the model geometry is simple and the necessary load at stress freezing is calculated simply based on expected fringe order at a point of interest, here one needs more information on the temperature dependent behavior of epoxy. Tests reveal that the Young's modulus E and F_σ at room temperature (30°) and at stress freezing temperature (70°) are, respectively, 1500 MPa, 10 N/mm/fringe and 17 MPa, 0.27 N/mm/fringe. As polymers can sustain large strains before yielding, the linear limit of this epoxy is estimated to be 11 000 µε and the coefficient of thermal expansion of epoxy is 160×10^{-6} °C. Using this information, estimate the permissible tensile stress that can be applied to the representative rectangular cross-section for stress freezing. Deliberate and compare the cases when the (i) thermal strains are ignored and (ii) thermal strains are considered.

18. In the previous problem, can you also find the permissible load to avoid buckling during stress freezing when the compressive load is applied? The representative specimen to base your calculations on can be assumed as a fixed-fixed column with a cross sectional area of 14×2.5 mm^2 and a length of 45 mm. Consider the behaviour of epoxy being similar in both tension and compression and the maximum linear strain which the material can withstand in compression before yielding is also 11 000 µε.

References

[1] Ramesh K 2008 Photoelasticity *Springer Handbook of Experimental Solid Mechanics* ed J William and N Sharpe (New York: Springer) pp 701–42

[2] Aben H 2007 On the role of T. J. Seebeck in the discovery of the photoelastic effect in glass *Proc. Est. Acad. Sci., Eng.* **13** 283–94 https://kirj.ee/public/Engineering/2007/issue_4/eng-2007-4-3.pdf

[3] Frocht M M 1941 *Photoelasticity* (New York: Wiley)

[4] Frocht M M 1948 *Photoelasticity* (New York: Wiley)

[5] Ramesh K 2000 *Digital Photoelasticity Advanced Techniques and Applications* (Berlin: Springer)

[6] Dally J W and Riley W F 1978 *Experimental Stress Analysis* (New York: McGraw-Hill)

[7] Ramesh K 2009 *e-book on Experimental Stress Analysis* (Chennai, India: IIT Madras) https://home.iitm.ac.in/kramesh/ESA.html

[8] Ramesh K 2015 Digital photoelasticity *Digital Optical Measurement Techniques and Applications* ed P Rastogi (Boston, MA: Artech House) pp 289–344

[9] Ramesh K 2011 Lec-12 ordinary and extraordinary rays *Video Lectures on Experimental Stress Analysis* (Chennai, India: NPTEL and MHRD) https://www.youtube.com/watch?v=s2HOSfwNoUE&list=PL21BB25670CDC2AEB&index=12

[10] Jones R C 1941 A new calculus for the treatment of optical systems I. Description and discussion of the calculus *J. Opt. Soc. Am.* **31** 488–93

[11] Ramesh K 2017 P_Scope® – a virtual polariscope *Photomechanics Lab.* (IIT Madras) https://home.iitm.ac.in/kramesh/p_scope.html (accessed Apr. 06, 2021)

[12] Ramesh K 2011 Lec-22 fringe ordering in photoelasticity *Video Lectures on Experimental Stress Analysis* (Chennai, India: NPTEL and MHRD) https://www.youtube.com/watch?v=eZ4KWB6uhqw&list=PL21BB25670CDC2AEB&index=22

[13] Durelli A J 1973 Log-art as a means of humanization *Exp. Mech.* **13** 232–7

[14] Sanford R J 1980 Application of the least-squares method to photoelastic analysis *Exp. Mech.* **20** 192–7

[15] Aben H 1979 *Integrated Photoelasticity* (New York: McGraw-Hill International Book Company)

[16] Aben H and Guillemet C 1993 *Photoelasticity of Glass* (Berlin: Springer)

[17] Swain D, Philip J, Pillai S A and Ramesh K 2016 A revisit to the frozen stress phenomena in photoelasticity *Exp. Mech.* **56** 903–17

[18] Ju Y, Wang L, Xie H, Ma G, Mao L, Zheng Z and Lu J 2017 Visualization of the three-dimensional structure and stress field of aggregated concrete materials through 3D printing and frozen-stress techniques *Constr. Build. Mater.* **143** 121–37

[19] Heywood R B 1952 *Designing by Photoelasticity* (New York: Springer)

[20] Hariprasad M P, Ramesh K and Prabhune B C 2018 Evaluation of conformal and non-conformal contact parameters using digital photoelasticity *Exp. Mech.* **58** 1249–63

[21] Ramesh K, Yadav A K and Pankhawalla V A 1995 Plotting of fringe contours from finite element results *Commun. Numer. Methods Eng.* **11** 839–47

[22] Pathak P M and Ramesh K 1999 Validation of finite element modelling through photoelastic fringe contours *Commun. Numer. Methods Eng.* **15** 229–38

[23] Ramesh K and Pathak P M 1999 Role of photoelasticity in evolving discretization schemes for FE analysis *Exp. Tech.* **23** 36–8

[24] Karthick Babu P R D and Ramesh K 2006 Development of photoelastic fringe plotting scheme from 3D FE results *Commun. Numer. Methods Eng.* **22** 809–21

[25] Babu K and Ramesh K 2006 Role of finite elements in evolving the slicing plan for photoelastic analysis of three-dimensional models *Exp. Tech.* **30** 52–6

[26] Ravichandran M, Karthick Babu P R D, Rao V and Ramesh K 2007 New initiatives on comparison of whole-field experimental and numerical results *Strain* **43** 119–24

[27] Neethi Simon B and Ramesh K 2010 A simple method to plot photoelastic fringes and phasemaps from finite element results *J. Aerosp. Sci. Technol.* **62** 174–82 http://www.aerojournalindia.com/2010 contents/Content -Aug-10.html

[28] Ramesh K 2011 Lec-04 physical principle of strain gauges, photoelasticity and Moiré *Video Lectures on Experimental Stress Analysis* (Chennai, India: NPTEL and MHRD) https://www.youtube.com/watch?v=kzWbdP5gqb0&list=PL21BB25670CDC2AEB&index=4

[29] Dora Pallicity T, Ramesh K, Mahajan P and Vengadesan S 2016 Numerical modeling of cooling stage of glass molding process assisted by CFD and measurement of residual birefringence *J. Am. Ceram. Soc.* **99** 470–83

[30] Pallicity T D, Vu A-T, Ramesh K, Mahajan P, Liu G and Dambon O 2017 Birefringence measurement for validation of simulation of precision glass molding process *J. Am. Ceram. Soc.* **100** 4680–98

[31] Brinson H F 1970 The ductile fracture of polycarbonate *Exp. Mech.* **10** 72–7

[32] Burger C P 1980 Nonlinear photomechanics *Exp. Mech.* **20** 381–9

[33] Ito K 1962 New model materials for photoelasticity and photoplasticity *Exp. Mech.* **2** 373–6

[34] Dally J W and Mulc A 1973 Polycarbonate as a model material for three-dimensional photoplasticity *J. Appl. Mech. Trans. ASME* **40** 600–5

[35] Shimamoto A and Takahashi S 1990 Calibration of materials for photoelastoplastic models *Exp. Mech.* **30** 114–9

[36] Subramanyam Reddy M and Ramesh K 2018 Study of photoplastic behaviour of polycarbonate using digital image correlation *Exp. Mech.* **58** 983–95

[37] Parsons E, Boyce M C and Parks D M 2004 An experimental investigation of the large-strain tensile behavior of neat and rubber-toughened polycarbonate *Polymer (Guildf).* **45** 2665–84

[38] G'Sell C, Hiver J M and Dahoun A 2002 Experimental characterization of deformation damage in solid polymers under tension, and its interrelation with necking *Int. J. Solids Struct.* **39** 3857–72

[39] Subramanyam Reddy M 2018 *Study of Fracture and Photoplastic Behviour of Polycarbonate Using Digital Image Correlation* (Chennai, India: IIT Madras)

[40] Raghava R, Caddell R M and Yeh G S Y 1973 The macroscopic yield behaviour of polymers *J. Mater. Sci.* **8** 225–32

[41] Buisson G and Ravi-Chandar K 1990 On the constitutive behaviour of polycarbonate under large deformation *Polymer (Guildf).* **31** 2071–6

IOP Publishing

Developments in Photoelasticity
A renaissance
K Ramesh

Chapter 2

Fringe multiplication, fringe thinning and carrier fringe analysis

Methods for digital fringe multiplication and digital fringe thinning are discussed. A comparative evaluation of the efficacy of various thinning methodologies is then presented. Applicability of these methodologies to evaluate the stress field parameters in the neighbourhood of interacting cracks in conjunction with over-deterministic non-linear least squares method is demonstrated. The importance of crack-tip refinement and strategies for data collection are discussed. Evaluation of residual stresses with the use of carrier fringes and fringe thinning is discussed for polycarbonate and glass. Calibration of glass using the carrier fringes method for both annealed and stressed float glass are then discussed. The influence of flow induced residual stress on the crack-tip stress field parameters is presented finally[1].

2.1 Introduction

Developments in phase shifting techniques (PSTs) and processing of colour information have revolutionised photoelastic analysis. However, in the early development of digital photoelasticity, digital image processing (DIP) systems were essentially used to automate the procedures that were done either optically or manually. Techniques for fringe multiplication are now possible without any additional optics and in fringe thinning methodologies, fringe skeletonization is effected by identifying the minimum intensity points using the DIP hardware and software. These methodologies are still relevant in a certain class of problems. The application areas chosen deal with important and difficult problems in stress analysis.

Extracting the data along the fringe skeleton is quite useful in carrier fringe analysis, and in problems dealing with the evaluation of stress intensity factors

---

[1] Parts of this chapter have been adapted with permission of Elsevier from the author's own work in [29], [38], [43].

(SIFs) in fracture mechanics, contact stress parameters etc. If the fringes are not sufficient in either dark or bright fields, then one must adopt fringe multiplication techniques to augment the information. One can also augment the information by using carrier fringes that find wide applications in glass stress analysis.

In this chapter, the techniques for fringe multiplication are discussed followed by methods for fringe thinning. The use of these techniques to evaluate stress field parameters for interacting cracks is discussed. The problems selected are such, the fringe patterns are sufficiently complex and only a multi-parameter solution is appropriate. The configurations of interacting cracks that are collinear, stacked, asymmetric and at generic orientations are considered under both uniaxial and biaxial loadings. A brief introduction to fracture mechanics is presented to appreciate the complexity of the problems and the elegance of photoelasticity to address these.

Evaluation of residual stresses is the most challenging problem in engineering analysis. Polycarbonate is now widely used as a structural material and these sheets have residual stresses introduced during manufacturing. The use of carrier fringes for quantitative evaluation of residual stress is illustrated for this example. Glass is birefringent and has residual stresses intentionally introduced during manufacturing to enhance its use in structural applications. Evaluation of residual stresses in glass that are classified as thickness and edge stresses is discussed using carrier fringes and fringe thinning. Evaluation of the photoelastic constant for glass is discussed with the help of carrier fringes to augment the retardation. It would be of interest to see how the residual stresses would affect the crack-tip stress field parameters, which is discussed at the end for flow induced residual stresses in a polycarbonate specimen with cracks parallel and perpendicular to the flow stress.

2.2 Digital fringe multiplication

Fringe multiplication is easily accomplished using DIP hardware. A simple digital subtraction of bright and dark-field images could result in fringe multiplication by an order of two. In such a case, one gets the grey scale value of the pixel $g(x, y)$ as a function of the retardation δ as [1],

$$g(x, y) \approx I_a \cos \delta \qquad (2.1)$$

In equation (2.1), the extinction of light will occur when $\delta = (2n + 1)\, \pi/2$ where, $n = 0$, 1, 2, ... The resultant image is termed as a *mixed-field* image and the fringe orders are labelled as $N = 0.25, 0.75, 1.25, ...$ Figure 2.1 shows the fringe multiplication obtained for the problem of interacting cracks that are at an arbitrary orientation.

Using PSTs or by twelve fringe photoelasticity (TFP) (chapters 3 and 4), one can get the fringe order at every pixel in the image. However, for certain applications such as extraction of SIFs or contact stress parameters, it is desirable to get the data from fringe skeletons so that the geometric shape of the fringe field is captured reasonably and fed into the data processing software for easy convergence to the results. From the whole field fringe order results, one can extract mixed-field fringe orders and hence, can be considered as fringe multiplication for the purpose of data analysis. Figure 2.2(a) shows the colour isochromatic fringe field of asymmetric

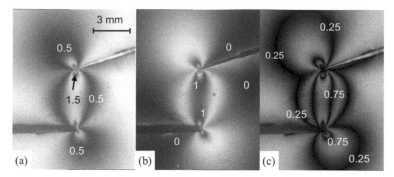

Figure 2.1. Fringe multiplication of interacting cracks by image subtraction of bright and dark field images: (a) bright field image, (b) dark field image, and (c) mixed-field image.

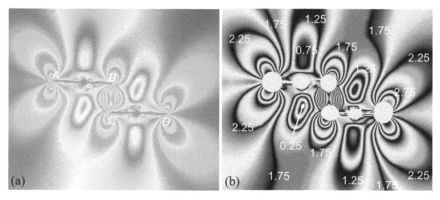

Figure 2.2. (a) Dark field colour isochromatics of asymmetric cracks recorded in white light, (b) mixed-field image obtained from the coloured isochromatics by extracting the total fringe order using colour image processing (section 4.9).

interacting cracks and figure 2.2(b) shows the mixed-field image extracted from colour information processing using a single isochromatic image.

2.3 Digital fringe thinning

One of the simplest methods for fringe thinning is to treat the fringe patterns as a binary image and the fringe centre lines are determined by a process of erosion [2]. However, in general, the centrelines may not be the actual fringe skeletons. This is because, only those points where the intensity of light transmitted is zero depict a fringe. Due to experimental difficulties, it is difficult to extract the fringe skeletons by collecting only those points with zero intensity. Identifying the fringe areas in each image greatly simplifies the development of fringe thinning methodologies. As photoelastic images have high contrast, a simple global thresholding operation can identify the fringe areas. In this the pixels above a threshold are made into white and below the threshold are made into black resulting in a binary image. The choice of the threshold is problem dependent. Figure 2.3(a–c) shows the fringe areas for

Figure 2.3. Fringe areas identified for various threshold values: (a) 60, (b) 80, (c) 110, (d) same as (b) but semi-thresholded.

different threshold values for asymmetric interacting cracks. If the thresholding operation is such that the intensity variation below the threshold is retained, then it is known as *semi-thresholding*. Figure 2.3(d) shows a semi-thresholded image. Such images are useful for further processing of the image based on intensity variation.

Over the years, several techniques have been proposed for fringe thinning. The early methods used a binary image corresponding to the fringe pattern [2]. Methods based on the variation of intensities within the fringe were developed later [3–6]. These intensity-based methods can be classified into mask-based methods and global methods.

2.3.1 Binary based method

In this method, the grey scale image of the fringe pattern is converted into a binary image by applying a suitable global threshold. The skeleton is determined by a process of erosion by progressively removing the outer layers of the fringe until only the skeleton is left. The fringe image is scanned sequentially from left to right, right to left, top to bottom and bottom to top to identify and eliminate the border pixels forming the fringe band. A 3×3 pixels mask is considered to identify the edge pixels (figure 2.4). As an example, when the scan direction is from left to right, there can be four combinations to identify whether the pixel $(0, 0)$, represents a fringe skeleton point:

    a. $(0, 1)$ and $(0, -1)$ are both fringe points;
    b. $(0, -1)$ is a fringe point and $(0, 1)$ is not a fringe point;
    c. $(0, -1)$ is not a fringe point and $(0, 1)$ is a fringe point;
    d. $(0, 1)$ and $(0, -1)$ are not fringe points.

The point $(0, 0)$ is removed if any of the three conditions 'a–c' is satisfied. In the case where the fourth condition is satisfied, then the point is retained. The conditions

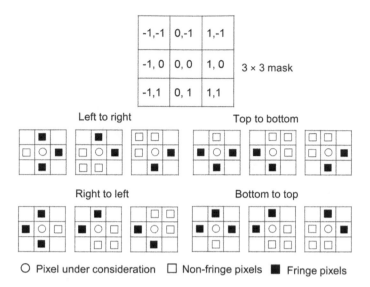

Figure 2.4. Pictorial representation of the elimination conditions for the four scanning directions in the Chen and Taylor algorithm [2].

for elimination for all the four scanning directions are pictorially shown in figure 2.4. The process of erosion is repeated until only a single pixel representing the fringe skeleton remains.

2.3.2 Mask-based methods

The mask-based method proposed by Yatagai *et al* [3] uses a 5×5 mask to detect the local fringe minima. The scanning is performed in four directions, as shown in figure 2.5(a). For the mask shown in figure 2.5(b), the minima conditions to be satisfied for two typical directions 1 and 3 (figure 2.5(c)), respectively, are the following:

$$I(0, 0) + I(0, -1) + I(0, 1) < I(-2, 1) + I(-2, 0) + I(-2, -1) \text{ and}$$

$$I(0, 0) + I(0, -1) + I(0, 1) < I(2, 1) + I(2, 0) + I(2, -1)$$

$$I(0, 0) + I(-1, -1) + I(1, 1) < I(-2, 1) + I(-2, 2) + I(-1, 2) \text{ and}$$

$$I(0, 0) + I(-1, -1) + I(1, 1) < I(1, -2) + I(2, -2) + I(2, -1)$$

Similar conditions also exist for directions 2 and 4. A point is identified as a point on the fringe skeleton if the minima conditions are satisfied for any two or more directions.

Umezaki *et al* [4] proposed a more exhaustive scanning involving eight directions (figure 2.5(c)) for the extraction of fringe skeleton. For row-wise (direction 1), column-wise (direction 2) and diagonal (directions 3 and 4) scanning, the intensity

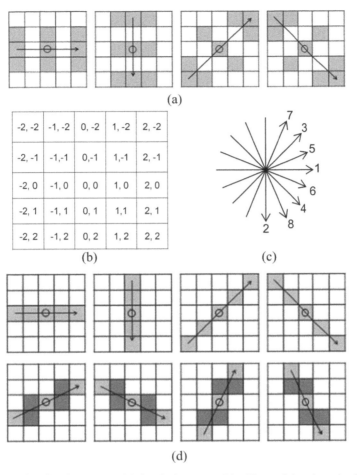

Figure 2.5. (a) Scanning directions along with the pixels processed for fringe minima detection in Yatagai *et al* method [3], (b) a 5 × 5 pixel mask, (c) scanning directions with labels, (d) eight direction scanning scheme for fringe minima detection in Umezaki *et al* method [4].

values are directly considered for the identification of fringe skeleton. For the other four scanning directions (directions 5, 6, 7 and 8), the averages of the pixels shown as dark grey (figure 2.5(d)) are considered. The point under consideration (0, 0) is identified as the fringe skeleton if it is the point of minimum intensity for at least two of the eight scanning directions. The 5 × 5 mask is moved till all the pixels in the image are processed.

2.3.3 Global fringe thinning method

The global fringe thinning method involves two steps—fringe edge detection and skeletonization. Since photoelastic fringe patterns are generally of remarkably high contrast, fringe areas are identified by semi-thresholding. If the background intensity varies over the domain, one may use dynamic thresholding in which the threshold value is selected differently for various regions based on certain criteria automatically or segment the image using tiles and assign appropriate threshold values. One may also use normalization schemes (section 4.10) to stretch the dynamic range of the image and reduce the non-uniform variation of intensities over the image.

Once the fringe areas are identified, the next step is to locate the minimum intensity points (rather than points of zero intensity) within the fringe. To account for varied fringe shapes and curvatures, the image is scanned row-wise, column-wise, diagonal, and cross-diagonal directions. Logical operators are suitably employed to get a fringe skeleton free of noise and discontinuities. The scheme is illustrated in figure 2.6. In each of these scans, the minimum intensity points in the fringe areas (zones between the fringe edges) are identified.

Figure 2.7 summarises the skeletons obtained for various scans. One can see that in each scan, when the scan direction is tangential to a fringe, information is lost, and noise is generated in addition. These are scan direction dependent and a unique process of logical operations is used to get a continuous fringe skeleton free of noise. It can be noted that the 'OR' operation of orthogonal scans removes the gaps in the fringe skeleton, while the final 'AND' operation removes the noise.

Figure 2.8 shows the fringe-skeleton superimposed on the original fringe patterns obtained by various methods. The binary based method is shown in figure 2.8(a). In view of progressive thinning, the approach is basically iterative in nature and the result shown here is obtained after 70 iterations. Spurious skeletons are identified in zones where the fringes are thicker.

Since the mask-based methods of Yatagai and Umezaki do not identify fringe edges, fringe points are identified even in non-black regions. Further, these two schemes do not guarantee fringe skeleton of one pixel width. In dense fringe zones, the skeletons look nice and brighter because of this reason. Only the global fringe thinning algorithm of Ramesh and Pramod [5] guarantees fringe skeleton of one pixel width and its performance stands out. If one uses, fringe normalized image (section 4.10) for processing, its performance is observed to be even better, as shown in figure 2.8(e).

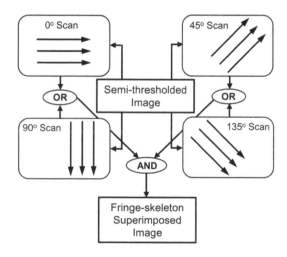

Figure 2.6. Global fringe thinning algorithm of Ramesh and Pramod [5].

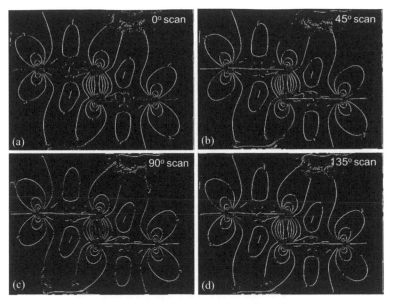

Figure 2.7. Fringe skeleton obtained for various scan directions: (a) 0-deg scan, (b) 45-deg scan, (c) 90-deg scan, (d) 135-deg scan.

2.4 Need of fracture mechanics to quantify cracks

In the beginning of the nineteenth century, many structures failed in service even though they were designed based on the best design practices of that time. During World War II, several Liberty ships that were made by welding broke into two in the North Atlantic sea. This along with several other catastrophic failures prompted scientists to look back and see what could have caused such failures [7].

A study by Inglis for a small elliptical hole in an infinite plate subjected to uniaxial tension (σ_{xx}) has shown that for an elliptical hole having semi-major axis of length a and semi-minor axis of length b, the maximum stress developed is given by,

$$\sigma_{\max} = \sigma_{xx}\left(1 + \frac{2a}{b}\right) \tag{2.2}$$

Equation (2.2) indicates that when one has a crack, i.e., as b tends to zero, the maximum stress could theoretically become infinite! The result of this study proves how cracks could be extremely dangerous; more than what the scientists thought of! Further, the simple representation of discontinuities contributing to an increase in stress concentration is no longer sufficient, and a new field of mechanics known as *fracture mechanics* came into existence through research done by several investigators across the world [8]. With such new developments, the experimentalists have to determine the stress field parameters in the neighbourhood of the crack-tip. Calculation of the stress concentration factors (SCFs) from photoelastic results is quite simple, as seen in chapter 1. In a similar vein, photoelasticity is also

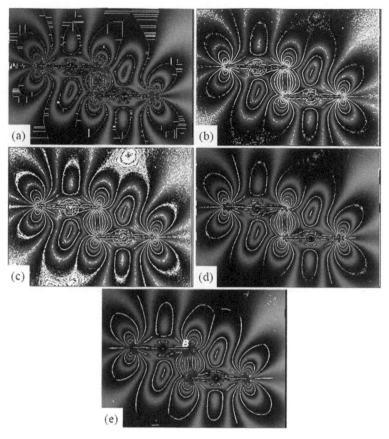

Figure 2.8. Fringe-skeleton superimposed image obtained by various fringe thinning methodologies: (a) binary-based method, (b) Yatagai *et al* method, (c) Umezaki *et al* method, (d) global fringe thinning by Ramesh and Pramod on original image, (e) global fringe thinning using the normalized image.

well suited to evaluate the crack-tip stress field parameters though it is slightly involved.

Welding has evolved over the years and, with close quality control, even critical structures such as bridge trusses are made through welding these days due to its ease of manufacturing. It is also a common practice now to have in-built crack arresters in any structural design. All these have been possible through the development of fracture mechanics. Photoelasticity has made a significant contribution in the development of the governing equations of fracture mechanics. The development of stress field equations in fracture mechanics is discussed next.

2.5 Development of the stress field equation in the neighbourhood of the crack-tip

Westergaard in 1939 [9] proposed the complex stress function method for solving the opening mode problem, which played a significant role in the development of linear

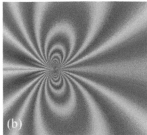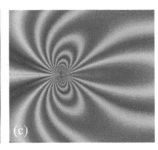

Figure 2.9. Theoretically simulated crack-tip isochromatics using P_Scope® [12]: (a) classical Westergaard solution (equation (2.4)). The figure also shows the convention that the crack-tip is taken as the origin and x-axis aligns with the crack; (b) Irwin's modification of the Westergaard's equation (equation (2.5)); (c) generalised Westergaard equation (equation (2.7)). Note that in (a) along the crack axis the fringe order is zero, in (b) it is a constant and only in (c) does the fringe order vary, indicating that maximum shear stress can vary along the crack-axis. Courtesy: P_Scope® software, IIT Madras [12].

elastic fracture mechanics. Westergaard in his original paper, demonstrated that 'in a restricted but important group of cases the normal stresses and the shearing stress in the directions x and y can be stated in the form'

$$\begin{Bmatrix} \sigma_x \\ \sigma_y \\ \tau_{xy} \end{Bmatrix} = \begin{Bmatrix} \mathrm{Re}\, Z - y\mathrm{Im}Z' \\ \mathrm{Re}\, Z + y\mathrm{Im}Z' \\ -y\, \mathrm{Re}\, Z' \end{Bmatrix} \tag{2.3}$$

where $Z = Z(z) = Z(x + iy) = \mathrm{Re}\, Z + i\, \mathrm{Im}\, Z$ and $Z' = Z'(z) = \mathrm{Re}\, Z' + i\, \mathrm{Im}\, Z'$

Figure 2.9(a) shows the representative coordinate system that takes the crack-tip as the origin and crack axis as x-axis for defining the stress as well as displacement field equations. In 1958, Wells and Post [10] performed the very first test of photoelastic studies on running cracks using a Cranz–Schardin camera. The experiments revealed that the disturbance from a running crack is proportional to crack length when the external boundaries are sufficiently far away. It is also observed that in the absence of inelastic effects, the stress distribution in the vicinity of the running crack approximates to that of an equivalent statically loaded cracked plate extended by a fixed displacement.

Irwin [11] observed that the concept of SCF is meaningless for a sharp crack and introduced the concept of stress intensity factor (SIF) that linked the crack length and the stress in a novel way to characterise the influence of the crack. In all, he classified three modes of loading: opening mode, where the crack faces open (Mode-I), in-plane shearing mode where the crack faces slide in the plane (Mode-II) and out-of-plane shear where the crack faces slide out of plane (Mode-III). In a generic problem, combination of these modes in various degrees can be witnessed. Of these, Mode-I is quite common and considered as the most important. Unlike the SCF, which is dimensionless, SIF has the units of MPa $\sqrt{\mathrm{m}}$. The classical Westergaard equations in terms of SIF (K_I) for Mode-I loading are,

$$\left\{\begin{matrix} \sigma_x \\ \sigma_y \\ \tau_{xy} \end{matrix}\right\} = \frac{K_I}{\sqrt{2\pi r}} \cos\frac{\theta}{2} \left\{\begin{matrix} 1 - \sin\dfrac{\theta}{2} \sin\dfrac{3\theta}{2} \\ 1 + \sin\dfrac{\theta}{2} \sin\dfrac{3\theta}{2} \\ \sin\dfrac{\theta}{2} \cos\dfrac{3\theta}{2} \end{matrix}\right\} \tag{2.4}$$

The non-interacting crack-tips 'A' and 'D' shown in figure 2.2(a) are loaded in Mode-I and show that the isochromatics are symmetric about the crack axis and they have a prominent tilt about the vertical axis. Irwin [13] in 1958 reported that the photoelastic fringes observed by Wells and Post [10] on centre cracked panels did not match the isochromatic contours predicted by Westergaard's solution (figure 2.9(a)), in which the fringes are symmetric about the crack axis as well as perpendicular to the crack axis. He suggested that a correction needs to be applied to the stress field equation for a finite body and introduced the constant non-singular stress term (σ_{ox}) to the σ_x term. Thus, the modified Westergaard's equation is,

$$\left\{\begin{matrix} \sigma_x \\ \sigma_y \\ \tau_{xy} \end{matrix}\right\} = \frac{K_I}{\sqrt{2\pi r}} \cos\frac{\theta}{2} \left\{\begin{matrix} 1 - \sin\dfrac{\theta}{2} \sin\dfrac{3\theta}{2} \\ 1 + \sin\dfrac{\theta}{2} \sin\dfrac{3\theta}{2} \\ \sin\dfrac{\theta}{2} \cos\dfrac{3\theta}{2} \end{matrix}\right\} + \left\{\begin{matrix} -\sigma_{ox} \\ 0 \\ 0 \end{matrix}\right\} \tag{2.5}$$

This simple correction remarkably changed the appearance of the theoretically simulated isochromatics and they show a tilt (fringes are forwarded tilted in this case) as observed in the experiments (figure 2.9(b)). On scrutiny, one can observe that the angle of tilt (the angle of the line joining the crack-tip and the farthest point on a fringe measured from the crack axis) increases and becomes 90° (figure 2.9(b)) close to the crack-tip as observed in the experiment. Thus, photoelasticity has been invaluable in the early development of fracture mechanics as the fringe geometry prompted the researchers to look for answers. Continuing this spirit, in 1979, Sanford [14] brought out the inadequacy of Irwin's modifications and other modified Westergaard equations for solving problems in which the crack-tip is closer to a free boundary or situated in a stress gradient field. In such cases, several fringe orders can occur along the crack axis indicating a variation of maximum shear stress along the crack axis. In developing fracture mechanics equations, one usually assumes that the in-plane shear stress along the crack axis is zero for Mode-I loading due to symmetry. The contribution of Sanford was that the condition that is usually imposed to make the in-plane shear stress zero along the crack axis due to symmetry also restricted the maximum shear stress to be zero along the crack axis unknowingly! This needs to be corrected and he proposed the Generalized Westergaard equation, which is given as,

$$\begin{Bmatrix} \sigma_x \\ \sigma_y \\ \tau_{xy} \end{Bmatrix} = \begin{Bmatrix} \mathrm{Re}\, Z - y\mathrm{Im}Z' - y\mathrm{Im}\, Y' + 2\,\mathrm{Re}\, Y \\ \mathrm{Re}\, Z + y\mathrm{Im}Z' + y\mathrm{Im}\, Y' \\ -y\,\mathrm{Re}\, Z' - y\,\mathrm{Re}\, Y' - \mathrm{Im}\, Y \end{Bmatrix} \qquad (2.6)$$

where $Z(z) = \sum_{j=0}^{J} C_{2j} z^{(2j-1)/2}$ and $Y(z) = \sum_{j=0}^{J} C_{2j+1} z^{j}$.

The equations when used to calculate the maximum shear stress along the crack axis predicted a variation (figure 2.9(c)) but the in-plane shear stress remained zero satisfying the basic requirement [15]. Here again, the geometry of the fringe patterns has prompted Sanford to look for answers. This idea is well exploited in case studies discussed in chapter 5 to use the geometry of fringe patterns observed in photoelasticity as a guide to seek answers for modelling the problem correctly.

In 1987, Atluri and Kobayashi [16] reported another set of multi-parameter stress field equations for the combined loading case (Mode-I and Mode-II) in a general form. However, these equations had a few typographical errors which were later corrected by Ramesh *et al* [17]. The corrected Atluri and Kobayashi multi-parameter stress field equation [17] is given as,

$$\begin{Bmatrix} \sigma_x \\ \sigma_y \\ \tau_{xy} \end{Bmatrix} = \sum_{n=1}^{\infty} \frac{n}{2} A_{\mathrm{I}n} r^{\frac{n-2}{2}} \begin{Bmatrix} \left[\left(2 + \frac{n}{2} + (-1)^n\right)\cos\left(\frac{n}{2}-1\right)\theta - \left(\frac{n}{2}-1\right)\cos\left(\frac{n}{2}-3\right)\theta \right] \\ \left[\left(2 - \frac{n}{2} - (-1)^n\right)\cos\left(\frac{n}{2}-1\right)\theta + \left(\frac{n}{2}-1\right)\cos\left(\frac{n}{2}-3\right)\theta \right] \\ \left[\left(\frac{n}{2}-1\right)\sin\left(\frac{n}{2}-3\right)\theta - \left\{\frac{n}{2}+(-1)^n\right\}\sin\left(\frac{n}{2}-1\right)\theta \right] \end{Bmatrix}$$
$$- \sum_{n=1}^{\infty} \frac{n}{2} A_{\mathrm{II}n} r^{\frac{n-2}{2}} \begin{Bmatrix} \left[\left(2 + \frac{n}{2} - (-1)^n\right)\sin\left(\frac{n}{2}-1\right)\theta - \left(\frac{n}{2}-1\right)\sin\left(\frac{n}{2}-3\right)\theta \right] \\ \left[\left(2 - \frac{n}{2} + (-1)^n\right)\sin\left(\frac{n}{2}-1\right)\theta + \left(\frac{n}{2}-1\right)\sin\left(\frac{n}{2}-3\right)\theta \right] \\ \left[-\left(\frac{n}{2}-1\right)\cos\left(\frac{n}{2}-3\right)\theta - \left\{-\frac{n}{2}+(-1)^n\right\}\cos\left(\frac{n}{2}-1\right)\theta \right] \end{Bmatrix} \qquad (2.7)$$

The coefficients $A_{\mathrm{I}1}$, $A_{\mathrm{I}2}$, ... and $A_{\mathrm{II}1}$, $A_{\mathrm{II}2}$, ... are the unknown Mode-I and Mode-II parameters, respectively. The '-$\sigma_{\mathrm{o}x}$' stress term introduced by Irwin is also embedded in this equation and it is labelled as T-stress in numerical literature of fracture mechanics. The SIFs and T-stress can be computed from the coefficients as $K_{\mathrm{I}} = A_{\mathrm{I}1}\sqrt{2\pi}$, $K_{\mathrm{II}} = -A_{\mathrm{II}1}\sqrt{2\pi}$ and T-stress, $T = 4A_{\mathrm{I}2}$.

Fracture theories that predict crack growth or failure, require the calculation of SIFs from the field. Irwin [13], while discussing the work of Wells and Post, also proposed a simple one-point data analysis to determine both Mode-I SIF (K_{I}) as well as $\sigma_{\mathrm{o}x}$. This is termed as a two-parameter solution and dominated the evaluation of SIF by photoelasticity in the early stages. As noted earlier, when several fringes cross the crack axis, this solution is insufficient, and one needs to use a multi-parameter equation even if the focus is on the first few parameters. Ramesh *et al* [17] have shown the elegance of corrected Atluri and Kobayashi multi-parameter stress field equations in the estimation of crack-tip stress field parameters using isochromatic data from photoelasticity. The advantage of the corrected Atluri and Kobayashi multi-parameter equation is that the terms are generalized. It is relatively easier to include as many terms as required [17]. Thus, the accuracy of the stress field solution in a body with finite boundaries can be improved significantly. Ramesh *et al* [18] have noted that the equations developed by Williams, Sanford, and Atluri and Kobayashi are similar and established an interrelationship between the coefficients.

2.6 Study of interacting cracks

Structural components develop cracks at multiple locations during service. Though in early developments of fracture mechanics, one is concerned with only a single crack, the need for understanding the interaction effects of cracks has become a necessity now. The simplest of these studies is what can happen if two cracks come close. This effect can be estimated through changes in the SIF at the individual crack-tips.

Analytical and experimental works to study different types of crack interaction problems are reported in the literature [19–23]. A detailed study on multiple crack interaction under uniaxial loading was done by Kachanov [19] using an analytical method based on the superposition principle. Various representative geometries including two collinear, stacked, and asymmetric cracks are examined and the effects produced by these configurations are brought out in that study. Of these, asymmetric cracks are found to produce maximum amplification in SIF. These observations are easy to visualize if one does a photoelastic analysis of these crack configurations (figure 2.10). Even a simple visualization of isochromatic contours gives a qualitative idea of the level of stress intensification. Among the various configurations, only the asymmetric configuration has the highest intensity of fringes in both uniaxial and biaxial loading. Another interesting aspect is that stress shielding (reduction in SIFs) occurs when one moves from uniaxial to biaxial loading for the same crack-configuration.

In many practical applications, the stress field is usually biaxial, and it is important to study the interaction effect of cracks under biaxial loading. Cruciform specimen has been commonly used for biaxial testing and for fatigue life assessment of components under biaxial loading [24–27]. Early studies by researchers were confined to the study of the influence of biaxial loading only for a single crack [27, 28]. Zakeri *et al* [27] used photoelasticity to study the lateral load effect on a centre cracked cruciform specimen using a two-parameter stress field equation. Theocaris and Spyropoulos [28] determined the SIF for a slant crack under biaxial loading with the help of a series form of stress field equation using photoelasticity. The crack interaction effects under biaxial loading have been studied comprehensively by Vivekanandan and Ramesh [29]. In the subsequent sections (sections 2.7–2.10), the methodology for quantitative evaluation of crack-tip stress field parameters is discussed and is applied to the study of interacting cracks.

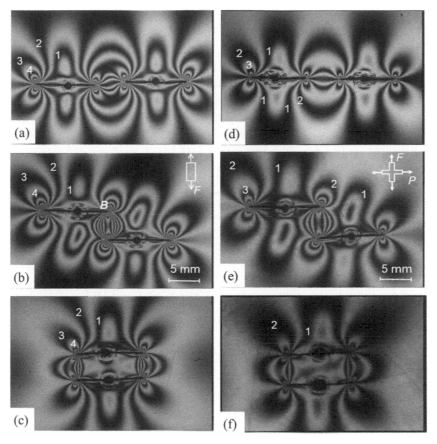

Figure 2.10. The dark field isochromatic fringe pattern: uniaxial tensile load of 1186 N, (a) collinear, (b) asymmetric, (c) stacked crack configurations; biaxial ratio of 0.25 ($P = 294$ N, $F = 1186$ N): (d) collinear, (e) asymmetric, (f) stacked crack configurations.

2.7 Evaluation of stress-field parameters using non-linear least squares analysis

2.7.1 Problem formulation

The stress-optic law relates the fringe order N and the in-plane principal stresses σ_1 and σ_2 as,

$$\frac{NF_\sigma}{t} = \sigma_1 - \sigma_2 \tag{2.8}$$

where F_σ is the material stress fringe value and t is the thickness of the specimen. For a plane stress problem, the stress components σ_x, σ_y and τ_{xy} are related to the principal stresses as,

$$\sigma_1, \sigma_2 = \frac{\sigma_x + \sigma_y}{2} \pm \sqrt{\frac{(\sigma_x - \sigma_y)^2}{4} + \tau_{xy}^2} \tag{2.9}$$

Substituting equation (2.9) in (2.8), one can define an error function g for the mth data point given as,

$$g_m = \left\{ \frac{\sigma_x - \sigma_y}{2} \right\}_m^2 + (\tau_{xy})_m^2 - \left\{ \frac{N_m F_\sigma}{2t} \right\}^2 \tag{2.10}$$

Equation (2.10) is a non-linear equation in terms of the unknown parameters A_{I1}, A_{I2}, ... and A_{II1}, A_{II2}, ... Initial estimates are made for these unknown parameters and substituted in equation (2.10) and most often the error will not be zero since the estimates are not accurate. The estimates are then corrected using an iterative process based on Taylor's series expansion of g_m. Finally, one can arrive at solving a simple matrix problem represented as,

$$\{g\}_i = -[b]_i \{\Delta A\}_i \tag{2.11}$$

where $\{g\}_i$, $[b]_i$ and $\{\Delta A\}_i$ represent the following matrices

$$\{g\}_i = \begin{Bmatrix} g_1 \\ g_2 \\ \vdots \\ g_m \\ \vdots \\ g_M \end{Bmatrix}_i \quad [b]_i = \begin{bmatrix} \frac{\partial g_1}{\partial A_{I1}} & \frac{\partial g_1}{\partial A_{I2}} & \cdots & \frac{\partial g_1}{\partial A_{Ik}} & \frac{\partial g_1}{\partial A_{II1}} & \frac{\partial g_1}{\partial A_{II2}} & \cdots & \frac{\partial g_1}{\partial A_{IIl}} \\ \frac{\partial g_2}{\partial A_{I1}} & \frac{\partial g_2}{\partial A_{I2}} & \cdots & \frac{\partial g_2}{\partial A_{Ik}} & \frac{\partial g_2}{\partial A_{II1}} & \frac{\partial g_2}{\partial A_{II2}} & \cdots & \frac{\partial g_2}{\partial A_{IIl}} \\ \vdots & \vdots & & \vdots & \vdots & \vdots & & \vdots \\ \frac{\partial g_M}{\partial A_{I1}} & \frac{\partial g_M}{\partial A_{I2}} & \cdots & \frac{\partial g_M}{\partial A_{Ik}} & \frac{\partial g_M}{\partial A_{II1}} & \frac{\partial g_M}{\partial A_{II2}} & \cdots & \frac{\partial g_M}{\partial A_{IIl}} \end{bmatrix}_i \quad \{\Delta A\}_i = \begin{Bmatrix} \Delta A_{I1} \\ \Delta A_{I2} \\ \vdots \\ \Delta A_{Ik} \\ \Delta A_{II1} \\ \Delta A_{II2} \\ \vdots \\ \Delta A_{IIl} \end{Bmatrix}_i$$

In the above equations, i represents the ith iteration and M is the total number of data points. The number of Mode-I and Mode-II parameters are represented by k and l, respectively. From equation (2.11), one gets the following equation,

$$[b]_i^T \{g\}_i = -[b]_i^T [b]_i \{\Delta A\}_i \tag{2.12}$$

Solving this, one can get $\{\Delta A\}_i$. Using this, one can update the coefficients as,

$$\{A\}_{i+1} = \{A\}_i + \{\Delta A\}_i$$

2.7.2 Convergence criteria

The iteration must be stopped by an appropriate convergence criterion. A fringe order error minimization criterion is used in which the fringe orders corresponding to the selected data points are calculated theoretically using the newly calculated values of $\{A\}_{i+1}$ during every iteration step and are compared with the experimentally obtained fringe orders. The convergence criterion is satisfied if,

$$\frac{\sum |N_{\text{theory}} - N_{\text{exp}}|}{\text{Total no. of data points}} \leqslant \text{convergence error} \tag{2.13}$$

where N_{theory} is the theoretical fringe order and N_{exp} is the actual fringe order observed in the experiment. *A priori* one does not know how many terms are required to model the stress field for a given problem. It is always recommended to start from a minimum number of parameters and to progressively increment them by one as the problem demands. The initial value of convergence error considered is 0.5, which is progressively decreased until it is 0.05 or less. Below this value, the data points usually match well with the reconstructed image. The parameters thus obtained are used to recalculate $(\sigma_1-\sigma_2)$ using equations (2.7) and (2.9) and fringe order N at every point in the data field is evaluated using equation (2.8). Using the intensity equation for the respective dark, bright or mixed-fields, the fringe patterns are theoretically reconstructed with the data points echoed back and compared with the experimental image. Only when this comparison is good is the solution accepted. This makes the SIF evaluation procedure self-consistent.

2.8 Subtleties in the evaluation of crack-tip stress field parameters

The over-deterministic method utilises the data from the field to evaluate the parameters in a least squares sense. In principle, one should be able to take data randomly from the field and if they are properly conditioned statistically, they should lead to a unique solution. As data collection from the field was quite cumbersome before the advent of digital photoelasticity, the two-parameter method of Irwin was widely used as it requires data from just one point closer to the crack-tip. However, with the development of whole field data extraction methods like PST or by processing the data from colour image processing, which provided data at every pixel, the focus of the researchers shifted using them as such, randomly for evaluation of the crack-tip stress field parameters. However, such random data collection does not always guarantee the correct solution. The key understanding developed over the years has revealed that the numerical solution needs to be guided with proper choice of data that reflect the geometric features of the fringes precisely in the close vicinity of the crack-tip. In many instances, the geometric features of the photoelastic fringes have guided theoretical development including that of fracture mechanics. Here, the fringe skeletonization plays a useful role and even if data is available over the field from PST or colour image processing, it is profitable to extract data from the fringe skeletons such as a mixed-field image (figure 2.2(b)).

Another subtle aspect is that in Mode-II, the second term is zero in equation (2.7). Unless this aspect is appropriately handled in coding, it can lead to division by zero and a mixed-mode evaluation of the fracture parameters is not possible. In fact, such issues have hampered several groups working on photoelasticity to be satisfied with the two-parameter solution of Irwin and not adopting to use multi-parameter solution for fracture mechanics. The paper by Ramesh *et al* [17] has clearly prompted more researchers to adopt the use of multi-parameter solution in recent times. This adaptability is further simplified by the development of the software PSIF [30] dedicated to the photoelastic evaluation of the SIF.

2.8.1 Data collection module of PSIF

The PSIF software has an excellent data collection module, which by the click of relevant features can extract data from the fringe thinned image as well as along fringe contours from whole field data collection methods that provide fringe data (.frn file) over the field (figure 2.11). The data collection module is common for a

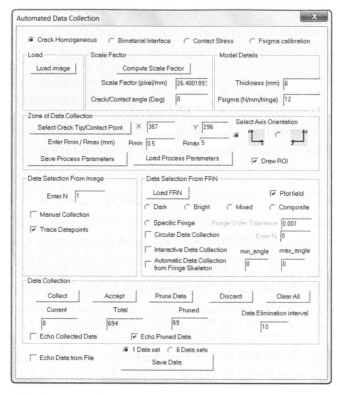

Figure 2.11. Data collection menu of the software PSIF [30] showing various features. Courtesy: PSIF software, IIT Madras [30].

class of problems that require positional coordinates and the corresponding fringe order for further processing, and these can be selected from the menu. One essentially gets the image as an assembly of pixels and for processing the data, one must have them in physical coordinates, and this is achieved by evaluating the scale factor that is decided by the set of lenses and extenders used for image grabbing. As the zone of data collection is problem dependent, they can also be specified suitably, and so can the relevant coordinate system depending on the orientation of the crack. Then the data can be collected from the fringe-skeletonised image or extracted from whole field fringe data provided as .frn file by PST or TFP (chapter 4). In these by the selection of the respective field, fringe skeleton data is extracted automatically. Many of the menu items shown in figure 2.11 are self-explanatory.

2.8.2 Crack-tip refinement

The origin is the crack-tip and it is usually picked interactively by the user and hence prone to error. The accuracy of the evaluated SIF depends on the careful selection of the crack-tip location. Neethi and Ramesh [31] had observed for a bi-material interface crack problem that even a variation by 1 or 2 pixels in the crack-tip coordinates can alter the solution and the convergence of the iteration scheme. For correcting the crack-tip error, one can consider the crack-tip also as an unknown and determine it as part of the least squares solution. Although, this appears straight forward, it adds unnecessary computational complexities.

Following, the success of the simple grid search methodology proposed by Neethi and Ramesh to detect the crack-tip location, it is now customary to use it for all crack problems as well as in contact stress analysis where origin needs to be identified interactively [32]. In this method, the initial origin (crack-tip location) is picked by the user from the image interactively. With that as origin, non-linear least squares analysis (or linear least squares for Hertzian contact) is performed to identify the desired number of parameters required as discussed before, for correctly modelling the stress field near the crack-tip.

Once the number of parameters is frozen, a 5×5 pixels mask surrounding the initially estimated origin (crack-tip) is considered, and the convergence errors are evaluated at each position of the origin in the 5×5 mask. The next location of the possible origin is identified as the pixel location based on the least convergence error within the 5×5 pixels mask. If the identified pixel coordinate is not at the centre of the 5×5 mask, keeping this identified pixel as the origin (crack-tip), another 5×5 pixels mask is considered and the convergence error is computed for the pixel mask. This process is continued, until the estimated origin with the least convergence error becomes the centre of the pixel mask. A flow chart for the grid search method for identifying crack-tip coordinates is shown in figure 2.12. This can be done in PSIF with the click of a menu option.

2.9 Experimental evaluation of stress field parameters for interacting cracks

In this section, the interaction effect of two cracks embedded in a finite plane homogeneous region under uniaxial as well as biaxial tensile loads is investigated using photoelasticity. The crack tip stress fields in such cases are usually overly complex and one must use a large number of parameters in the stress field equations to get an accurate estimate of the SIFs.

2.9.1 Uniaxial and biaxial loadings

Three different crack configurations viz., collinear, asymmetric, and stacked are considered in uniaxial and biaxial loadings (figure 2.10). For uniaxial loading, a load of 1186 N is applied and for biaxial loading, a cruciform specimen is loaded at a biaxial ratio of 0.25 with $F = 1186$ N and $P = 294$ N. The epoxy used has been tested for its elastic properties and the Young's modulus is found to be 2650 MPa and the Poisson's ratio is 0.43. The material stress fringe value is obtained as 12.5 N/mm/ fringe. Stress intensity factors are estimated from the crack tip isochromatic fringe patterns obtained from photoelasticity (figure 2.10). It is to be observed and noted that at interacting tips, the fringes exhibit mixed-mode behaviour, whereas non-interacting tips remain predominantly in Mode-I. This gives the first qualitative information about the interaction effect. Corrected Atluri–Kobayashi stress field

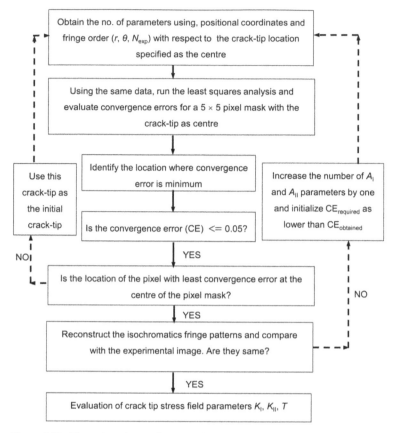

Figure 2.12. Flow chart for semi-automatic identification of crack-tip coordinates.

equations in conjunction with over-deterministic least squares approach [16, 17] are used for SIF evaluation. Advanced digital image processing techniques are used for processing the experimental images of isochromatic fringe patterns. The solution is verified by theoretically reconstructing the fringe patterns and comparing them with the experimental fringe patterns. For one of the crack-tips (figure 2.10(b), tip B), the utility of the non-linear least squares method is presented in some detail. The fringe skeletons obtained from the fringe thinning process for figure 2.10 is shown in figure 2.8(e), which is used for data collection.

Figure 2.13 shows the close-up of experimental fringe pattern for crack-tip B (interacting crack tip) along with the theoretically reconstructed fringe patterns from 5 to 9 parameters each for Mode-I and Mode-II with the data points echoed back. The number of data points collected is 152 and the zone of data collection is $0.18 < r/a < 1.02$. Below the crack line, due to crack-tip interaction, the fringes are similar to what is seen in a circular disc under diametral compression (figure 1.8). The data is collected only up to the half of the disc like region. From the reconstructed patterns, it is observed that an eight-parameter solution is appropriate for this problem. The summary of the SIF values obtained for each of these cases is given in table 2.1.

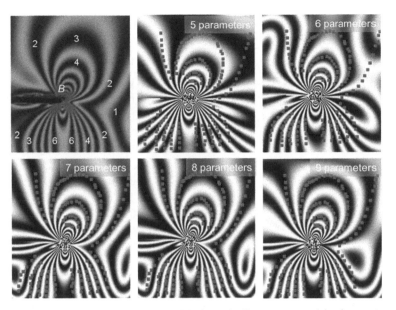

Figure 2.13. Experimental fringe pattern along with theoretically reconstructed isochromatics with experimental data points in red echoed for various numbers of stress field parameters for interacting tip (tip B) of the uniaxial asymmetric configuration, load = 1186 N.

For the evaluation of the interaction effect, the SIFs of the interacting cracks are compared with the SIF values of a single central crack under pure Mode-I loading ($K_0 = \sigma\sqrt{\pi a}$) for the same applied loads. The normalized SIF values for the three crack configurations under uniaxial and biaxial loads are shown in table 2.2. It is to be understood that under no interaction effect, $K_I/K_0 = 1$ and $K_{II}/K_0 = 0$. Deviation from these values represents either amplification or shielding in SIFs. The stacked configuration has a shielding effect and the second crack acts like a stress relieving hole. Data provided in table 2.2 indicate that in collinear and asymmetric crack configurations under Mode-I loading, the interaction causes amplification in Mode-I SIFs (K_I) at the crack-tips. In biaxial loading, there is a shielding effect for the asymmetric configuration. This is an interesting result, but there can be amplification too, which is brought out with an exhaustive study of crack interaction with improved centre-crack, as discussed next.

2.9.2 Study of angularly oriented cracks subjected to biaxial loading

Among the various specimens for biaxial testing, cruciform specimens are convenient to load, and the biaxial ratios can be changed easily. For various types of cruciform specimens, Rajiv and Ramesh [33] obtained the stress field numerically and postprocessed the results to get photoelastic fringe patterns. As photoelasticity directly gives the principal stress difference, for equi-biaxial loading, the zeroth fringe order (that appears as black in colour) represents the zone of biaxiality in the specimen. The focus is to choose a specimen that has the largest size of this equi-

Table 2.1. Summary of stress field parameters evaluated from photoelastic data for tip B of uniaxial asymmetric crack configuration for a load of 1186 N.

Parameters	5 terms	6 terms	7 terms	8 terms	9 terms
K_I, MPa(m)$^{1/2}$	1.05	0.907	0.811	0.748	0.750
K_{II}, MPa(m)$^{1/2}$	0.055	0.061	0.035	0.066	0.066
σ_{ox}, MPa	−0.0184	2.412	4.068	5.148	5.148
A_{I1}, MPa(mm)$^{1/2}$	13.274	11.440	10.227	9.434	9.460
A_{I2}, MPa	0.001	−0.603	−1.017	−1.287	−1.287
A_{I3}, MPa(mm)$^{-1/2}$	−0.723	−0.527	−0.180	0.293	0.318
A_{I4}, MPa(mm)$^{-1}$	0.025	0.125	0.125	0.118	0.111
A_{I5}, MPa(mm)$^{-3/2}$	0.013	−0.082	−0.108	−0.091	−0.086
A_{I6}, MPa(mm)$^{-2}$		0.011	0.020	0.010	0.010
A_{I7}, MPa(mm)$^{-5/2}$			−0.004	0.001	0.000
A_{I8}, MPa(mm)$^{-3}$				−0.001	−0.000
A_{I9}, MPa(mm)$^{-7/2}$					−0.000
A_{II1}, MPa(mm)$^{1/2}$	−0.692	−0.768	−0.446	−0.831	−0.834
A_{II2}, MPa	0.000	0.000	0.000	0.000	0.000
A_{II3}, MPa(mm)$^{-1/2}$	−0.214	−0.261	−0.362	−0.296	−0.284
A_{II4}, MPa(mm)$^{-1}$	0.012	0.071	0.150	0.057	0.045
A_{II5}, MPa(mm)$^{-3/2}$	−0.005	−0.007	0.009	0.029	0.029
A_{II6}, MPa(mm)$^{-2}$		0.000	−0.016	−0.033	−0.034
A_{II7}, MPa(mm)$^{-5/2}$			0.003	0.007	0.007
A_{II8}, MPa(mm)$^{-3}$				−0.001	−0.001
A_{II9}, MPa(mm)$^{-7/2}$					0.000
Convergence error	0.422	0.203	0.107	0.084	0.085

Table 2.2. Normalized SIFs for interacting cracks.

	Uniaxial		Biaxial ($\beta = 0.25$)	
Configuration	K_I/K_0	K_{II}/K_0	K_I/K_0	K_{II}/K_0
Collinear	1.49	0.05	–	–
Stacked	0.93	0.18	0.67	0.05
Asymmetric	1.50	0.13	0.98	0.19

biaxial zone. The study suggested the use of a simple cut type cruciform specimen with a suitable fillet.

For this study, a specimen of total length 300 mm and an arm width of 90 mm as shown in figure 2.14(a) is found to be suitable for biaxial tests with the loading system available at the laboratory. In this, P and F are the forces applied in the horizontal and

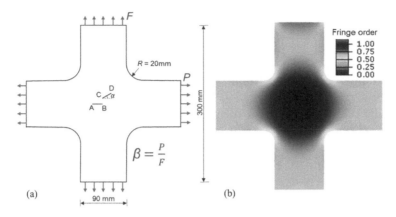

Figure 2.14. (a) Overall dimensions of the cruciform specimen with cracks of lengths 20 mm. The crack-tip 'C' is at a distance of 5 mm perpendicular to the crack-tip 'B'. (b) Post processed FE results of isochromatics for equi-biaxial load showing a large area of biaxial loading at the centre of the specimen. Reprinted from [29] with permission from Elsevier.

the vertical directions, respectively; β is the biaxial ratio which is the ratio of the horizontal load to the vertical load and α is the inclination angle of the second crack with respect to the horizontal crack. The crack tips are labelled as A, B, C and D. Figure 2.14(b) shows the photoelastic fringe pattern for the cruciform specimen under equi-biaxial load ($P = 211.5$ N, $F = 211.5$ N) obtained from numerically evaluated stress fields. Amongst the various trial fillet radii, for a fillet radius of 20 mm, the biaxial zone is found to be the maximum and a circle of 90 mm diameter can fit in that zone. Studies by Ebrahim *et al* [34] confirm the importance of modifying the geometric parameters in making the central zone of the specimen to be the failure zone.

A commercially available high density polyurethane sheet of 5 mm thickness is used to make the specimens. The material is linear until a stress value of about 2 MPa and the corresponding Young's modulus in this zone is 50.9 MPa. Using a circular disc under diametral compression as the model, the material stress fringe value is found to be 8 N/mm/fringe. Two fine cracks each of 20 mm length at different orientation angles (α) are cut in the biaxial zone using sharp industrial razor blades. One of the advantages of using this specimen is that one can easily insert an industrial razor blade by which a fine crack of width 0.1 mm can be obtained. The crack thus introduced is straight and is also quite sharp compared to what was achievable by a fine hack saw, as in figure 2.10.

The loading system has a provision to accommodate different loads in the horizontal and vertical directions. Thus, the biaxial ratio β, can be easily varied with the loading system. The loaded specimens are viewed in a circular polariscope arrangement. The loads are selected in such a way that the crack opening displacements are small and there are sufficient fringes for data collection by digital photoelastic methods. A stress value of 0.47 MPa is found to meet this requirement for all biaxial ratios. This corresponds to an axial load of 211.5 N. Appropriate load is applied in the horizontal direction to alter the force value P to achieve biaxial ratios from 0 to 1 in steps of 0.25. Initially, a single centre crack is introduced, and the tests are conducted for various biaxial ratios. Then the second crack is introduced at an angle and similar specimens are made for various angles to perform the tests. In a normal course, the load applied is quite small and fringes are not sufficient for data reduction. However, by employing fringe multiplication, the data has been enhanced, as shown in figure 2.1, for a sample case of the angular crack at 15° for a biaxial ratio of 0.5.

It has been mentioned that it is desirable to collect data along the fringe skeleton, which has been done for the previous example (figure 2.13) in one fashion and in this example, a more systematic study has been carried out. Modern data collection with uncertainty calculations is implemented in PSIF to process the fringe patterns. The trace option in PSIF software allows one to collect all the points forming the fringe (figure 2.15(a)). One can also instantly extract fringe

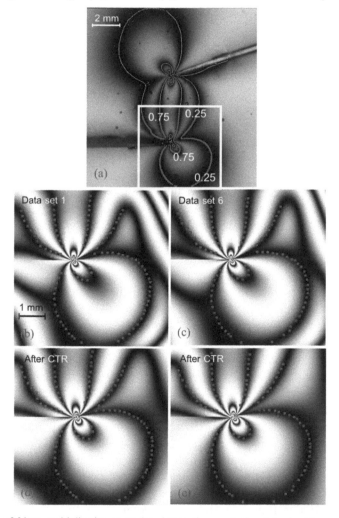

Figure 2.15. Use of fringe multiplication to analyse interacting crack-tips: (a) mixed-image showing zone of analysis with continuous data extracted. Reconstructed fringe pattern with 9 Mode-I and 9 Mode-II parameters with data points echoed: (b) data set 1, (c) data set 6, (d) data set 1 after CTR, (e) data set 6 after CTR.

skeleton information from the whole field data of fringe patterns using advanced digital photoelastic techniques.

The results must be independent of the choice of data. A novel approach is implemented in PSIF that allows skipping of data at regular intervals from the continuous data (figure 2.15(a)); thus, it can provide a distinct data set yet preserve the complete geometry of the fringe patterns. Six data sets are generated and the reconstructed pattern with nine parameters each of Mode-I and Mode-II for the initial crack-tip ((315, 429) in pixels) selected is shown in figure 2.15(b) and (c) for two datasets 1 and 6. The Mode-I SIF is $K_I = 0.078$ MPa$\sqrt{m}$ for the data set 1 and the convergence error obtained is 0.013, which is the least for all the six data sets processed. Crack-tip refinement (CTR) [31] is done for dataset 1, which showed a small shift of the order of two to three pixels in the crack-tip, which in physical units is about 0.05 mm for the problem. For all the data sets, with the new crack-tip (318, 427), the least squares analysis is done, and the results show improved convergence and a slight increase in the values of Mode-I SIF was observed. The values obtained for data set 6 is remarkably close to the mean of the values evaluated for the six datasets after CTR. The results of the reconstruction with CTR for data sets 1 and 6 are shown in figure 2.15(d and c). The convergence errors obtained for both these cases are small and are 0.010 and 0.011, respectively. The result from data set 6 with CTR (figure 2.15(e)) is taken to be the final result as the reconstruction is much closer to the experimental fringes compared to other data sets. Noticeably a triangular fringe feature ahead of the crack is seen in figures 2.15(b–d), which is absent in the experimental image (figure 2.15(a)) as well as in figure 2.15(e). The values obtained are $K_I = 0.0823$ MPa$\sqrt{m}$; $K_{II} = 0.0153$ MPa$\sqrt{m}$ and T-stress $= 0.194$ MPa. Uncertainty is defined as the ratio of the standard deviation divided by the square root of the number of data sets. The uncertainties are estimated to be 0.0005, 0.0004 and 0.0008 for K_I, K_{II} and T-stress, respectively. This example has brought out the importance of using CTR—though the changes are small in the crack-tip location. The reconstruction is proven to be much better, and an improved estimation of the parameters is obtained, as otherwise it may have been underestimated, that is not desirable.

A study by Abhishek Gandhi in 2017 [35] on the possible accuracy of SIF evaluation by various experimental methods such as strain gauges, transmission photoelasticity, coherent gradient sensing (transmission and reflection modes) and digital image correlation established that among all these methods, photoelasticity guarantees the most accurate evaluation of SIF. The use of PSIF software [30] has been demonstrated to solve even complex mixed-mode problems very convincingly. This will go a long way in helping researchers to evaluate the SIFs reliably by photoelasticity in problems of practical interest.

The experiments are repeated for various crack angles and biaxial ratios. As discussed for the sample crack, the stress-field parameters are meticulously evaluated for all these configurations. Among the crack tips, tips B and C are the interacting crack tips (figure 2.14(a)). The SIF values thus obtained are normalized with the SIF of a single centre crack of 20 mm length (0.074 MPa$\sqrt{m}$) in the cruciform specimen subjected to uniaxial loading ($\beta = 0$). Introduction of the lateral loads causes a shielding effect at the tips of a single crack. When two cracks interact, both amplification and shielding effects are seen depending on the crack tip under consideration, crack interaction angle and biaxiality ratio. Figure 2.16 shows the normalized SIF values at tip C as a function of α for different values of β.

A maximum amplification of about 155% is observed at the interacting tips B and C. Tip B has an amplification in the Mode-I SIF for all biaxial ratios. For tip C, under uniaxial tension, decreasing amplification is seen up to an angle (α) of 30° and after that shielding effect is seen, as shown in figure 2.16(a).

At tip C there is an amplification in Mode-I SIF followed by shielding and the transition occurs between $\alpha = 32.5°$ to 45° when the biaxial ratio is increased from 0 to 1 (figure 2.16). The Mode-II SIF is quite significant and comparable to Mode-I SIF value for tip C. The value of Mode-II SIF increases and then decreases, and the transition occurs at about $\alpha = 52.5°$ for tip C.

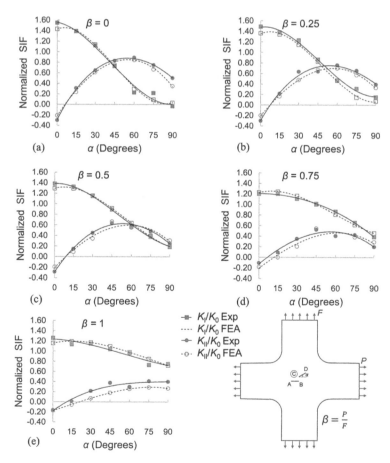

Figure 2.16. Normalized SIF at crack tip C as a function of crack angle α for different biaxial ratios β. Reprinted from [29] with permission from Elsevier.

2.10 Empirical relations for estimating normalized SIF under biaxial loading

Empirical equations are evaluated in such a manner that, the normalized SIF values $(K_I/K_0)_\beta$ under different biaxial ratios are estimated when α, β and the normalized SIF value under uniaxial tension at $\alpha = 0$, $(K_I/K_0)_{\alpha,\beta=0}$ are known. Two types of normalizations are carried out. In the first case, the SIF for a single crack in a cruciform specimen subjected to Mode-I loading (0.074 MPa$\sqrt{m}$) is used. The empirical relations that can be used to estimate the values of normalized SIF at tips B and C are:

$$\left(\frac{K_I}{K_0}\right)_\beta^{\text{Tip B}} = \left(-0.0033\alpha + \left(\frac{K_I}{K_0}\right)_{\alpha,\,\beta=0}^{\text{Tip B}}\right) + (0.0012\alpha - 0.3312)\beta \qquad (2.14)$$

The value of α is to be used in degrees for this and other empirical equations. For tip C, two empirical relations are necessary for two different ranges of interaction angles α and are given as:

$$\left(\frac{K_I}{K_0}\right)_\beta^{\text{Tip C}} = \left(-0.0119\alpha + \left(\frac{K_I}{K_0}\right)_{\alpha,\,\beta=0}^{\text{Tip C}}\right) + (0.0107\alpha - 0.3763)\beta \qquad (2.15)$$

$$(\text{for } \alpha \text{ up to } 30°)$$

$$\left(\frac{K_I}{K_0}\right)_\beta^{\text{Tip C}} = \left(-0.018\alpha + \left(\frac{K_I}{K_0}\right)_{\alpha,\,\beta=0}^{\text{Tip C}}\right) + (0.0104\alpha - 0.1823)\beta \qquad (2.16)$$

$$(\text{for } \alpha \text{ from } 30° \text{ to } 90°)$$

For tip B, the maximum error is 12%, the minimum error is 0.16% and the average absolute error is 4.7% using the empirical relations. For tip C, the equations provide an estimate with higher deviation only for $\alpha = 60°$, but it is on the conservative side. The error at $\alpha = 90°$ shows a higher value up to 57% pertaining to exceedingly small values of normalized SIF which are insignificant. Excluding these cases, the maximum error for estimates using the empirical relations at tip C is 19%, minimum error is 0.5% and the average absolute error is 4.5%.

In the second case only the analytical SIF value of a crack in a uniaxial tension specimen is used ($K_0 = \sigma\sqrt{\pi a}$) for normalization as it can be more beneficial for the designers.

$$\left(\frac{K_I}{K_0}\right)_\beta^{\text{Tip B}} = \left(-0.0029\alpha + \left(\frac{K_I}{K_0}\right)_{\alpha,\,\beta=0}^{\text{Tip B}}\right) + (0.001\alpha - 0.2905)\beta \qquad (2.17)$$

For tip C, two empirical relations are necessary for two different ranges of interaction angles α and are given as:

$$\left(\frac{K_{\mathrm{I}}}{K_0}\right)_\beta^{\mathrm{Tip\,C}} = \left(-0.0181\alpha + \left(\frac{K_{\mathrm{I}}}{K_0}\right)_{\alpha,\,\beta=0}^{\mathrm{Tip\,C}}\right) + (0.0093\alpha - 0.329)\beta \tag{2.18}$$

(for α upto $30°$)

$$\left(\frac{K_{\mathrm{I}}}{K_0}\right)_\beta^{\mathrm{Tip\,C}} = \left(-0.0184\alpha + \left(\frac{K_{\mathrm{I}}}{K_0}\right)_{\alpha,\,\beta=0}^{\mathrm{Tip\,C}}\right) + (0.0101\alpha - 0.2061)\beta \tag{2.19}$$

(for α from $30°$ to $90°$)

2.11 Use of carrier fringes in photoelasticity

Single image acquisition based digital photoelastic methods are important as they give whole field results using only a single image of the photoelastic fringe pattern. This makes it suitable for solving problems in an industrial scenario and it is adaptable for time varying phenomenon too. Carrier fringes have been used in conventional photoelasticity for the measurement of retardation (isochromatic fringe order) at a point in the model [36, 37]. To use the carrier fringe method (CFM) for quantitative analysis at a point, one must know the isoclinic angle (principal stress orientations) at that point *a priori*.

Polycarbonate is widely used in structural applications. It is birefringent and hence amenable for direct photoelastic analysis. One of the important issues to recognize is that the commercial sheets of polycarbonate have residual stresses and if they are not quantified, the final design using them would be compromised. Flow induced residual stress that is uniform can be measured using CFM as the principal stress direction is known.

The use of carrier fringes is specifically advantageous for the analysis of components made of glass that is weakly birefringent. In this, CFM is used to amplify the retardation in the model for measurement purposes. In many problems involving the stress analysis of glass, the principal stress orientations are already known. Thus, CFM is ideally suited for the evaluation of stresses and it is also quite useful even to calibrate the glass to find the stress-optic coefficient.

2.12 Residual stresses in a commercial polycarbonate sheet

2.12.1 Nature of the residual stresses

Polycarbonate (PC) sheets are manufactured through the extrusion process. In this process, initially, the plastic raw material is heated using a twin-screw mechanism. The hot plastic melt is then forced through a die to make the sheet of required thickness. As the plastic melt flows through the die, the polymer molecules are stretched in the flow direction. If the material is cooled before the molecules are fully relaxed, the molecular orientation is locked into the sheet, resulting in plastic flow induced residual stresses. After coming out of the die, the plastic sheet is cooled to room temperature. If the sheet is cooled non-uniformly, then thermally induced residual stresses are introduced. Thus, residual stresses in a PC sheet are in general, a combination of flow and thermal induced stresses.

Figure 2.17(a) shows the dark field isochromatic fringe pattern observed in a commercially available polycarbonate sheet of thickness 1.2 mm using white light. The constant colour of the isochromatics indicates that residual stress in that region is uniform. Figure 2.17(b) shows the isochromatics after making a thin slit in the sheet [38]. It is observed that the fringe pattern does not show any distortion even after a slit is cut across it. Broutman and Krishnakumar [39], have reported that no distortion indicates that the residual stress induced is due to plastic flow. Thus, the polycarbonate sheet used in the current study has only plastic flow induced residual stresses.

2.12.2 Measurement of plastic flow induced residual stresses using CFM

The residual stress in the specimen (figure 2.17) can be determined only by measuring the total fringe order at that point. Due to uniform nature of the residual isochromatics, it is not possible to find the residual stress using the conventional Tardy's method of compensation (section 1.11) or a digital polariscope using phase shifting conveniently (chapter 3). Even colour code is not sufficient to identify the integer part of the fringe order.

In CFM [36, 37], the residual birefringence in the PC sheet is superimposed by carrier fringes resulting in the formation of composite fringes. The carrier fringes are introduced by a stress frozen C-specimen (figure 2.18) of 50 mm width and 5.7 mm thickness, which was subjected to eccentrically loaded axial tension while stress freezing. The carrier fringes are vertical straight lines caused by the superposition of bending and tensile stresses. The optical arrangement for measuring residual stress of the polycarbonate sheet in a conventional circular polariscope using the carrier is shown in figure 2.18. Figure 2.19 shows the superposition of the residual stresses in a small region of the PC sheet with the carrier fringes. In figure 2.19, region *A* is the

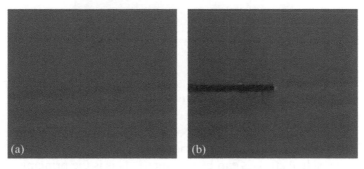

Figure 2.17. Isochromatic fringe patterns in (a) original PC sheet, (b) after making the slit. Reprinted from [38] with permission from Elsevier.

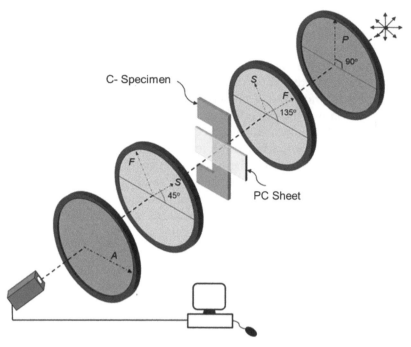

Figure 2.18. Schematic diagram of the circular polariscope arrangement used in CFM. Reprinted from [38] with permission from Elsevier.

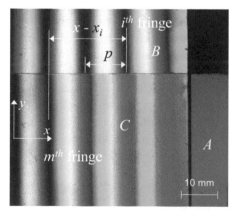

Figure 2.19. Measurement of residual stresses in the PC sheet using CFM.

residual birefringence observed in the PC sheet, region B is the carrier fringes in the C-specimen and region C shows the composite fringes. Due to the residual birefringence, a horizontal shift is observed in the composite fringes, which is constant. This shift in the fringes is proportional to the residual stresses in the PC sheet [38]. Since the retardation in the carrier is already known, the retardation in the PC sheet due to residual stresses can be measured. This is the principle behind CFM and the methodology is discussed next.

For a monochromatic light source, the intensity of light emerging from a dark field circular polariscope with the superposition of the carrier fringes is given as [36, 37],

$$I = I_b + \frac{I_a}{2}\Big[1 - \sin^2(\theta_{pc} - \theta_c) \cos 2\pi(N_{pc} - N_c)$$
$$- \cos^2(\theta_{pc} - \theta_c)\cos 2\pi(N_{pc} + N_c)\Big] \tag{2.20}$$

where I_a is the amplitude of light intensity, I_b accounts for the background intensity, N_c and N_{pc} are the fringe orders in the carrier and the PC sheet, respectively, and θ_c and θ_{pc} are the principal stress orientations in the carrier and the PC sheet.

From equation (2.20), it is evident that the light intensity depends on two parameters, fringe order N and isoclinic angle θ. Under special conditions like $\theta_{pc} - \theta_c = 0$ or $\pm 90°$, the intensity of light depends on the fringe order only and is given as,

$$I = I_b + \frac{I_a}{2}[1 - \cos 2\pi(N_{pc} \pm N_c)] \tag{2.21}$$

The total fringe order of the composite fringes is given as,

$$N_t = |N_{pc} \pm N_c| \tag{2.22}$$

In the current investigation, $\theta_{pc} - \theta_c = \pm 90°$ as the PC sheet and the carrier are pre-aligned that way. For the reference directions shown in figure 2.19, the stresses in the carrier vary linearly along the 'x' direction but are constant in the 'y' direction. Since the retardation is proportional to the stresses in the carrier, it is a function of 'x' only. In the PC sheet, as observed in region A of figure 2.19, the retardation is constant. Hence, the total retardation in equation (2.22) can be written as,

$$N_t(x) = |N_c(x) \pm N_{pc}| \tag{2.23}$$

The pitch 'p' of the carrier is defined as the distance between two adjacent integral fringe orders. The retardation at a point x on the carrier (N_c), with respect to an ith carrier fringe can be written as,

$$N_c(x) = N_c^i \pm \frac{x - x_i}{p} \tag{2.24}$$

The fringe order of the mth composite fringe matches with the ith carrier fringe. Now considering the point at which $N_t^m(x) = N_c^i$ on the composite fringes, equation (2.23) becomes,

$$N_t^m(x) = N_c^i = \left| N_{pc} \pm N_c^i \pm \frac{x - x_i}{p} \right| \tag{2.25}$$

Solving equation (2.25), the residual fringe order in the PC sheet can be obtained as,

$$N_{pc} = \left| \frac{x - x_i}{p} \right| \tag{2.26}$$

where $(x-x_i)$ is the deviation of the fringes at an ordinate y, as shown in figure 2.19. The residual fringe order for this PC sheet is found to be 1.85 fringe orders. Using this, $(\sigma_1-\sigma_2)$ is evaluated from the stress-optic law and is estimated to be 9.99 MPa.

2.13 Nomenclature of stresses in glass

Glass is one of the oldest man-made materials and the history of glass making can be traced back to the third century B.C. One of the main concerns in the applicability of glass is its brittle nature and failure under tensile stress. Over the years, researchers have understood that the structural behaviour of glass can be regulated to a large extent by controlling the residual stress in it [40–43]. Hence, its measurement and control are crucial in glass industries. Photoelasticity has been in use for stress measurement in glass since it exhibits stress-induced birefringence. The residual stress in glass varies from 0–1 MPa in a moulded glass lens, 70–120 MPa in thermally tempered glass plates to as high as 1000 MPa in chemically tempered glass plates. The last decade has seen rapid advancements in glass stress analysis using digital photoelasticity [43].

Thermal residual stresses in plate glass are generally divided into two—thickness and membrane stresses [40]. The thickness stresses are the stresses induced due to the thermal gradients across the thickness of the glass plates. Membrane stresses are introduced due to the thermal gradients along the surface of the plate. Membrane stresses near the edge of the glass plate are termed as edge stress. In glass literature [40, 44], one would also find a term called surface stress which denotes the combined effect of the thickness and membrane stresses on the glass surface. Nomenclature of residual stress as thickness, membrane and edge stress is convenient as the reason for their formation can be identified and controlled separately. Further, in measurement, the optical methods lend themselves to measure these separately.

Figure 2.20 illustrates the nature of residual stresses in a heat-treated glass plate in typical blocks taken at selected locations [43]. For a block at A that is taken away from the edges, the variation of thickness stress components on the x and y planes is shown. Tensorially, these are σ_y and σ_x components for the coordinate system shown in figure 2.20. Variation of σ_x and σ_y across the thickness of the plate is parabolic in nature with compression near the surface and tension in the central region. The tensile stress is usually half the magnitude of the compressive stress [45]. The compressive stress near the surface increases the bending strength of glass, whereas the tension in the mid-plane affects its fragmentation properties. The magnitude of thickness stress is found to depend on the cooling process [46, 47] and the dimensions of the glass plate [48]. It is reported that the stress state is hydrostatic ($\sigma_x \approx \sigma_y$) at zones away from the edges and cut-outs [40, 41].

Membrane stresses are created due to the non-uniformity in cooling across the surface of the glass plate. They are constant throughout the thickness of the plate [40]. Among the membrane stresses, the stresses near the edge are of interest to the glass manufacturers. Block B (figure 2.20) is taken along the edge parallel to the y-direction and the edge stress component for a typical section in the $-x$-plane is illustrated. Tensorially, it is σ_y and is compressive in nature. Similarly, block C

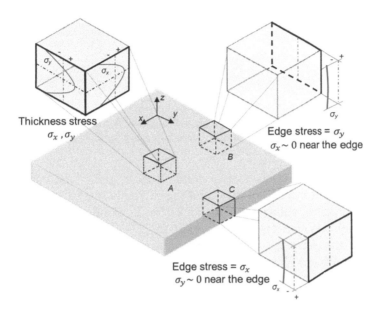

Figure 2.20. Schematic illustration of the nature of residual stress in a heat-treated glass plate. A rectangular block A is considered at a zone away from the edges to illustrate the nature and variation of the thickness stress. Blocks B and C located at the edges illustrate the nature of edge stresses. Reprinted from [43] with permission from Elsevier.

(figure 2.20) is taken along the edge parallel to the x-direction and σ_x is the edge stress for a typical section in the y-plane. The σ_z component is usually neglected owing to the small thickness.

Edge stresses are created since the edges of the glass plate act as additional cooling surfaces and cool down faster compared to the central region. They are generally compressive in nature and are beneficial to arrest crack growth and improve glass strength. Aben *et al* [49, 50] reported a correlation between the edge and surface stresses in tempered glass plates. This helps to determine one type of stress when the other is known.

2.14 Thickness stress evaluation of commercially annealed float glass

Thickness stresses can be viewed in a circular polariscope when the glass is placed in such a way that the light passes along the x- or y-directions (figure 2.20). When the light is sent along the x-direction, the fringes are horizontal (figure 2.21(a)) which indicates that the difference in principal stresses is constant along the y-direction. These fringes qualitatively indicate the presence of residual stress in float glass. Since only the stress components lying in a plane perpendicular to the light direction contribute to the formation of photoelastic fringes and since $\sigma_z = 0$, one observes only the fringes due to σ_y. As glass is a weakly birefringent material, the retardations observed are generally small, which make them difficult to measure directly. In such cases, a carrier is used to amplify the retardation for its easier measurement. Aben and Guillemet [40] have reported the use of carrier fringes (also called test fringes) to

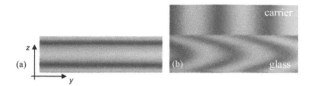

Figure 2.21. (a) Dark field isochromatics in a sample of float glass observed such that the light passes along +ve x-direction. (b) Isochromatic fringes obtained by superimposing glass with carrier fringes. Reprinted from [51] with permission from John Wiley and Sons.

qualitatively check the presence of residual stress in glass. The composite fringes obtained by the superposition of glass and the carrier (stress frozen C-specimen) are parabolic in shape, as shown in figure 2.21(b). It is interesting to note that the variation of thickness stress is also parabolic [51].

Quantitative measurement of the residual stresses is possible with the CFM in conjunction with digital photoelasticity. The dimensions of float glass specimen used for the experiment are $150 \times 50 \times 8$ mm^3 [51]. Once the glass is cut, the sides are carefully ground and buffed. The float glass sample is immersed in a bath of immersion liquid of refractive index, $n = 1.52$ (the same as that of glass) supplied by Cargille Laboratories, USA. The use of immersion liquid eliminates any reflection/refraction of light at the glass surface and is found to enhance the quality of the fringe pattern. The float glass sample and carrier are aligned such that the principal stresses in the glass beam and the carrier are perpendicular to each other [52]. The experiments are performed using sodium vapour lamp (589.3 nm), and the images are grabbed using a Sony Donpisha XC-003P CCD camera. To reduce the noise due to electrical fluctuations, time averaging is performed using 16 frames. The experiment is repeated three times to ensure repeatability of the method.

To improve the accuracy of calculations, it is desirable to extract the fringe skeletons using digital fringe thinning. It has been demonstrated earlier that among the various thinning methodologies, the global fringe thinning algorithm of Ramesh and Pramod [5] guarantees fringe skeleton of high quality even in complex crack-tip isochromatics, and hence it is used to extract the fringe skeleton (figure 2.22). For the coordinate systems shown in figures 2.20 and 2.21, and following section 2.12.2, the residual fringe order of glass can be expressed as

$$N_g(z) = \frac{|y - y_i|}{p} \tag{2.27}$$

In the derivation of equation (2.27), the key point to note is that the mth composite fringe corresponding to the ith fringe order of the carrier fringe has to be identified first. To determine this in monochromatic light, one must use certain basic information of the nature of stress variations, which is discussed next.

2.14.1 Identification of composite fringes and stress calculations

One can easily identify the composite fringes corresponding to each carrier fringe under white light using the colour information, which is directly evident in

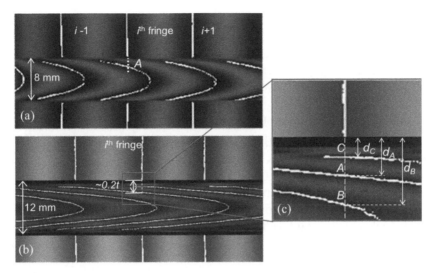

Figure 2.22. Fringe skeleton obtained using global fringe thinning algorithm (a) 8 mm thick glass (b) 12 mm thick glass. (c) Indicates how to identify the composite fringe corresponding to the ith carrier fringe. Reprinted from [51] with permission from John Wiley and Sons.

figure 2.21, figure 2.19 or in figure 2.22(a) but is not that evident in figure 2.22(b). The thickness stress is compressive on the surface and becomes tensile at mid-thickness. From equation (2.27), the residual stress will become zero when the extended carrier fringe intersects the composite fringe (A in figure 2.22(a)). In this case, the extended carrier fringe intersects only one composite fringe. Figure 2.22(b) shows the composite fringes in a float glass slice of dimensions $150 \times 100 \times 12$ mm^3. Here, the thickness of the specimen is more and further, the length of the light path is doubled from 50 mm to 100 mm, which increases the retardation considerably. It can be seen from the zoomed portion of the tile shown in figure 2.22(c) that the ith carrier fringe intersects three composite fringes at points C, A and B at depths of d_C, d_A and d_B from the surface, respectively.

To resolve the ambiguity, one has to use the knowledge that the depth of the compression layer from the surface of the glass is approximately 0.2 times the thickness [40]. The point A, depicting the zero stress must be within 0.18–0.22 times the thickness. For figure 2.22(a) this is straightforward and for figure 2.22(b), points C, A, B are at depths of 0.11, 0.20, and 0.39 times the thickness of the glass and hence the middle point is correctly labelled as A [51].

A distance of 1 mm is excluded from the top and bottom edges of the float glass for data extraction owing to irregularities due to polishing and buffing. The skeleton points of mth composite fringe corresponding to the ith carrier fringe are extracted. The exercise is repeated for the $(i + 1)^{th}$ and $(i-1)^{th}$ carrier fringes. The deviation of the composite fringe from the corresponding carrier fringe is measured and this is used to calculate the residual fringe order using equation (2.27). Average of these measurements is used as the final distribution. Next, a parabola is fit to the retardation data extracted and is extrapolated to the surfaces of glass.

The measurement process in CFM is quite simple as it amounts to finding only the deviation, which is easy to obtain by fringe thinning. It is established in photoelastic analysis that phase shifting techniques (PSTs) can give the most accurate estimate of fringe orders (chapter 3). Hence, it is desirable to compare the efficacy of CFM with PST, which is discussed next.

2.14.2 Comparison of thickness stresses measured using CFM and PST

Figure 2.23(a) shows the retardation measured using the CFM and six-step PST [51] for the glass sample of thickness 8 mm. The retardation obtained from automatic polariscope AP-07 (red light, $\lambda = 627$ nm) is normalized to 589.3 nm for comparison. The average deviation in the retardation data obtained by CFM is only 0.022, which is less than 4% of the retardation at the surface.

The photoelastic constant (C) of the float glass is necessary to compute the stresses. It is determined to be 2.6 TPa^{-1} by the method that will be discussed in

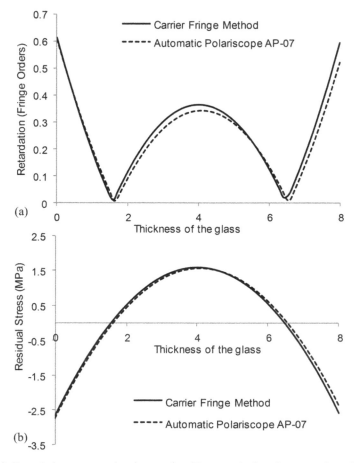

Figure 2.23. (a) Retardation measured using carrier fringe method and automatic polariscope AP-07 (Normalized to $\lambda = 589.3$ nm). (b) The corresponding residual stress values. Reprinted from [51] with permission from John Wiley and Sons.

section 2.15. Figure 2.23(b) shows the residual stress variation along the thickness of the 8 mm float glass sample measured using CFM and PST. The tensile stress at the mid-plane and the average compression at the surface are obtained as 1.55 and −2.55 MPa, respectively, using AP-07. The average tensile stress at the mid-plane is obtained as 1.65 MPa using CFM with an uncertainty of 0.012 MPa. The average surface compression is obtained as −2.74 MPa with an uncertainty of 0.011 MPa.

The utility of the method using carrier fringes is also demonstrated by measuring the residual stresses in glass specimens of different thicknesses [51]. The averages of residual stresses on both surfaces are −1.32, −1.62, −2.23, −2.74, and −3.9 MPa for float glass of thicknesses of 4, 5, 6, 8, and 12 mm, respectively. The corresponding midplane tensions are 0.69, 1.03, 1.39, 1.65, and 2.28 MPa, respectively.

2.15 Calibration of glass

Photoelastic calibration of glass, i.e., the determination of the photoelastic constant of glass, is important. Unlike in conventional photoelasticity where material stress fringe value (F_σ, which has the unit of N/mm/fringe) is widely used for quantifying the photoelastic behaviour; in glass literature, one finds the use of photoelastic constant C (TPa^{-1}). For most glasses, the value of C usually lies in the range of 2–4 TPa^{-1} [40], whereas for commonly used photoelastic materials like epoxy, it is in the order of 50 TPa^{-1}. Hence, the birefringence in glass is weak and its accurate determination poses a challenge. Calibration is performed on simple specimens with known stress state. The specimens used in general are glass fibre under tension, four-point bending, concentrated force at a point, uniaxial tension/compression and cylinder/disc under diametral compression [43]. Considering the ease of specimen preparation and accuracy of loading, the use of beam under four-point bending is preferable.

A rectangular beam (120 × 20 × 6.05 mm^3) made of glass material (LB2000) is subjected to four-point bending. The glass specimen to be calibrated should be free from any residual stresses as per ASTM standards. When the unloaded beam is viewed in a dark field circular polariscope, the specimen is completely dark indicating that the glass specimen is free of residual stresses (figure 2.24(a)). When the carrier fringes are superimposed, they appear undistorted and remain vertical (figure 2.24(b)). When a load of 47 N is applied over the beam to bend it, the edges of the beam become slightly bright and the neutral axis at the centre is black as in figure 2.24(c), indicating only a small value of retardation. The maximum bending stress introduced at this load is 1.79 MPa and the corresponding fringe order at the outermost fibre is 0.203 only. When the loaded beam is viewed with the carrier, the carrier fringes within the beam rotate (figure 2.24(d)). This rotation is perceptible, easily measurable, and thus, the carrier fringes have successfully helped in amplifying the retardation in the beam for accurate measurement. This rotation increases, as the applied load is increased.

It is interesting to observe that the introduction of the carrier also results in the composite fringe shape that is akin to the stress variation! This is an interesting coincidence—it was parabolic in measuring thickness stresses and linear for a beam

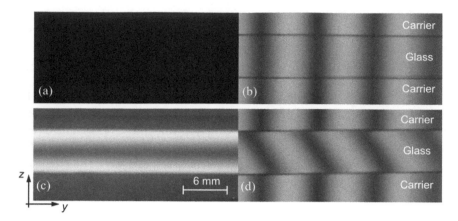

Figure 2.24. Dark field isochromatics in the LB2000 glass specimen under (a) no load, (b) no load with carrier superimposed, (c) a load of 47 N, (d) a load of 47 N with carrier superimposed. Reprinted from [52] with permission from Elsevier.

under pure bending. In fact, this aspect of carrier fringes is used to develop optical comparators to monitor the process parameters for glass tempering by comparing the shape of the parabola. Thus, CFM is quite useful in an industrial scenario as a simple qualitative tool to monitor the manufacturing process of glass.

The deviation from the carrier fringe and the pitch of the carrier fringes can be utilized to determine the fringe orders over the depth of the beam.

$$\sigma_1 - \sigma_2 = \frac{N_g \lambda}{Ct} \tag{2.28}$$

where, $(\sigma_1 - \sigma_2)$ is the principal stress difference, N_g is the retardation in fringe order, t is the thickness of the glass beam and λ is the wavelength of the light used (589.3 nm). In this case, since the glass beam is subjected to bending stress only, σ_2 is zero and the only unknown in equation (2.28) is the photoelastic constant C. Hence, equation (2.28) can be recast as,

$$C = \frac{N_g \lambda}{\sigma_y t} \tag{2.29}$$

By substituting the beam bending equation, equation (2.29) becomes,

$$C = \frac{2 N_g \lambda I_{xx}}{z W a t} \tag{2.30}$$

where, W is the total load applied on the four-point bent beam and a is the distance between the loads $W/2$ to effect the bending moment. Since the edges of the glass beam are chamfered, an ordinate was chosen at a distance of 0.5 mm below the top edge for measurements. Once the retardation (fringe order) is measured for several data points along the height of the beam, a plot is made between fringe order versus distance from the neutral axis of the beam. Let S_1 $(= N_g/z)$ be the slope of the least squares line obtained for this data for a particular load. The experiment is repeated for a few loads. For each of these loads, slope S_1 is calculated. Next, another plot is made between these slopes (S_1) and the respective loads. As the beam is loaded within its elastic limit, the plot between the slope S_1 and the corresponding load is expected to be a straight line. A least squares line is drawn for this plot and its slope ($S_2 = N_g/zW$) is used for obtaining the value of C. Equation (2.30) can be written in terms of slope S_2 as,

$$C = \frac{2S_2\lambda I_{xx}}{at} \tag{2.31}$$

Equation (2.31) gives the photoelastic constant of the glass specimen in TPa^{-1} considering maximum data points from the experiments. The average value of C obtained is 2.79 TPa^{-1} with an uncertainty of 0.026. The uncertainty value achievable using automated phase shifting is 0.025 [52]. Thus, the use of multiple data points obtained for various load values guarantees high accuracy of evaluation of the photoelastic constant C using CFM.

2.15.1 Calibration of commercial float glass

The float process for the manufacturing of float glass has gained popularity owing to its unique advantages like excellent surface finish, uniform thickness, superior optical quality, and bright appearance. Manufacturing of float glass involves heating raw materials to about 1500 °C, cooling them to about 600 °C over a bed of molten tin followed by annealing [53]. The cooling process generates residual stresses due to the thermal gradients across the thickness of the glass. Owing to mass production, the temperature cycles are usually chosen such that the production time is minimal, which in turn may compromise the level of annealing. Hence, some residual stresses will be retained in glass whose magnitude depends on the cooling cycle adopted.

It is reported that, apart from the glass composition, the photoelastic constant value of glass depends even on the thermal history. Fontana [54] has found that the stress optic coefficient of annealed soda lime glass is lower than that of un-annealed glass. Hence, it is desirable to develop a method of calibration that can handle specimens with residual stress.

A rectangular beam ($150 \times 23 \times 8$ mm^3) of glass is sliced from a float glass panel of thickness 8 mm. The cut surfaces of the glass beam are then ground, buffed, and polished to make them transparent to light. The residual stress due to improper thermal cooling would result in thickness stresses that vary parabolically over the depth revealed by the carrier fringes, as shown in figure 2.25(a).

Figure 2.25(b) shows the fringes as seen in a loaded float glass beam with skeletons extracted by the global fringe thinning algorithm of Ramesh and Pramod [5]. In each of these cases, the individual stress variations can be evaluated based on the deviations $|y_i-y_u|$ and $|y_i-y_\ell|$, respectively. As the focus is on finding the calibration constant, which requires the variation of the bending stress due to external loads, it can be obtained based on the deviation $|y_\ell-y_u|$. If these are obtained for various loads, then using equation (2.31), the calibration constant can be obtained and is found to be 2.72 TPa^{-1}. To verify the evaluation of the calibration constant, the fringes are

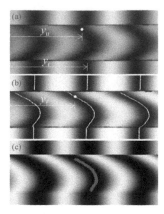

Figure 2.25. (a) Float glass beam showing parabolic variation of thickness stresses. (b) Float glass beam is subjected to four-point bending showing the result of superposition of the thickness stress and the linear variation of the bending stress. (c) Theoretically reconstructed carrier and composite fringes of (b) with experimentally collected data shown as red points. Adapted from [55] with permission from Elsevier.

reconstructed, and the data used for the data reduction is echoed (figure 2.25(c)). If the float glass beam is carefully annealed, then the calibration constant is found to be 2.61 TPa^{-1} [55]. This is deviating by 4% and confirms the observation of Fontana [54] that the stress-optic coefficient of the annealed glass is lower.

2.16 Edge stress analysis in tempered glass panels

Edge stresses can be viewed in a circular polariscope when the glass is placed in such a way that the light passes along the z-direction (figure 2.20). The carrier and the toughened glass plate ($600 \times 600 \times 6$ mm^3) are oriented in such a way that their principal stress directions are perpendicular to each other. The composite fringes obtained are shown in figure 2.26(a) in which the skeletons obtained by the global fringe thinning algorithm of Ramesh and Pramod [5] are superimposed. Near the glass edge, the composite fringes deviate with respect to the horizontal carrier fringes and the deviation is quite high if the retardation is high due to increased thickness as in figure 2.26(b). This deviation qualitatively indicates the presence of the edge stress in the plate.

Now, if one identifies the fringe in the composite fringe pattern corresponding to the reference i^{th} carrier fringe ($N_t = N_i$), one gets the residual retardation in the glass plate as,

$$N_g(x) = \frac{|y - y_i|}{p} \tag{2.32}$$

where, p is the pitch of the carrier. As in the determination of thickness stresses, here again, the identification of the corresponding m^{th} composite fringe is challenging using the monochromatic light. The extended i^{th} carrier fringe may intersect more than one composite fringe (fringe 3 and fringe 4 in figure 2.26(a)) for a 6 mm thick and three fringes (3, 4 and 5) for a 10 mm thick tempered glass plates. Which one of these needs to be used is the question.

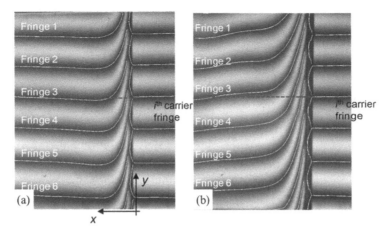

Figure 2.26. Illustration of the identification of composite fringe pattern in (a) 6 mm glass (b) 10 mm glass.

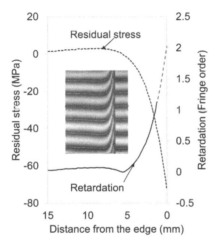

Figure 2.27. Variation of retardation with distance from the edge of the glass plate. The identified fringes (reference carrier fringe—blue, corresponding composite fringe—red) are shown in the inset. The figure also shows the variation of residual stress with distance from the edge of the glass plate.

From the mechanism of residual stress generation in glass, the integrated retardation is maximum at the edge and drops drastically with distance from the edge. Typically, the fringe orders at distances of about 10–15 mm from the edge would be close to 0 ($N < 0.3$) [56]. The mth composite fringe corresponding to i^{th} carrier fringe is identified by searching for the composite fringe that is far away from the edge, typically at about 15 mm in this problem. The composite fringe is identified as fringe 3 for both the cases, which is clear from figure 2.26. Once a point on the composite fringe is identified, the data from the entire fringe is extracted using appropriate fringe tracking schemes to get the deviations (see inset in figure 2.27).

Figure 2.27 shows the variation of residual fringe order as a function of distance from the edge of the glass plate. The fringe order is extrapolated to the edge of the glass plate using polynomial extrapolation and is obtained as 1.97. Figure 2.27 also shows the residual stress plot obtained using the stress-optic law. The edge stress in the tempered glass plate is obtained as -70.97 MPa. Hence, the glass plate can be classified as 'fully toughened'. It is to be noted that CFM is applicable only for the measurement along a straight edge.

The merit of equation (2.26) proposed by Ajovalasit et al [37], is that the retardation near the edge of the glass plate is expressed in terms of just the spatial deviation of the composite fringes. However, it assumes that the retardation along the edge of the glass plate is constant. It is seen that in some practical situations, the retardation along the edge of a tempered glass plate is not always uniform. This is because of the non-uniform heat transfer along the edges of the plate during the manufacturing process. The evaluation of residual stress in such cases is discussed by Vivek Ramakrishnan [57].

2.17 Influence of residual stress on crack-tip stress field parameters

It is seen that polycarbonate sheets have flow induced residual stress introduced during manufacturing. To study the influence of residual stress on the crack-tip stress fields, SEN (single edge notched) specimens with dimensions of $150 \times 50 \times 1.2$ mm^3 are cut from two different zones such that the residual stresses are uniform over the specimen size and also another specimen that is totally free of residual stresses by a careful process of annealing is taken. Depending on the flow stress direction (loading axis— along the 150 mm side), the specimens are labelled with subscripts as longitudinal (L) or transverse (T). When viewed in a white light circular polariscope, the annealed specimen is black and the other two specimens appear with uniform colour and their respective fringe orders are mentioned in figures 2.28(a–c). The determination of flow induced residual stresses has been discussed in section 2.12.2 and for the L and T specimens these are 11.25 MPa and 9.99 MPa, respectively [38].

A crack length of 10 mm is cut in these specimens such that the residual stresses are perpendicular (SEN$_L$) and parallel (SEN$_T$) to the crack axis. To aid the comparison, the annealed specimen is also cut with a crack and labelled as SEN$_A$. These are subjected to Mode-I loading and viewed in a dark field circular polariscope with a monochromatic illumination. The yield stress (σ_{ys}) of polycarbonate is 67.5 MPa. The specimens are loaded in such a way that the ratio of applied stress (σ) to yield stress (σ_{ys}) is varied from 0 to 0.312 in steps of 0.0195. The isochromatic fringe patterns for SEN$_A$, SEN$_L$, and SEN$_T$ specimens at different selected σ/σ_{ys} ratios are shown in figure 2.28. The fringes in the SEN$_A$ and SEN$_L$ specimen are forward tilted for all the values of σ/σ_{ys} ratio. The fringes in the SEN$_T$ specimen are initially backward tilted and as the σ/σ_{ys} ratio increases they become forward tilted.

Observing the fringe order variation in figure 2.28, it can be observed that for SEN$_L$ specimen, residual stresses and external loads add up while for SEN$_T$ specimen, they try to cancel each other. The crack tip stress field is affected by the direction and magnitude of the flow induced residual stress near the crack tip.

To quantify the effect of residual stresses on fracture parameters, the captured isochromatic fringe pattern images are post-processed. Fringe order and pixel

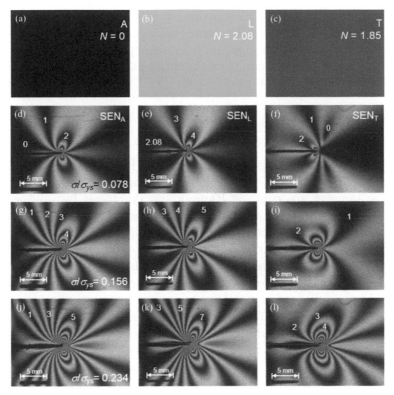

Figure 2.28. (a)–(c) Dark field residual fringe pattern in the three specimens seen in white light at $\sigma/\sigma_{ys} = 0$. (d)–(l) Dark field isochromatics for the three specimens under monochromatic light at different selected σ/σ_{ys} ratios. Adapted from [38] with permission from Elsevier.

coordinates data along the fringes having integer fringe orders are collected. The data is then used in an over-deterministic least squares method in conjunction with corrected Atluri–Kobayashi crack-tip stress field equation to evaluate the fracture parameters. Beyond σ/σ_{ys} ratio of 0.156, yielding is observed in the specimen and the conditions of small-scale yielding (SSY) are no longer valid and only up to this load are the SIFs evaluated. The difference between SIF values for the SEN_L and SEN_A specimens is 1.24%–7.08%, whereas for the SEN_T and SEN_A specimens, it is 0.2%–6.09%. Overall, these variations are small and SIF values are unaffected by the presence of the residual stresses [38].

However, the residual stresses had a noticeable effect on the value of T-stress. In the presence of tensile residual stresses perpendicular to the crack axis, as in the case of the SEN_L specimen, T-stress is negative for all σ/σ_{ys} ratios. In the presence of tensile residual stresses parallel to the crack axis, as in the case of the SEN_T specimen, T-stress is initially positive, and it decreases as the σ/σ_{ys} ratio increases. The T-stress for the SEN_T specimen eventually changes its sign from positive to negative as evidenced by the formation of forward tilted fringes at higher σ/σ_{ys} ratios in figure 2.28 [38]. Thus, the residual stresses have a stronger influence on the T-stress values.

The simple crack growth models involving only the SIFs are not sufficient to predict crack growth if the T-stress becomes significant. In such a case, one needs to use the generalized maximum tangential stress theory (GMTS) (section 5.5). Hence, when residual stresses are present, from the current study, it indicates that for crack growth predictions, one must employ only GMTS theory as T-stress becomes significant.

2.18 Closure

In this chapter, the use of fringe multiplication and fringe thinning techniques have been discussed for a range of problems. Among the fringe thinning methods, the global fringe thinning algorithm of Ramesh and Pramod is fast and assures high quality fringe skeletons of one pixel width in the domain. It has been shown that the use of fringe normalization (chapter 4), further improves the performance of the algorithm and its use is recommended for all fringe thinning requirements.

Both fringe multiplication and fringe thinning find immense application in the extraction of fringe data from crack-tip fringe fields. Problems that would be brushed aside as insufficient in data have been made to reveal relevant information by simple digital fringe multiplication. With sophisticated data extraction methods, crack-tip stress field parameters are evaluated for interacting crack-tips, whose fringe fields are quite complex in both fringe geometry and mode mixity. The need for collecting data along fringe geometry for effective data processing is emphasised. Refinements on improving the determination of crack-tip origin and use of multiple data sets to evaluate the uncertainty in the evaluation of SIFs and T-stress has been brought out. The use of colour information as another method for fringe multiplication is indicated in this chapter and is used for data extraction in chapter 4. This approach has an additional advantage that it uses only one colour image, hence could be used to advantage in time varying phenomenon.

The various uses of the carrier fringes method are brought out for the estimation of flow induced residual stress in polycarbonate, thickness, and edge stresses in float glass as well as calibration of glass. Finally, how flow induced residual stresses influence the magnitudes of SIFs and T-stress is discussed.

Exercises

1. While doing the analysis of a bimaterial specimen containing a crack lying on one of the materials, Ajith multiplied the fringes using the isochromatic fringes recorded in dark (figure (a)) and bright field (figure (b)) arrangements of a circular polariscope. The resulting mixed field fringes are shown in figure (c). Label the fringe orders in figure (c).

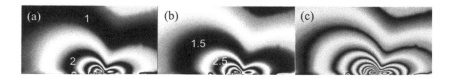

2. Explain the basic principle involved in digital image processing (DIP). List at least five applications where DIP is extensively employed these days.

3. What is meant by thresholding an image? What are the different criteria that are usually employed to determine the threshold value? What is meant by semi-thresholding and where is it used?

4. How is fringe thinning followed by fringe ordering still relevant even after the development of fully automated procedures like phase shifting methods?

5. What are the various methods for fringe thinning? Explain with neat sketches, the functioning of a binary based algorithm in fringe thinning.

6. What is the basic criterion used in intensity-based fringe thinning algorithms? How is this aspect incorporated in mask-based and global fringe thinning algorithms?

7. Mention the various steps involved in the global fringe thinning algorithm. What is the basic difficulty in handling fringes of arbitrary orientation? In what way do the logical operators help in achieving fringe thinning for a generic problem? How can fringe skeleton extraction be improved in high density fringe zones?

8. In optical methods of stress analysis, one needs the data of fringe order and its positional co-ordinates in S.I units for further processing. In DIP, the image is converted as an assembly of pixels. Explain how you would find the positional co-ordinates from the digitised image and mention the precautions to be taken.

9. Explain the role of photoelasticity in the development of fracture mechanics.

10. To study the crack on a turbine blade, Jithin conducted a photoelastic analysis using a cracked specimen made of polycarbonate. While doing least squares analysis, he found that the theoretical fringe patterns match well with the experimental fringes when the number of Mode-I and Mode II parameters is eight and the convergence error is found to be 0.02. The obtained values of parameters are given in the table below. Find out the values of K_I, K_{II} and the T-stress. Also, write software code to reconstruct these fringe patterns if the model thickness is 6 mm and the material stress fringe value is 8 N/mm/fringe.

Mode I parameters	Value	Mode II parameters	Value
A_{I1}, MPa(mm)$^{1/2}$	1.1024	A_{II1}, MPa(mm)$^{1/2}$	−0.1659
A_{I2}, MPa	−0.0698	A_{II2}, MPa	0.0000
A_{I3}, MPa(mm)$^{-1/2}$	0.0336	A_{II3}, MPa(mm)$^{-1/2}$	0.0458
A_{I4}, MPa(mm)$^{-1}$	0.0119	A_{II4}, MPa(mm)$^{-1}$	0.0171
A_{I5}, MPa(mm)$^{-3/2}$	0.0101	A_{II5}, MPa(mm)$^{-3/2}$	0.0023
A_{I6}, MPa(mm)$^{-2}$	0.0011	A_{II6}, MPa(mm)$^{-2}$	0.0059
A_{I7}, MPa(mm)$^{-5/2}$	0.0000	A_{II7}, MPa(mm)$^{-5/2}$	0.0014
A_{I8}, MPa(mm)$^{-3}$	0.0001	A_{II8}, MPa(mm)$^{-3}$	0.0002

11. The figure given below shows the isochromatic fringes (figure a) observed while loading two spur gears which are in mesh. One gear is made of aluminium and the other is made of epoxy. The numerically plotted isochromatic fringes with different numbers of Mode I and Mode II parameters are also shown in figures (b–d). Arrange figures (b–d) in the increasing order of the number of parameters used for the reconstruction.

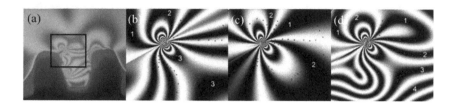

12. For a crack configuration as shown in figure 2.14 in a specimen subjected to biaxial loading, determine the normalised SIF at crack tip B for a biaxial ratio (β) of 0.25 and an interacting crack orientation angle (α) of 45° using the empirical relation provided in the chapter. The normalisation is done using the SIF for a single crack in a cruciform specimen subjected to Mode-I loading ($K_0 = 0.074$ MPa$\sqrt{m}$). The normalized SIF value under uniaxial tension at $\alpha = 0$ is $(K_I/K_0)_{\alpha,\beta=0} = 1.55$. Plot a graph between the normalised SIF and the biaxial ratios and see how the variation is for different interacting crack orientation angles.

13. In the calibration experiment of a glass using a four-point bending test, the dimension of the specimen is $120 \times 20 \times 6.05$ mm^3 and the wavelength of the light used is 589.3 nm. The distance from the load to effect the bending moment (moment arm) is 12.68 mm. In the experiment, the fringe orders are measured at several points along the height of the specimen and a plot is made between fringe order versus distance from neutral axis of the beam in mm. Let S_1 be the slope of this plot. The experiment is repeated for different loads and the slope S_1 is obtained for each load. Then, another plot is made between these slopes (S_1) and the respective loads in N. The slope of the least squares line plotted for this data is obtained as 0.0015. Determine the calibration constant C of the specimen.

14. The residual stresses in a polycarbonate sheet can be classified into plastic flow induced residual stresses and thermal induced stresses. Explain how photoelasticity can be employed to distinguish these stresses?

15. To measure flow induced residual stresses in a polycarbonate sheet of thickness 1.5 mm using CFM, a circular polariscope arrangement is used as shown in figure 2.18. The polycarbonate sheet and the carrier are placed in such a manner that the principal stress directions are perpendicular to each other. The carrier fringes are in vertical direction and the distance between adjacent fringes is 28 mm. The composite fringes are also in the vertical direction with a horizontal shift in the fringe order compared to the carrier

fringes. It is observed that the shift between corresponding fringe orders in the carrier and the composite fringe is 54 mm. Determine the residual stresses in the polycarbonate specimen if the material stress fringe value is 8 N/mm/fringe.

16. The thickness stresses are the stresses introduced due to the thermal gradients across the thickness of the glass plates which can be determined by means of CFM in photoelasticity. Figure 2.22(a) shows the fringe skeleton of the carrier and composite fringes obtained for a specimen of dimensions $150 \times 50 \times 8 \text{ mm}^3$. The experimental arrangement to obtain these is as discussed in section 2.14. It is observed from the fringe patterns that at the middle of the specimen, the distance between the composite fringe to the corresponding carrier fringe is 0.35 times the pitch of the carrier fringes. The calibration constant C of the glass is 2.65 TPa^{-1} and the wavelength of the light used is 589.3 nm. Determine the thickness stress at the centre of the specimen.

17. A beam of depth 40 mm, thickness 6 mm is subjected to pure bending of 1 Nm. The beam is viewed through carrier fringes and the fringes are observed as shown. Using the reference axes as shown, determine the stresses in the beam at the top surface and at the centre of the beam using the CFM. The material stress fringe value of the model material is 12 N/mm/fringe. Verify your calculations with the beam theory. First solve this problem by taking a copy of this image and extract fringe skeletons by hand sketch. If you can, extract the fringe skeleton by digital means and compare your results.

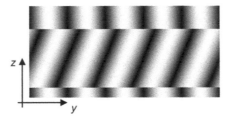

18. To determine the edge stresses on a glass plate using the CFM, the composite fringes are obtained as discussed in section 2.16. The composite fringes and carrier fringes for a 6 mm thick glass plate are shown in figure 2.26(a). It is observed that the ith carrier fringe intersects more than one composite fringe. What criteria can be used to determine the composite fringe corresponding to a carrier fringe? The distance between the composite fringe and the corresponding carrier fringe at a point 2 mm from the edge of the glass plate is found to be 17 mm and the pitch of the carrier fringe is 26 mm. The calibration constant C of the glass is 2.6 TPa^{-1} and the wavelength of the light used is 589.3 nm. Determine the value of edge stresses at this point.

References

[1] Toh S L, Tang S H and Hovanesian J D 1990 Computerized photoelastic fringe multiplication *Exp. Tech.* **14** 21–3

[2] Chen T Y and Taylor C E 1989 Computerized fringe analysis in photomechanics *Exp. Mech.* **29** 323–9

[3] Yatagai T, Nakadate S, Idesawa M and Saito H 1982 Automatic fringe analysis using digital image processing techniques *Opt. Eng.* **21** 432–5

[4] Umezaki E, Tamaki T and Takahashi S 1989 Automatic stress analysis of photoelastic experiment by use of image processing *Exp. Tech.* **13** 22–7

[5] Ramesh K and Pramod B R 1992 Digital image processing of fringe patterns in photomechanics *Opt. Eng.* **31**

[6] Ramesh K and Singh R K 1995 Comparative performance evaluation of various fringe thinning algorithms in photomechanics *J. Electron. Imaging* **4** 71–83

[7] Ramesh K 2007 *e-book on Engineering Fracture Mechanics* (Chennai: IIT Madras) https://home.iitm.ac.in/kramesh/EFM.html

[8] Ramesh K 2012 *Video Lectures on Engineering Fracture Mechanics* (Chennai: NPTEL and MHRD) https://youtube.com/playlist?list=PLbMVogVj5nJTTGvVQL5ks-woI_vqDTKXF

[9] Westergaard H M 1939 Bearing pressures and cracks *J. Appl. Mech.* **44** A49–53

[10] Wells A A and Post D 1958 The dynamic stress distribution surrounding a running crack: a photoelastic analysis *SESA Proc.* (Washington, DC: Naval Research Laboratory) 69–92

[11] Irwin G R 1957 Analysis of stresses and strains near the end of a crack traversing a plate *J. Appl. Mech.* **24** 361–4

[12] Ramesh K 2017 P_Scope®-a virtual polariscope *Photomechanics Lab.* (IIT Madras) https://home.iitm.ac.in/kramesh/p_scope.html accessed 06.04.21

[13] Irwin G R 1958 Discussion of the dynamic stress distribution surrounding a running crack-a photoelastic analysis *SESA Proc.* **16** 93–5

[14] Sanford R J 1979 A critical re-examination of the Westergaard method for solving opening-mode crack problems *Mech. Res. Commun.* **6** 289–94

[15] Ramesh K 2012 Lec-22 multi-parameter stress field equations *Video Lectures on Engineering Fracture Mechanics* 1st edn (Chennai: NPTEL and MHRD) https://youtu.be/ZRjxa29Sy1o?list=PLbMVogVj5nJTTGvVQL5ks-woI_vqDTKXF

[16] Atluri S N and Kobayashi A S 1987 Mechanical response of materials *Handbook on Experimental Mechanics* (Englewood Cliffs, NJ: Prentice-Hall) pp 1–37

[17] Ramesh K, Gupta S and Kelkar A A 1997 Evaluation of stress field parameters in fracture mechanics by photoelasticity—revisited *Eng. Fract. Mech.* **56** 25–45

[18] Ramesh K, Gupta S and Srivastava A K 1996 Equivalence of multi-parameter stress field equations in fracture mechanics *Int. J. Fract* **79** R37–41

[19] Kachanov M 1993 Elastic solids with many cracks and related problems *Advances in Applied Mechanics* vol 30 ed J W Hutchinson and T Y Wu (New York: Academic) pp 259–445 https://doi.org/10.1016/S0065-2156(08)70176-5

[20] Yan X 2004 Interaction of arbitrary multiple cracks in an infinite plate *J. Strain Anal. Eng. Des* **39**

[21] Lange F F 1968 Interaction between overlapping parallel cracks; a photoelastic study *Int. J. Fract. Mech.* **4** 287–94

[22] Phang Y and Ruiz C 1984 Photoelastic determination of stress intensity factors for single and interacting cracks and comparison with calculated results. Part I: two-dimensional problems *J. Strain Anal. Eng. Des* **19** 23–34

[23] Mehdi-Soozani A, Miskioglu I, Burger C P and Rudolphi T J 1987 Stress intensity factors for interacting cracks *Eng. Fract. Mech.* **27** 345–59

[24] Breitbarth E, Besel M and Reh S 2018 Biaxial testing of cruciform specimens representing characteristics of a metallic airplane fuselage section *Int. J. Fatigue* **108** 116–26

[25] Shlyannikov V N, Tumanov A V and Zakharov A P 2014 The mixed mode crack growth rate in cruciform specimens subject to biaxial loading *Theor. Appl. Fract. Mech.* **73** 68–81

[26] Bastun V N 1994 Fracture of thin-walled bodies with crack under biaxial loading *Eng. Fract. Mech.* **48** 703–9

[27] Zakeri M, Ayatollahi M R and Nikoobin A 2007 Photoelastic study of a center-cracked plate—the lateral load effects *Comput. Mater. Sci.* **41** 168–76

[28] Theocaris P S and Spyropoulos C P 1983 Photoelastic determination of complex stress intensity factors for slant cracks under biaxial loading with higher-order term effects *Acta Mech.* **48** 57–70

[29] Vivekanandan A and Ramesh K 2019 Study of interaction effects of asymmetric cracks under biaxial loading using digital photoelasticity *Theor. Appl. Fract. Mech.* **99** 104–17

[30] Ramesh K 2000 PSIF software—photoelastic SIF evaluation *Photomechanics Lab* (IIT Madras) https://home.iitm.ac.in/kramesh/psif.html

[31] Simon B N and Ramesh K 2009 Effect of error in crack tip identification on the photoelastic evaluation of SIFs of interface cracks *Fourth Int. Conf. on Experimental Mechanics (Singapore)* vol 7522 75220D https://doi.org/10.1117/12.852519

[32] Hariprasad M P and Ramesh K 2018 Analysis of contact zones from whole field isochromatics using reflection photoelasticity *Opt. Lasers Eng* **105** 86–92

[33] Singh R K and Ramesh K 2013 Comparative study of various cruciform specimen geometry using numerically obtained photoelastic fringe contours *1st Indian Conf. Applied Mechanics (Chennai)* pp 126–7

[34] Lamkanfi E, Van Paepegem W and Degrieck J 2015 Shape optimization of a cruciform geometry for biaxial testing of polymers *Polym. Test.* **41** 7–16

[35] Gandhi A 2017 *A Study of Errors in the Stress Intensity Factor, Estimated Using Different Experimental Techniques, due to Three Dimensional and Corner Singularity Effects* (IIT Kanpur)

[36] Ramesh K 2000 *Digital Photoelasticity Advanced Techniques and Applications* (Berlin: Springer) https://doi.org/10.1115/1.1483353

[37] Ajovalasit A, Petrucci G and Scafidi M 2012 Photoelastic analysis of edge residual stresses in glass by automated 'test fringes' methods *Exp. Mech.* **52** 1057–66

[38] Subramanyam Reddy M and Ramesh K 2016 Photoelastic study on the effect of flow induced residual stresses on fracture parameters *Theor. Appl. Fract. Mech.* **85** 320–7

[39] Broutman L J and Krishnakumar S M 1974 Cold rolling of polymers 2. Toughness enhancement in amorphous polycarbonates *Polym. Eng. Sci.* **14** 249–59

[40] Aben H and Guillemet C 1993 *Photoelasticity of Glass* (Berlin: Springer) https://doi.org/10.1007/978-3-642-50071-8

[41] Le B E 2007 Glass rheology *Glass: Mechanics and Technology* (New York: Wiley) ch 6 https://onlinelibrary.wiley.com/doi/book/10.1002/9783527617029

[42] Shelby J E 2005 *Introduction to Glass Science and Technology* (Cambridge: Royal Society of Chemistry) https://doi.org/10.1039/9781847551160

[43] Ramesh K and Vivek R 2016 Digital photoelasticity of glass: a comprehensive review *Opt. Lasers Eng.* **87** 59–74

[44] Kishii T 1981 Laser biascope for surface stress measurement of tempered glasses *Opt. Laser Technol.* **13** 261–4

[45] Redner A S and Hoffman B 2001 Detection of tensile stresses near edges of laminated and tempered glass *Proc. of Glass Processing Days* pp 589–91

[46] Chen Y, Lochegnies D, Defontaine R, Anton J, Aben H and Langlais R 2013 Measuring the 2D residual surface stress mapping in tempered glass under the cooling jets: The influence of process parameters on the stress homogeneity and isotropy *Strain* **49** 60–7

[47] Lochegneis D, Romero E, Anton J, Errapart A and Aben H 2005 Measurement of complete residual stress fields in tempered glass plates *Proc. of Glass Processing Days (Tampere, Finland)* pp 88–91

[48] Haldimann M, Luible A and Overend M 2008 *Structural Use of Glass* (Zurich: International Association for Bridge and Structural Engineering)

[49] Aben H, Lochegnies D, Chen Y, Anton J, Paemurru M and Õis M 2015 A new approach to edge stress measurement in tempered glass panels *Exp. Mech.* **55** 483–6

[50] Aben H, Anton J, Paemurru M and Õis M 2013 A new method for tempering stress measurement in glass panels *Est. J. Eng.* **19** 292–7

[51] Vivek R and Ramesh K 2015 Residual stress analysis of commercial float glass using digital photoelasticity *Int. J. Appl. Glas. Sci.* **6** 419–27

[52] Ramesh K, Vivek R, Tarkes Dora P and Sanyal D 2013 A simple approach to photoelastic calibration of glass using digital photoelasticity *J. Non. Cryst. Solids* **378** 7–14

[53] Nielsen J H 2009 Tempered glass: bolted connections and related problems (Lyngby: Technical University of Denmark) https://orbit.dtu.dk/en/publications/tempered-glass-bolted-connections-and-related-problems

[54] Fontana E H 1985 Stress-optical coefficients for glasses in their annealing range *Am. Ceram. Soc. Bull.* **64** 1456–8

[55] Vivek R and Ramesh K 2016 A novel method for the evaluation of stress-optic coefficient of commercial float glass *Meas. J. Int. Meas. Confed.* **87** 13–20

[56] Naveen Y A, Ramesh K and Ramakrishnan V 2012 Use of carrier fringes in the evaluation of edge residual stresses in a glass plate by photoelasticity *Joint Int. Conf. of the 2nd ISEM—11th ACEM—SEM Fall Conference (Taiwan)* pp 26–8

[57] Vivek R 2016 *Single image based isochromatics evaluation in digital photoelasticity and its applications* (IIT Madras)

Chapter 3

Phase shifting techniques in photoelasticity

Phase shifting techniques (PSTs) have truly revolutionised all optical techniques including photoelasticity. In photoelasticity, one must evaluate the isoclinic and isochromatic phases, which are interdependent. These are evaluated by processing intensity data using inverse trigonometric functions, which are multivalued, hence posing certain difficulties. In the wrapped isoclinic phasemap, one observes *inconsistent* zones, which translate into *ambiguous* zones in the isochromatic phasemap. Why such zones are formed and how these can be resolved is discussed in this chapter. How to calibrate the polariscope used in PST so that the optical alignment of the elements matches with the theoretical development is presented. In the case of isoclinic phasemaps, unwrapping refers to the process of obtaining the direction of either σ_1 or σ_2 consistently over the domain, while in the case of isochromatic phasemaps, unwrapping refers to suitable addition of the integral value to the fractional retardation values to get continuous fringe order data. Methods of adaptive quality guided phase unwrapping and adaptive multi-directional progressive smoothing to avoid progression of noise and handling the features of an isoclinic phasemap like the presence of isotropic points, π-jumps are discussed. Phase shifting in colour domain is demonstrated showing its improved performance in evaluating the isoclinics. Parallel unwrapping using multiple seed points and its application to a multi-segment industrial component to evaluate both isoclinic and isochromatic phasemaps is illustrated[1].

3.1 Introduction

Phase shifting technique (PST) has been applied to many classical interferometric techniques [1]. This has been achieved using intensity information for extracting the phase information. The advent of affordable high quality digital image acquisition

---

[1] Parts of this chapter have been adapted with permission of Elsevier from the author's own work in [62], with the permission of SAGE from the author's own work in [6], with the permission of SPIE from the author's own work in [49] and with the permission of Springer from the author's own work in [2].

systems has made this transformation possible. In most classical interferometers, the path length of the two interfering beams is distinct and separate. In general, phase differences can be introduced by altering the optical path length of any one of the light beams. Usually, the phase of the reference light beam is altered in known steps. Photoelasticity falls into a special category, in that the two light beams cannot be treated separately but always go together. The use of an external mirror as used in other interferometric techniques to introduce a phase shift is not necessary here. In practice, a change in phase between the beams involved is achieved by appropriately rotating the optical elements of the polariscope.

In fact, in the early development of digital photoelasticity, numerous papers have been written on using generic polariscope arrangements of various hues. If the optical elements are kept at arbitrary positions, a great degree of flexibility in utilising the intensity information for proposing newer methodologies for data reduction is possible. The research has come through a full circle, and now the focus has been on how to use even conventional photoelastic optical arrangements for PST. Nevertheless, a study on the intensity of light transmitted for arbitrary orientations of the optical elements is desirable to appreciate the development of the methodology of phase shifting in photoelasticity.

3.2 Intensity of light transmitted for generic arrangements of plane and circular polariscopes

In the subsequent discussions, δ is used to represent the retardation (in radians) introduced by the model, θ is the orientation of the slow axis of the model (one of the principal stress directions at the point of interest) with respect to the horizontal, $ke^{i\omega t}$ is the incident light vector, E_β and $E_{\beta+\pi/2}$ are the components of exit light vector along the analyser axis and perpendicular to the analyser axis, respectively. The angular orientations of the various optical elements are referred to with respect to the x-axis.

Figure 3.1(a) shows a photoelastic specimen kept in a plane polariscope with the polariser and the analyser kept at arbitrary angles of α and β, respectively. Using Jones calculus, the components of the exit light vector along the analyser axis and perpendicular to the analyser axis are obtained as [2–5]

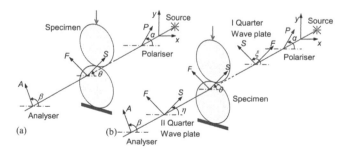

Figure 3.1. Generic arrangements of (a) plane polariscope, (b) circular polariscope.

$$\left\{ \begin{array}{c} E_\beta \\ E_{\beta+\pi/2} \end{array} \right\} = \left[\begin{array}{cc} \cos\beta & \sin\beta \\ -\sin\beta & \cos\beta \end{array} \right] \left[\begin{array}{cc} \cos\dfrac{\delta}{2} - i\sin\dfrac{\delta}{2}\cos 2\theta & -i\sin\dfrac{\delta}{2}\sin 2\theta \\ -i\sin\dfrac{\delta}{2}\sin 2\theta & \cos\dfrac{\delta}{2} + i\sin\dfrac{\delta}{2}\cos 2\theta \end{array} \right]$$

$$\times \left\{ \begin{array}{c} \cos\alpha \\ \sin\alpha \end{array} \right\} k e^{i\omega t} \qquad (3.1)$$

The intensity of light transmitted I_p (the subscript 'p' denotes that the incident light is plane polarised) is obtained as

$$I_p = I_a\left[\cos^2\frac{\delta}{2}\cos^2(\beta - \alpha) + \sin^2\frac{\delta}{2}\cos^2(\beta + \alpha - 2\theta) \right] \qquad (3.2)$$

where I_a accounts for the amplitude of the incident light vector and the proportionality constant. It is interesting to note that equation (3.2) is not altered when the model slow axis orientation is changed from θ to $\theta + \pi/2$. In other words, θ can represent either the major or minor principal stress direction at the point of interest.

If $\alpha = \beta + \pi/2$, equation (3.2) reduces to

$$I_p = I_a\left[\sin^2\frac{\delta}{2}\sin^2 2(\theta - \beta) \right] \qquad (3.3)$$

Equation (3.3) represents the intensity of light transmitted for crossed positions of polariser-analyser combination.

In most PSTs that use a circular polariscope, only the second quarter-wave plate and the analyser are kept at arbitrary positions of η and β, respectively (figure 3.1(b)). The polariser is kept at 90-deg and the quarter-wave plate-I is kept at either $\xi = 135$-deg or 45-deg. Using Jones calculus, the components of the exit light vector along the analyser axis and perpendicular to the analyser axis (for $\xi = 135$-deg) are obtained as [2–5],

$$\left\{ \begin{array}{c} E_\beta \\ E_{\beta+\pi/2} \end{array} \right\} = \frac{1}{2}\left[\begin{array}{cc} \cos\beta & \sin\beta \\ -\sin\beta & \cos\beta \end{array} \right] \left[\begin{array}{cc} 1 - i\cos 2\eta & -i\sin 2\eta \\ -i\sin 2\eta & 1 + i\cos 2\eta \end{array} \right]$$

$$\times \left[\begin{array}{cc} \cos\dfrac{\delta}{2} - i\sin\dfrac{\delta}{2}\cos 2\theta & -i\sin\dfrac{\delta}{2}\sin 2\theta \\ -i\sin\dfrac{\delta}{2}\sin 2\theta & \cos\dfrac{\delta}{2} + i\sin\dfrac{\delta}{2}\cos 2\theta \end{array} \right] \left[\begin{array}{cc} 1 & i \\ i & 1 \end{array} \right] \left\{ \begin{array}{c} 0 \\ 1 \end{array} \right\} k e^{i\omega t} \qquad (3.4)$$

The combination of polariser at 90-deg and the quarter-wave plate-I at $\xi = 135$-deg produces a left circularly polarised light. Denoting the intensity of light transmitted as I_ℓ (the subscript 'ℓ' denotes that the incident light is left circularly polarised), the intensity expression is obtained as

$$I_\ell = \frac{I_a}{2} + \frac{I_a}{2}[\sin 2(\beta - \eta)\cos\delta - \sin 2(\theta - \eta)\cos 2(\beta - \eta)\sin\delta] \qquad (3.5)$$

where I_a accounts for the amplitude of light vector and the proportionality constant. If in figure 3.1(b), the model fast axis is kept at θ (i.e., slow axis at $\theta + \pi/2$), then the intensity of light transmitted is,

$$I_\ell = \frac{I_a}{2} + \frac{I_a}{2}[\sin 2(\beta - \eta) \cos \delta + \sin 2(\theta - \eta) \cos 2(\beta - \eta) \sin \delta] \qquad (3.6)$$

Unlike for the generic arrangement of a plane polariscope (equation (3.2)), inspection of equations (3.5) and (3.6) shows that the intensity of light transmitted depends on the orientation of the model fast or slow axes for the cases when the optical elements after the model are kept at arbitrary orientations. As the model fast or slow axes are not known *a priori*, this leads to an ambiguity in the sign of fractional retardation δ. Hence, this aspect needs special attention in automated techniques based on intensity measurement.

In figure 3.1(b), if the quarter-wave plate-I is rotated such that the slow axis is kept at $\xi = 45$-deg, then one gets a right circularly polarised light being incident on the model. The intensity of light transmitted I_r (the subscript 'r' denotes that the incident light is right circularly polarised), including the effect of model slow or fast axis orientation for this case is,

$$I_r = \frac{I_a}{2} - \frac{I_a}{2}[\sin 2(\beta - \eta) \cos \delta \mp \sin 2(\theta - \eta) \cos 2(\beta - \eta) \sin \delta] \qquad (3.7)$$

where the upper sign $(-)$ corresponds to the case for which the model slow axis is at θ and the lower sign $(+)$ corresponds to the case for which the model fast axis is at θ. The intensity of light transmitted depends on the orientation of the model fast or slow axis in this case as well.

Many investigators have explicitly added a term I_b to the intensity expression to account for the stray light/background illumination. In view of the optical inhomogeneity and non-uniform characteristics of the light source and the diffuser, the quantities I_a and I_b are functions of the spatial co-ordinates viz., $I_a(x, y)$ and $I_b(x, y)$. With this explicit understanding, these will be represented as I_a and I_b in further text.

3.3 Development of phase shifting techniques

The various phase-shifted images required for PST are obtained in photoelasticity by suitably rotating the optical elements in a polariscope [6–8]. Ramesh [2] clearly brought out that rotation of an optical element is akin to providing a phase-shift, which has been seen in the example of the Tardy's method of compensation (section 1.11). By recording a few phase-shifted images as intensity distributions, and subsequently processing them, one could obtain the photoelastic parameter distributions over the model domain. Although many researchers reported techniques that employ this principle, the credit for calling such methodologies as phase shifting techniques goes to Asundi [9].

The concept of phase shifting in photoelasticity was first introduced by Hecker and Morche [10] in 1986. They recorded five phase-shifted images and demonstrated the possibilities for the determination of isochromatic parameter over the whole domain using a circular polariscope arrangement. Later, Kihara [11] reported an eight-step method using a mixed polariscope arrangement for data acquisition and examined the exit light using a quarter-wave plate and an analyser combination. Patterson and Wang [12] extended the work of Hecker and Morche [10] and proposed a six-step PST for the determination of both isoclinic and isochromatic parameters, which gave a new impetus to digital photoelasticity. As only fringe order N and principal stress direction θ need to be determined, Sarma et al [13] focused on the least number of optical arrangements to evaluate these and proposed a three-step method using a plane polariscope. Although their approach is conceptually good, true whole-field determination of the parameters was not possible [14]. Asundi [9] extended Tardy's method of compensation using four phase-shifted images for the evaluation of isochromatics lying on a particular isoclinic contour.

The challenges in the initial days of development of digital photoelasticity were the identification of the right combination of optical arrangements and the efficient processing of the recorded intensity data. Ramesh and Ganapathy in 1996 [14] performed a Jones calculus analysis of the six-step phase shifting method of Patterson and Wang [12] and simplified the evaluation of the expression for intensity of light transmitted. This approach opened the possibility of exploring newer optical arrangements conveniently. Once the basic methodology had been established for whole-field parameter determination, later research focused on refining these methodologies for better accuracy. Mangal and Ramesh [15] have experimentally recorded the characteristic intensity variations in the high stress gradient zones due to poor monochromaticity of the light source using modern day CCD cameras. To enhance the accuracy of data reduction in stress concentration zones, Ramesh and Sreedhar [16] reported the use of optically enhanced tiling (OET) and Xue-Feng et al [17] extended this for phase unwrapping.

In 1998, Ajovalasit et al [18] proposed a slight but important modification of the basic six-step phase shifting algorithm that uses both left and right circularly polarized lights incident on the model judiciously. Prashant and Ramesh [19] observed that the choice of optical arrangement (equations (3.5–3.7)) has an influence on processing the experimentally recorded images as the role of quarter-wave plate error is dependent on the choice of optical arrangements. When the input handedness is changed, in the presence of quarter-wave plate error ε (equation (4.4)), the term $\sin\varepsilon$ gets negated as $-\sin\varepsilon$ in the intensity equations [19]. This offers the possibility of judiciously combining left and right circularly polarised lights to minimise the quarter-wave plate error. Prashant and Ramesh [19] showed that six sets of such arrangements are possible (table 3.1). In this table, the subscripts 1–6 denote the intensity equations for various optical arrangements in a six-step circular polariscope PST. The final intensity equations remain the same (corresponding to intensity equations 5–10 in table 3.2) but the way the optical arrangements have been aligned to get these has been the subject of research to minimise the quarter-wave plate error. Of these, Ajovalasit et al [18] proposed set 6 in their paper in 1998 for PST. It is to be noted that the studies on the influence of quarter-wave plate error implicitly assume that both the quarter-wave plates (I and II) have a uniform and

Table 3.1. Combinations of intensity equations that give the least error in the presence of quarter-wave plate mismatch error.

Set 1	Set 2	Set 3	Set 4	Set 5	Set 6
I'_1	I'_1	I'_1	I'_1	I'_1	I'_1
I'_2	I'_2	I'_2	I'_2	I'_2	I'_2
I'_{3L}	I'_{5L}	I'_{3L}	I'_{5L}	I'_{3L}	I'_{3R}
I'_{3R}	I'_{5R}	I'_{3R}	I'_{5R}	I'_{4L}	I'_{4R}
I'_{4L}	I'_{6L}	I'_{6L}	I'_{4L}	I'_{5R}	I'_{5L}
I'_{4R}	I'_{6R}	I'_{6R}	I'_{6R}	I'_{6R}	I'_{6L}

The prime denotes intensity equation in the presence of quarter-wave plate error, while L and R denote optical arrangement with left and right circularly polarised light, respectively.

Table 3.2. Optical arrangement for ten-step PST.

α	ξ	η	β	Intensity equation
$\pi/2$	-	-	0	$I_1 = I_b + I_a \sin^2 \dfrac{\delta}{2} \sin^2 2\theta$
$5\pi/8$	-	-	$\pi/8$	$I_2 = I_b + \dfrac{I_a}{2} \sin^2 \dfrac{\delta}{2}(1 - \sin 4\theta)$
$3\pi/4$	-	-	$\pi/4$	$I_3 = I_b + I_a \sin^2 \dfrac{\delta}{2} \cos^2 2\theta$
$7\pi/8$	-	-	$3\pi/8$	$I_4 = I_b + \dfrac{I_a}{2} \sin^2 \dfrac{\delta}{2}(1 + \sin 4\theta)$
$\pi/2$	$3\pi/4$	$\pi/4$	$\pi/2$	$I_5 = I_b + \dfrac{I_a}{2}(1 + \cos \delta)$
$\pi/2$	$3\pi/4$	$\pi/4$	0	$I_6 = I_b + \dfrac{I_a}{2}(1 - \cos \delta)$
$\pi/2$	$3\pi/4$	0	0	$I_7 = I_b + \dfrac{I_a}{2}(1 - \sin 2\theta \sin \delta)$
$\pi/2$	$3\pi/4$	$\pi/4$	$\pi/4$	$I_8 = I_b + \dfrac{I_a}{2}(1 + \cos 2\theta \sin \delta)$
$\pi/2$	$\pi/4$	0	0	$I_9 = I_b + \dfrac{I_a}{2}(1 + \sin 2\theta \sin \delta)$
$\pi/2$	$\pi/4$	$3\pi/4$	$\pi/4$	$I_{10} = I_b + \dfrac{I_a}{2}(1 - \cos 2\theta \sin \delta)$

identical wavelength mismatch error. The mismatch error, in fact, is not uniform over the field. Further, this variation could be different for the second quarter-wave plate as compared to the first. All these variations influence the result from experimentally recorded images.

In conventional photoelasticity, it is well documented that the influence of quarter-wave plate error is the least for those optical arrangements in which the quarter-wave plates are crossed [4]. The first two optical arrangements used by

Patterson and Wang [12] in their six-step PST correspond to conventional bright and dark fields. Even for these, they chose generic optical arrangements in which the quarter-wave plates are not crossed. This is primarily because when PST development was at its initial stages, exploration of unconventional optical arrangements was the choice of the researchers to propose newer methods. Although theoretically it may not affect the calculation of the photoelastic parameters, experimentally the influence of quarter-wave plate error could be significant. Following the experience from conventional photoelasticity, Ramesh [5] further improved the six-step PST proposed by Ajovalasit *et al* [18] by incorporating the use of crossed quarter-wave plates wherever possible to minimize the error due to mismatch of quarter-wave plates.

Sai Prasad and Ramesh [20] did a systematic study on the role of background intensity on the performance of various PSTs for both plane and circular polariscope arrangements. The study has brought out the importance of accounting for the background light intensity explicitly in designing the algorithm for it to perform satisfactorily in processing experimentally recorded images. The study also suggested a simple way to assess any new PST.

One of the earliest methods for isoclinic evaluation by processing intensity data was proposed by Brown and Sullivan in 1990 [21]. They recorded four isoclinics, i.e., 0°, 22.5°, 45° and 67.5° using a monochromatic light source for whole-field isoclinic evaluation, which is termed as a polarisation stepping method. As isoclinics are not defined on isochromatic skeletons, they suggested that for recording the polarisation stepped images, the load must be minimized such that only fringe order less than or equal to 0.5 be present. Chen and Lin [22] proposed a polarisation stepping method in which both the polariser and the analyser are kept parallel. In 1999, Mangal and Ramesh [23] proposed a load-stepping method combined with PST to get the continuous isoclinics for the whole-field. The isoclinics obtained from two different loads are logically added to reduce the isochromatic interaction in the isoclinics. Petrucci [24] utilized the optical arrangement of Brown and Sullivan [21] and replaced the monochromatic light source with a white light source. In doing so, the noise due to isochromatic-isoclinic interactions is significantly reduced at moderate fringe orders (fringe order 2 at the centre of the disc) [21]. Although one can evaluate the isoclinics over the image domain, the need for a smaller number of isochromatics has been one of the issues in the development of digital photoelasticity. In this chapter, the examples chosen are such that they exhibit many isochromatics, yet it is possible to evaluate the parameters to sufficient accuracy for further processing to illustrate the current advancements in the field.

Nurse [25] proposed wavelength stepping and utilized a monochrome CCD camera with three different filters to get three different wavelengths using a white light source. He adopted an over-deterministic least squares approach to reduce the noise in the evaluation of isoclinics. Around the year 2002, researchers realized that although isoclinic evaluation is simple in conventional photoelasticity, it is quite involved in digital photoelasticity. To address this issue, composite six-step PSTs have been reported by Barone *et al* [26] and Ramji and Ramesh [27].

The various PSTs for isochromatic evaluation also provided isoclinics with variable accuracies. Ramji *et al* [28] performed a comparative study on various spatial domain algorithms for the evaluation of isoclinics. They examined the isoclinic values that are obtained using plane, circular, and mixed polariscope arrangements and suggested that the plane polariscope based algorithms are better for the isoclinic parameter evaluation. Zhenkun *et al* [29] have proposed a five-step PST and performed a comparative study between the use of monochromatic light and white light. They concluded that the isoclinics obtained using white light gives better results, except for the zones where the fringe order is zero.

Ramji and Ramesh in 2008 proposed a ten-step PST [30–32]. The methodology intelligently combines a four-step approach of Brown and Sullivan [21] for isoclinic data evaluation and a six-step circular polariscope based method for isochromatic data evaluation (table 3.2). The optical arrangements are carefully selected to minimize the influence of quarter-wave plate mismatch error. The quarter-wave plates are kept crossed wherever possible and the use of both left and right circularly polarised light (Set 6 of Prashant and Ramesh [19]) has also been accommodated judiciously. This has become a standard approach if one is interested in evaluating both isoclinics and isochromatics accurately [6, 33].

In 1997, Ramesh and Deshmukh [34] introduced PST in colour domain. They suggested that the green image plane ($\lambda \approx 546$ nm) in colour domain can act as an optical filter. As isochromatic unwrapping requires the value of total fringe order of at least one point in the domain, they suggested it could be obtained from the dark or bright field colour isochromatics of the six-step PST. Ji and Patterson [35] performed a study on the simulation of error in automated photoelasticity for their six-step PST [12]. They considered the effect of quarter-wave plate mismatch on photoelastic parameters in white light source and suggested that the median band filters, with centre wavelength of 550 nm, reduce quarter-wave plate mismatch error. Tamrakar and Ramesh [36] reported a more comprehensive simulation of quarter-wave plate mismatch error and optical misalignment in digital photoelasticity for a few PSTs by Jones calculus. A generic error simulation is reported by Ramji *et al* in 2009 [37]. Ajovalasit *et al* [38] studied the influence of spectral content of the light source, spectral response of the camera and quarter-wave plate error on a few PSTs using white light. With advancements in colour image processing methodologies (chapter 4), while using a 3CCD colour camera, the green plane image is used as the reference (546.1 nm) for identifying the total fringe order up to twelve fringes [39] from the dark field colour isochromatics. This opened the possibility of employing the ten-step method in colour domain [34]. The first four-plane polariscope images in colour assure high quality determination of isoclinics and the green plane of the next six images is then used to find the isochromatic data over the field. There are further variations of these methodologies and these will be discussed in chapter 4.

3.4 Evaluation of photoelastic parameters using intensity information

The number of unknowns at each point in a model are four, which are I_a, I_b, δ, and θ. In principle, just four optical arrangements should be sufficient to evaluate these. However, research in the past three decades has established that ten-images as in table 3.2 are needed for accurate evaluation of the parameters. This will become clear in the following sections.

Many of the methods reported earlier used fringe fields with only a few fringe orders, which showed less error in isoclinic evaluation. Such errors would increase if those methods are applied to fringe fields with dense fringe orders. If one needs an accurate evaluation of photoelastic parameters, the PST method should be robust so that the accuracy of parameters evaluation is not compromised due to higher retardations. Hence, consciously in the subsequent sections, models that are appropriately loaded to reveal a larger number of fringes are taken up for elucidating the performance of the ten-step method. Figure 3.2 shows the theoretically simulated ten phase-shifted images for the problem of a disc under diametral compression (dia = 60 mm). The diametral load is 1578 N, the thickness of the specimen is 6 mm, and the material stress fringe value is 12.88 N/mm/fringe. The fringe order at the centre of the disc is 5.2, that has been evaluated by the Tardy's method of compensation in chapter 1 (section 1.11), which is high, when compared

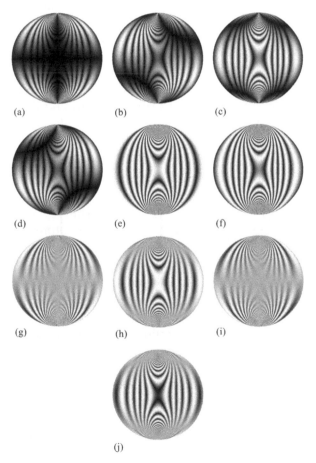

Figure 3.2. Ten-step phase-shifted images of a disc (diameter = 60 mm, load = 1578 N and F_σ = 12.88 N/mm/ fringe) under diametral compression. Courtesy: P_Scope® software, IIT Madras [40].

with the requirement for the value of maximum fringe order being less than 0.5 over the model domain as originally proposed by Brown and Sullivan in 1990 [21].

3.4.1 Calibration of polariscope

In conventional photoelasticity, the polariscope is generally aligned by inspecting the background light. In a dark field circular polariscope, the polariser and analyser should be kept crossed and the I and II quarter-wave plates must also be kept crossed. Although in theoretical development, for a quarter-wave plate, one will be naming the axes as fast or slow, it is not mandatory to check whether this alignment is satisfied in the experimental arrangement, as the intensity equations are independent of this (equation (1.15)). However, in digital photoelasticity absolute orientation of the optical elements plays a significant role (equations (3.5–3.7)). Hence, this must be addressed while aligning the polariscope.

The theoretically simulated phase-shifted images can be used as a guide to achieve the necessary calibration by means of comparison. The difference between figure 3.2(g and i) is that in one case, the quarter-wave plate slow axis is at 135° and in another case, it is at 45°. There is a definite change in the appearance of the intensity variation. As a circular disc is easily available, one can use these images as a reference to calibrate the polariscope so that the *fast* and *slow* axes of the quarter-wave plates are correctly identified. Every effort must be made to correctly align the polariscope such that one-to-one correspondence exists between the arrangement used for theoretical development of intensity equations and the arrangement used for experimental recording of intensities.

3.4.2 Evaluation of the photoelastic parameters

Using the first four equations in table 3.2, the isoclinic parameter is obtained as

$$\theta_c = \frac{1}{4}\tan^{-1}\left(\frac{I_4 - I_2}{I_3 - I_1}\right) = \frac{1}{4}\tan^{-1}\left(\frac{I_a \sin^2\frac{\delta}{2}\sin 4\theta}{I_a \sin^2\frac{\delta}{2}\cos 4\theta}\right) \text{for } \sin^2\frac{\delta}{2} \neq 0 \qquad (3.8)$$

where, the subscript 'c' refers to the principal value of the inverse trigonometric function in this equation as well as in subsequent equations. Equation (3.8) also gives the mathematical expressions of the intensity difference, which helps to visualise certain issues on parameter estimation. Representing equations in this manner, is adopted for some of the key equations discussed in this chapter. Equation (3.8) is valid only when $\sin^2(\delta/2) \neq 0$. This implies that the isoclinic parameter is undefined when $\delta = 0, 2\pi, 4\pi, \ldots$. In other words, the isoclinic contour is not defined at dark-field isochromatics and hence, it will not be continuous over the domain. When δ is not exactly equal to 0 or 2π or 4π etc, but close to these values, θ_c does not become indeterminate, but the value of θ_c determined will be unreliable and appears as noise in isoclinic fringe when plotted.

If white light is used as a source and the colour image is split into the respective image planes of red (R), green (G) and blue (B), then θ_c can be determined as

$$\theta_c \;=\; \frac{1}{4}\tan^{-1}\!\left(\frac{(I_{4,R}+I_{4,G}+I_{4,B})-(I_{2,R}+I_{2,G}+I_{2,B})}{(I_{3,R}+I_{3,G}+I_{3,B})-(I_{1,R}+I_{1,G}+I_{1,B})}\right) \tag{3.9}$$

The subscripts R, G, B refer to the red, green and blue colour planes of the image.

It is important to note that the first four intensity equations of table 3.2 are obtained in a crossed plane polariscope and the next six intensity equations are recorded in a generic circular polariscope. The isoclinics can also be determined by using only the intensity maps recorded in a generic circular polariscope as [2],

$$\theta_c = \frac{1}{2}\tan^{-1}\!\left(\frac{I_9-I_7}{I_8-I_{10}}\right) \;=\; \frac{1}{2}\tan^{-1}\!\left(\frac{I_a\sin\delta\sin2\theta}{I_a\sin\delta\cos2\theta}\right) \text{ for } \sin\delta\neq0 \tag{3.10}$$

Equation (3.10) shows that the isoclinic value is indeterminate at those points where $\delta=0$, π, 2π, 3π .... In principle, the noisy regions are doubled compared to those obtained by equation (3.8). This will become clear in section 3.8.1. Here again, the isoclinic contour will not be continuous over the domain. From the last six equations of table 3.2, the isochromatics can be obtained in two different ways as [2, 12],

$$\delta_c = \tan^{-1}\!\left(\frac{I_8-I_{10}}{(I_5-I_6)\cos2\theta_c}\right) \text{ for } \cos2\theta_c\neq0 \tag{3.11}$$

$$\delta_c = \tan^{-1}\!\left(\frac{I_9-I_7}{(I_5-I_6)\sin2\theta_c}\right) \text{ for } \sin2\theta_c\neq0 \tag{3.12}$$

Inspection of equations (3.11) and (3.12) shows that the evaluation of retardation (isochromatic fringe order) is interlinked to the evaluation of θ (isoclinic value). Further, in equations (3.11) and (3.12), δ_c has low-modulation regions when $\cos2\theta_c$ or $\sin2\theta_c$ is small. In these regions, the intensity differences (I_8-I_{10}) or (I_9-I_7) will also be small. The evaluation of δ_c will involve the ratio of two small discrete numbers and will be in considerable error in such regions. Despite this, many researchers used either equation (3.11) or equation (3.12) for retardation calculation, as there was a euphoria that one could get the retardation over the domain easily unlike the need for performing Tardy's method of compensation for every point in the domain. With advancements in research, this issue was ably addressed by Quiroga and Gonzàlez-Cano [41] and they suggested the following form of intensity processing:

$$\delta_c = \tan^{-1}\!\left(\frac{(I_9-I_7)\sin2\theta_c+(I_8-I_{10})\cos2\theta_c}{(I_5-I_6)}\right) \tag{3.13}$$

Albeit simple, it is a significant development noted by Ramesh [2] in 2000, and it took some time for the researchers to see its advantage and start using them.

It is to be noted that equations (3.8–3.10) and (3.13) involve inverse trigonometric operations and thus, one has multivalued solutions, and this poses a problem in correctly and uniquely evaluating δ and θ over the domain. To evaluate the inverse trigonometric function, one can use 'atan' function if one knows only the value of the ratio and it returns the value in the principal range $-\pi/2$ to $\pi/2$. If, however, the numerator and denominator can be independently evaluated including the sign, then, 'atan2' function is suitable for evaluation and it returns the value in the range $-\pi$ to π. Equations (3.8), (3.9), and (3.13) are amenable for using 'atan2' function and for equation (3.10), one can use only 'atan' function. Nevertheless, equations (3.8), (3.9) (using atan2) and (3.10) (using atan) give θ_c in the range of $-\pi/4$ to $\pi/4$. Physically, θ is in the range of $-\pi/2$ to $\pi/2$. The values of θ_c obtained from the trigonometric calculations need to be mapped to the actual physical range of θ and this process is called unwrapping of isoclinics in its simplest sense. More details regarding this will be discussed in sections 3.5–3.7. Use of atan2 function for δ_c (equation (3.13)) gives it in the range of $-\pi$ to π. For further processing, it is desirable that δ_c is expressed in the range of 0 to 2π. Ultimately, the retardation must be expressed as total fringe order (N) at a point, which requires suitable addition of the integer part of the fringe order. This is called unwrapping of the isochromatic parameter. With this simple introduction, in subsequent discussions the terms wrapped and unwrapped can be suitably interpreted for both isoclinics and isochromatics. More clarity will emerge after detailed discussions in sections 3.5–3.7.

The ten-step method advocates the use of unwrapped theta (θ) in evaluating the delta (equation (3.13)). The terms in equation (3.13) can be written as,

$$I_9 - I_7 = I_b + \frac{I_a}{2}(1 + \sin 2\theta \sin \delta) - \left(I_b + \frac{I_a}{2}(1 - \sin 2\theta \sin \delta)\right)$$

$$= I_a \sin 2\theta \sin \delta \qquad (3.14)$$

$$I_8 - I_{10} = I_b + \frac{I_a}{2}(1 + \cos 2\theta \sin \delta) - \left(I_b + \frac{I_a}{2}(1 - \cos 2\theta \sin \delta)\right)$$

$$= I_a \cos 2\theta \sin \delta \qquad (3.15)$$

$$I_5 - I_6 = I_b + \frac{I_a}{2}(1 + \cos \delta) - \left(I_b + \frac{I_a}{2}(1 - \cos \delta)\right) = I_a \cos \delta \qquad (3.16)$$

Substituting equations (3.14–3.16) in equation (3.13), one will get,

$$\delta_c = \tan^{-1}\left(\frac{I_a \sin 2\theta \sin \delta \sin 2\theta_c + I_a \cos 2\theta \sin \delta \cos 2\theta_c}{I_a \cos \delta}\right)$$

$$= \tan^{-1}\left(\frac{\sin \delta}{\cos \delta}(\sin 2\theta \sin 2\theta_c + \cos 2\theta \cos 2\theta_c)\right) \qquad (3.17)$$

Equation (3.17) clearly indicates that, if wrapped isoclinic value (θ_c) obtained from equations (3.8), (3.9) or (3.10) is substituted in equation (3.17), isoclinic-isochromatics interactions will be present. However, instead of wrapped isoclinic value, if unwrapped isoclinic values (instead of θ_c, use θ or more clearly $\theta_{\text{unwrapped}}$) are put in equation (3.17), it becomes,

$$\begin{aligned}
\delta_c &= \tan^{-1}\left(\frac{\sin\delta}{\cos\delta}(\sin 2\theta \sin 2\theta + \cos 2\theta \cos 2\theta)\right) \\
&= \tan^{-1}\left(\frac{\sin\delta}{\cos\delta}\right)
\end{aligned}$$
(3.18)

This indicates that the fractional retardation is not influenced by isoclinic angle calculation. The revised expression for δ_c is then,

$$\begin{aligned}
\delta_c &= \tan^{-1}\left(\frac{(I_9 - I_7)\sin 2\theta_{\text{unwrapped}} + (I_8 - I_{10})\cos 2\theta_{\text{unwrapped}}}{(I_5 - I_6)}\right) \\
&= \tan^{-1}\left(\frac{I_a \sin\delta}{I_a \cos\delta}\right)
\end{aligned}$$
(3.19)

3.5 Phasemaps in photoelasticity

One is normally accustomed to observing fringe patterns as results in optical methods. Instead, in PST one usually evaluates phasemaps from the intensity record of several images. Phasemap is a grey scale plot of the phase information. In photoelasticity, phasemaps can be plotted for isoclinics (θ) and isochromatics (δ). Isoclinic value physically lies in the range $-\pi/2$ to $+\pi/2$. A grey scale theta plot of isoclinic can be obtained by defining [4]:

$$g(x, y) = \text{INT}\left[\frac{255}{\pi}\left(\theta + \frac{\pi}{2}\right)\right] = \text{INT}[I]$$
(3.20)

where $g(x, y)$ is the grey level value at the point (x, y), INT $[I]$ is the nearest integer of I. Such a representation will plot 'black' for $\theta = -\pi/2$ and 'white' for $\theta = \pi/2$. As it plots the values for the complete range, it will give an unwrapped phasemap (figure 3.3(a)). The plot is theoretically simulated to show the major principal direction as unwrapped phasemap. If the same plotting scheme is adopted for the principal value of the isoclinic (θ_c) calculated in equation (3.8), which lies in the range $-\pi/4$ to $+\pi/4$, then it will give a wrapped phasemap. In this, one would see specific zones in the model domain (figure 3.3(b)) rather than a smooth variation of the grey scale values. The reason for these zones and their nature would be discussed in the next section.

Fractional retardation is obtained in the range $-\pi \leqslant \delta_c \leqslant \pi$. This must be first expressed in the range of 0 to 2π to plot it as a phasemap. This is then converted into grey levels between 0 and 255 for graphical plotting. The zero fractional retardation corresponds to pitch black and 2π fractional retardation corresponds to pure white, i.e., 255. The phasemap is plotted using the following relations [4]

$$\begin{aligned}
\delta_p &= \begin{cases} \delta_c & \text{for} \quad \delta_c > 0 \\ 2\pi + \delta_c & \text{for} \quad \delta_c \leqslant 0 \end{cases} \\
g(x, y) &= \frac{255}{2\pi}\delta_p
\end{aligned}$$
(3.21)

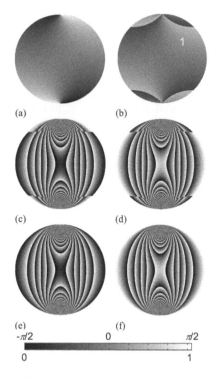

Figure 3.3. Theoretical simulation of phasemaps of a circular disc under diametral compression: (a) unwrapped isoclinics representing major principal stress direction consistently over the domain, (b) wrapped isoclinics, (c) wrapped isochromatics using equation (3.13) based on major principal direction, (d) wrapped isochromatics using equation (3.13) based on minor principal direction, (e) ambiguity free wrapped isochromatics based on major principal stress direction (equation (3.19)), (f) ambiguity free wrapped isochromatics based on minor principal stress direction (equation (3.19)). Courtesy: P_Scope® software, IIT Madras [40].

Thus, the phasemap obtained is only a wrapped phasemap as it has data of only fractional retardations over the domain. For getting the total fringe order, the wrapped phase must be unwrapped suitably by adding the integer value, dictated by the problem.

In wrapped isochromatics of figure 3.3(c), one can see similar zones as obtained in wrapped isoclinics of figure 3.3(b). Instead of the major principal direction, if one uses minor principal direction, the wrapped phasemap of isochromatics is shown in figure 3.3(d). One can see similar zones as in figure 3.3(b) and the zones observed in figures 3.3(c) and 3.3(d) are complimentary. Why such zones get formed and how these can be labelled are discussed in section 3.6. However, such zones are not seen in figures 3.3(e) and 3.3(f) as it uses unwrapped θ in calculating δ. Comparing the wrapped phasemap of figure 3.3(e) with that of the dark field image of figure 3.2(f), the features are remarkably similar. This can be used as an indirect check to ensure that the wrapped phasemap of isochromatics is free of ambiguities (section 3.6) suitable for effecting the phase unwrapping [42]. In complex problems involving multi bodies, its advantage will be seen. Unwrapping of isochromatics refers to the suitable addition of integral values to the fractional fringe orders to get the total fringe orders over the domain. Figure 3.3(f) is also free of ambiguities suitable for further processing. However, the unwrapping algorithm will be different for figure 3.3(e and f). In this chapter, examples are chosen to illustrate both the cases.

3.6 Intricacies in phasemaps of digital photoelasticity

In most optical techniques, the intensity information is related to a single physical information. For example, in holography, one gets out-of-plane displacements and in Moiré, one records either u or v displacement from a single experiment. On the other hand, in photoelasticity, the intensity information is affected by both the difference in principal stresses expressed as retardation (δ) and the orientation of the principal stress (θ). As the basic equations used for parameter estimation are inverse trigonometric functions, they are multivalued as noted before in section 3.4. Special care must be taken in finding the correct solution. If not handled properly, it can lead to certain *inconsistencies* in the isoclinic evaluation and results in *ambiguous* zones in isochromatic phasemaps. Comparing figure 3.3(b and c), the shape of the marked regions is identical in both. In isoclinic phasemap these are labelled as *inconsistent* zones and they lead to the formation of *ambiguous* zones in an isochromatic phasemap. This interdependence was first noted by Ramesh [2] in 2000. Why they are named as *inconsistent* and *ambiguous* will become clear in the subsequent discussions.

It is to be noted that the principal value of θ_c lies in the range of $-\pi/4 \leqslant \theta_c \leqslant \pi/4$. Since one does not know the principal stress direction *a priori*, there exists an inconsistency on whether θ_c corresponds to σ_1 or σ_2 direction over the domain. In view of this, the specific zones formed are labelled as *inconsistent* zones. In figure 3.3(b), the values of theta in Region 1 corresponds to maximum principal stress and in the other four small regions near the load application points, it corresponds to minimum principal stress. Associating these regions to specific principal stress direction is possible as these are theoretically generated. Once the concept is understood for a known problem, such regions in unknown problems can be identified and handled appropriately for processing the data.

Inconsistent zone is labelled in a relative perspective and at the boundaries of these zones, there is a jump in the value of isoclinic angle by $\pi/2$ degrees. If one wants to find only the orientation of the major principal stress over the domain, then the small four regions are not consistent with Region 1 and hence, these could be classified as *inconsistent* zones. One can also represent for the entire domain the orientation of the minimum principal stress and if done, then Region 1 becomes an *inconsistent* zone. The important aspect to note is, over the model domain, the isoclinic values in the range $-\pi/2 \leqslant \theta \leqslant \pi/2$ are to be obtained for one of the principal stresses consistently. This process is called unwrapping of isoclinic phasemap, which is discussed in section 3.7.

Calculation of fractional retardation requires first the evaluation of isoclinic angle. The use of *inconsistent* values of θ leads to the formation of *ambiguous* zones in isochromatic phasemap (figure 3.3(c), figure 3.4(a)). Unlike total fringe orders, which are invariably positive, fractional fringe orders have an associated sign. In these zones, there is an ambiguity on the sign of the fractional fringe order hence,

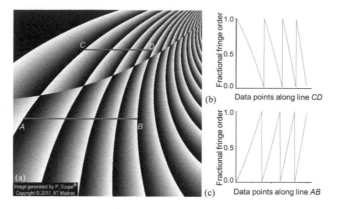

Figure 3.4. (a) Magnified view of the top left portion of figure 3.3(c) with lines *AB* and *CD* marked in normal and ambiguous zones, respectively. Gradient of intensity in the direction of increasing fringe order: (b) ambiguous zone, (c) normal zone. Courtesy: P_Scope® software, IIT Madras [40].

aptly named as *ambiguous* zones. Line *AB* is drawn in the normal zone, and line *CD* is drawn in the *ambiguous* zone (figure 3.4(a)). Figure 3.4(b and c) show the intensity variations along lines *CD* and *AB*, respectively. The slope of these lines is exactly the opposite. The sign of fractional retardation in *ambiguous* zones is opposite to that in the normal zone. For isochromatic unwrapping, the isochromatic phasemap has to be free of *ambiguous* zones [42, 43] (figure 3.3(e or f)). The *ambiguous* zone is again a relative term like an *inconsistent* zone.

3.7 Unwrapping methodologies

In the case of isoclinic phasemaps, unwrapping refers to the process of obtaining the direction of either σ_1 or σ_2 consistently over the entire domain, while in the case of isochromatic phasemaps, unwrapping refers to suitable addition of the integral value to the fractional retardation values to get continuous total fringe order data. The two focal issues are: how to avoid propagation of errors and how to handle complex geometries with cut-outs.

The choice of the scanning scheme for phase unwrapping influences its sensitivity to the propagation of errors. The simplest phase unwrapping methodology is to start from a seed point and scan the image horizontally, vertically or use these scans judiciously [44]. This algorithm can work well if it does not encounter a noise point, geometric discontinuity, or material discontinuity.

Another approach is to guide the path of unwrapping through a quality measure. In quality guided phase unwrapping, the pixels in the phasemap are graded depending on its quality where '1' refers to the highest quality and '0' denotes the lowest quality. Consequently, in the quality map, the black areas are of low quality, the white areas are of high quality and the grey areas are of intermediate quality. Unwrapping starts from the highest quality pixel and then proceeds towards the lowest quality pixels—thereby assuring the best possible results. It is computationally intensive, and several quality measures are reported in the literature. Among these, phase derivative variance is recommended for photoelasticity [44, 45].

Phase derivative variance gives the badness of the pixel as

$$\frac{\sqrt{\sum\left(\Delta^x_{i,j} - \overline{\Delta}^x_{m,n}\right)^2} + \sqrt{\sum\left(\Delta^y_{i,j} - \overline{\Delta}^y_{m,n}\right)^2}}{k^2} \qquad (3.22)$$

where for each sum, the indices (i, j) range over the $k \times k$ neighbourhood of each centre pixel (m, n) and $\Delta^x_{i,j}$ and $\Delta^y_{i,j}$ are the partial derivatives of the phase (i.e., the wrapped phase differences) along the x- and y-directions. The terms $\overline{\Delta}^x_{m,n}$ and $\overline{\Delta}^y_{m,n}$ are the averages of these partial derivatives in the $k \times k$ windows. Equation (3.22) is a root-mean-square measure of the variances of the partial derivatives in the x- and y-directions. The variance is negated consequently to represent goodness. As the scanning path is driven by quality, in principle, accumulation of error is minimized in this approach.

Phase unwrapping of complex model boundaries with cut-outs is a challenge by itself. Simply connected models are in general easy to handle. For multiply connected bodies, one can choose either domain delimiting or boundary masking approaches. In domain delimiting, the multiply connected models are unwrapped by dividing them into an assembly of simply connected domains. Madhu and Ramesh [46] proposed an approach to unwrap complex geometries having cut-outs. The boundary coordinates information required for delimiting is obtained using boundary extraction techniques.

In boundary masking approach, the image area is identified by a binary image. It is possible to handle any geometric shape using this approach. Geometric primitives to draw the boundary mask (figure 3.5) for even complicated geometries have been incorporated in digital photoelastic software such as *Digi*TFP® [47] and *Digi*Photo [48]. *Digi*TFP® is a comprehensive software for evaluating total fringe order and isoclinics at every point in the model using the colour information. It has the provision to handle phase shifting in colour domain. *Digi*Photo is a generic software covering various aspects of digital photoelasticity such as fringe multiplication, fringe thinning and PST using monochromatic light source. Both these software share many commonalities and have a provision to draw the boundaries of the model by simple drawing primitives. Using this drawing as a starting point, the domain is appropriately scanned from the user-specified seed points to generate the mask. It is essentially a binary image with white denoting the model domain and black denoting the background region (section 3.12.2). The features of the dialog box are self-explanatory.

Quality guided algorithm for phase unwrapping in conjunction with boundary masking would be a user-friendly phase unwrapping strategy. These are available as standard modules in *Digi*TFP® [47] and *Digi*Photo [48].

3.8 Evaluation of isoclinics

3.8.1 Origin of noise in isoclinic data

The choice of plane polariscope based isoclinic evaluation in ten-step PST is preferred because it has less noise when compared to other algorithms. Equation (3.10), which uses the intensity maps recorded in a circular polariscope is proposed

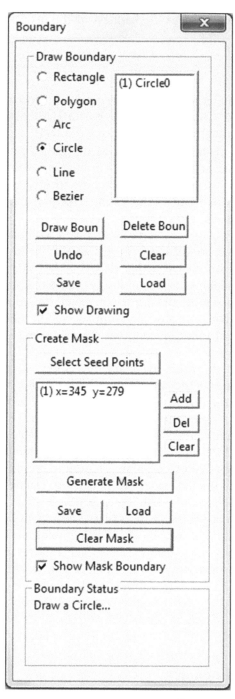

Figure 3.5. Figure showing the dialog box available in *Digi*TFP® [47] and *Digi*Photo [48] to create the boundary mask for identifying the model boundaries. Courtesy: [47, 48].

by Patterson and Wang [12]. Many researchers have used this until its issues were pointed out by Ramji *et al* [28]. Though Ramji *et al* [28] had shown that isoclinics obtained from circular polariscope based algorithm is noisy, a detailed study on what caused this noise is reported by Ramesh *et al* in 2015 [49]. The pertinent details are presented next to appreciate the supremacy of plane polariscope based methods for isoclinic evaluation by PST.

The experimental dark field image for the problem of a disc under diametral compression (diameter = 60 mm, load = 1578 N and F_σ = 12.88 N/mm/fringe) which corresponds to configuration 6 in table 3.2, is shown in figure 3.6(a). The isochromatic fringe order at the centre of the disc is 5.2 indicating that it is highly loaded. Experimentally obtained isoclinic phasemap using a plane polariscope is shown in figure 3.6(b) (i.e., using equation (3.8)) and a circular polariscope is shown in figure 3.6(c) (i.e., using equation (3.10)). Analytically obtained isoclinic phasemap for the disc is shown in figure 3.3(b).

Inspection of figures 3.6(b) and 3.6(c) shows that the isoclinic phasemaps are not continuous over the domain when compared to figure 3.3(b). This is due to noise and it is more in figure 3.6(c) than in figure 3.6(b) as already noted in section 3.4. The noise in figure 3.6(b) is because the isoclinic values are not defined on the isochromatics fringe skeleton as noted in section 3.4.

The origin of noise in the circular polariscope is different. The denominator of equation (3.10), to calculate isoclinics from a circular polariscope, is a function of the intensities of the 8th and 10th configurations shown in table 3.2. Pixels that exhibit a sign reversal in the denominator of equation (3.10) are identified and plotted as transitional pixelmap (figure 3.6(d)). There is a good comparison between

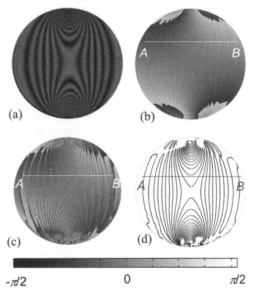

(a)　　(b)

(c)　　(d)

$-\pi/2$　　　0　　　$\pi/2$

Figure 3.6. Circular disc under diametral compression at a diametral load of 1578 N (a) experimental dark field image. Wrapped isoclinic phasemap: (b) experimental using equation (3.8), (c) experimental using equation (3.10), (d) transitional pixelmap for the denominator of equation (3.10). Adapted from [49] with permission of SPIE.

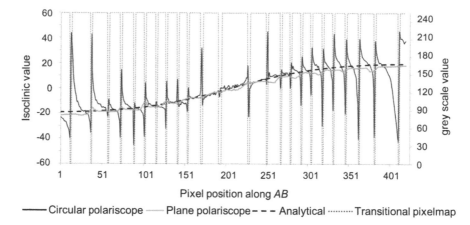

Figure 3.7. Wrapped isoclinic values and grey scale values along the line *AB* in figure 3.6. Reprinted from [49] with permission of SPIE.

the transitional pixelmap (figure 3.6(d)) and the noise in the isoclinic phasemap obtained using a circular polariscope (figure 3.6(c)).

Figure 3.7 shows a quantitative comparison of analytical and experimental isoclinic values obtained using plane as well as circular polariscope along with the grey scale values of the transitional pixelmap (figure 3.6(d)) for the line *AB* shown in Figs. 3.6(b–d). It can be observed from figure 3.7, that the θ values obtained using the circular polariscope based method have prominent noise when compared to θ obtained using a plane polariscope. Spikes of noise seen in figure 3.7 for the circular polariscope algorithm coincide with the jumps in the transitional pixelmap. From this understanding, the isoclinics obtained from a circular polariscope cannot be used for any further data processing due to its high noise. Hence, plane polariscope based algorithms must be used for isoclinics evaluation.

3.8.2 Unwrapping of isoclinics

To remove the inconsistencies in wrapped phasemap (figure 3.6(b)), the algorithm should first identify the boundaries of these zones. From the discussion in section 3.6, the boundaries of the *inconsistent* zones are characterized by a $\pi/2$-jump. Visually this boundary is smooth in a theoretically generated image (figure 3.3(b)) but in an experimental image, they are slightly different (figure 3.6(b)). The phase unwrapping procedure should be devised to identify this and in addition, it should also accommodate other features of the isoclinic phasemaps that are discussed next.

3.8.2.1 Isotropic points and pi-jumps

In a generic problem, unwrapped isoclinic phasemap will have isotropic points and π-jumps. At an isotropic point, all isoclinics merge, while a π-jump is a sudden jump of 180° in the isoclinic values, which starts from an isotropic point [2]. The challenge in devising a generic phase unwrapping algorithm is that it should accommodate the existence of such features, even if all of them may not be present in a specific problem. A visual appreciation of such features can help to understand it better.

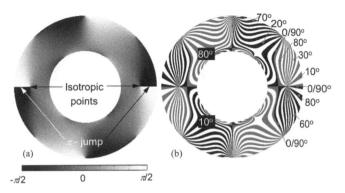

Figure 3.8. Theoretically generated isoclinics for ring under diametral compression: (a) unwrapped isoclinic phasemap of major principal direction with isotropic points and π-jumps indicated, (b) representation of all isoclinics in steps of 10°. Black corresponds to major principal stress; blue corresponds to minor principal stress directions. Courtesy: P_Scope® software, IIT Madras [40].

Phasemap of a ring under diametral compression exhibits both isotropic points and π-jumps. A theoretically generated unwrapped phasemap is shown in figure 3.8(a), which provides the major principal stress direction consistently over the domain. In conventional photoelasticity, one is conversant with the binary representation of isoclinics in steps of 10° (figure 3.8(b)). Binary representation of isoclinics from theoretical simulation can be done by the following code [4],

$$\text{If } ((-2.5° + \theta) < \theta < (2.5° + \theta))g_b(x, y) = 0, \text{ else } g_b(x, y) = 255 \qquad (3.23)$$

where, $\theta = -80°, -70°, \ldots 0° \ldots 10°, 20° \ldots 80°$ and g_b denotes the pixel intensity in the binary image. The features of the unwrapped phasemap can be appreciated from the binary representation of the isoclinics. The subtle aspect of the π-jump is brought out well by plotting the major and minor principal stress directions in different colours (figure 3.8(b)). The identification of isotropic points is straightforward.

In complex photoelastic models, many isotropic points/π-jumps may be present in the isoclinic phasemap. Such features are also a function of which principal stress direction is sought to be determined while unwrapping. By comparing the phasemap of disc shown in figure 3.3(a) to figure 3.8(a), one can simply say that handling a ring under diametral compression is more challenging. It is not so—even a disc under diametral compression can be quite challenging if one looks at the nature of unwrapped phasemap if it depicts the minor principal stress (figure 3.9(a)). It does show the presence of π-jumps. Figure 3.9(b) shows the unwrapped isoclinic phasemap of a ring under diametral compression corresponding to the minor principal direction, which is quite different from figure 3.8(a). So, special care needs to be taken to capture these features in the process of unwrapping. These are subtle concepts and are discussed threadbare here to appreciate the nuances in digital photoelasticity.

3.8.2.2 Adaptive quality guided phase unwrapping

Ramji and Ramesh [30–32] proposed an adaptive quality guided phase unwrapping (AQGPU) algorithm for unwrapping isoclinics which accommodates the presence of

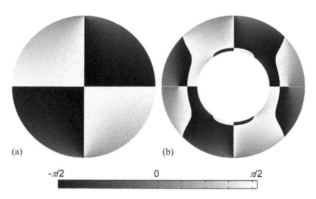

(a)   (b)

-π/2   0   π/2

Figure 3.9. Theoretically simulated unwrapped isoclinic phasemap corresponding to minor principal stress direction for the problems of: (a) disc under diametral compression, (b) ring under diametral compression. Courtesy: P_Scope® software, IIT Madras [40].

isotropic points as well as π-jumps suitably. A seed point of high quality is selected to initiate the phase unwrapping and the unwrapping path is dictated by the quality map. Phase unwrapping is done by applying suitable checking conditions for the immediate neighbourhood to the current pixel such as the left pixel, right pixel, bottom pixel, and top pixel positions. The checking conditions for left and right pixel positions are shown in table 3.3. The checking conditions autonomously take care of the isotropic point(s)/π-jumps occurring in the isoclinic phasemap while unwrapping. For identifying the *inconsistent* zones, a variable theta tolerance (*theta_tol*) is needed. A value of π/3 for *theta_tol* is found to be enough for a variety of problems. To explain the role of these conditions, in table 3.3 for the Left pixel position (LP), the checking conditions are marked as 1a, 2a, 3a and 4a, respectively. The conditions (1a & 3a) respectively take care of both the π-jumps and the presence of isotropic points while unwrapping. This also uses the *theta_tol* as an additional check. These conditions ensure that the isoclinic values lie in the range $-\pi/2 < \theta < +\pi/2$ so that at the interface of π-jump, it exactly reverses (from −π/2 to +π/2 or vice-versa). The conditions (2a & 4a) respectively take care of the π/2-jump denoting an *inconsistent* zone depending upon the sign of the gradient of the adjacent pixels.

Initially, from the seed point chosen, its four adjacent neighbours are unwrapped based on the checking conditions discussed previously. These unwrapped neighbours are sorted and stored based upon their quality values (highest to lowest) in a list called *adjoin list*. A minimum quality threshold decides up to which value of quality, the *adjoin list* would have and it changes as the unwrapping propagates. A starting value of 0.001 is generally used. Four neighbours of the pixel with the highest-quality values from the list are then unwrapped. If any of the neighbouring pixels have already been unwrapped, they are not considered for unwrapping. The highest quality pixel is removed from the list. The neighbours which are currently unwrapped and having quality value higher than the minimum quality threshold value are added to the *adjoin list*. They are sorted to find out the next pixel position for unwrapping. In this way, the path of unwrapping is decided by quality. It is essentially a region growing approach based upon flood filling algorithm.

Table 3.3. Checking conditions for isoclinic unwrapping based upon pixel position for AQGPU algorithm.

Pixel position	Checking condition
Left Pixel (LP)	$abs(\theta_{i,j} - \theta_{i-1,j}) > theta\_tol,\quad then$
	$\theta_{i-1,j} = \begin{cases} \theta_{i-1,j} + \dfrac{\pi}{2} & \theta_{i,j} <= \dfrac{-\pi}{2} \;\&\; \theta_{i-1,j} <= 0 \;\&\; (\theta_{i-1,j} - \theta_{i,j}) > theta\_tol \quad (1a) \\[2mm] \theta_{i-1,j} - \dfrac{\pi}{2} & \theta_{i,j} > \dfrac{-\pi}{2} \;\&\; (\theta_{i-1,j} - \theta_{i,j}) > theta\_tol \quad (2a) \end{cases}$
	$\theta_{i-1,j} = \begin{cases} \theta_{i-1,j} - \dfrac{\pi}{2} & \theta_{i,j} >= \dfrac{\pi}{2} \;\&\; \theta_{i-1,j} > 0 \;\&\; (\theta_{i-1,j} - \theta_{i,j}) < - theta\_tol \quad (3a) \\[2mm] \theta_{i-1,j} + \dfrac{\pi}{2} & \theta_{i,j} < \dfrac{\pi}{2} \;\&\; (\theta_{i-1,j} - \theta_{i,j}) < - theta\_tol \quad (4a) \end{cases}$
Right Pixel (RP)	$abs(\theta_{i,j} - \theta_{i+1,j}) > theta\_tol,\quad then$
	$\theta_{i+1,j} = \begin{cases} \theta_{i+1,j} + \dfrac{\pi}{2} & \theta_{i,j} <= \dfrac{-\pi}{2} \;\&\; \theta_{i+1,j} <= 0 \;\&\; (\theta_{i+1,j} - \theta_{i,j}) > theta\_tol \\[2mm] \theta_{i+1,j} - \dfrac{\pi}{2} & \theta_{i,j} > \dfrac{-\pi}{2} \;\&\; (\theta_{i+1,j} - \theta_{i,j}) > theta\_tol \end{cases}$
	$\theta_{i+1,j} = \begin{cases} \theta_{i+1,j} - \dfrac{\pi}{2} & \theta_{i,j} >= \dfrac{\pi}{2} \;\&\; \theta_{i+1,j} > 0 \;\&\; (\theta_{i+1,j} - \theta_{i,j}) < - theta\_tol \\[2mm] \theta_{i+1,j} + \dfrac{\pi}{2} & \theta_{i,j} < \dfrac{\pi}{2} \;\&\; (\theta_{i+1,j} - \theta_{i,j}) < - theta\_tol \end{cases}$

The *adjoin list* size is maintained as sum of L and M (L × M is the image size) to optimise the computational time. When the *adjoin list* exceeds the (L + M) size while unwrapping, the last half of the pixels are removed from the *adjoin list* and are labelled as postponed pixels. A new minimum quality threshold value is defined (quality value of the last pixel in the *adjoin list* after removal). The unwrapped values of the postponed pixels are maintained, and they are put back into the *adjoin list* at a later stage when the *adjoin list* gets empty. Once the list is filled, unwrapping process proceeds as discussed earlier.

Proceeding in this manner, unwrapping is done in a single step for the entire model domain (if simply connected) without any further human intervention. In problems that have multiple domains and those problems in industrial scenarios require advanced methods that are discussed later.

3.8.3 Importance of binary representation

The semblance of isochromatic skeleton on isoclinic phasemap (figure 3.10(a)) is actually noise and is clearly visible if the isoclinics are plotted over the domain as binary contours of 10° step. Binary representation of isoclinic data for figure 3.10(a) is given in figure 3.10(b). Noise due to isochromatic-isoclinic interaction is clearly visible as spikes along the lines of isochromatic skeleton in the isoclinic phasemap. Although theoretically, isoclinics for a given loading are independent of load, when the loading on the disc is different, while experimentally evaluating the isoclinics, the density of isochromatics has an influence on the result of isoclinics [49]. Thus, smoothing of isoclinics is essential for further use of this data. Different approaches used for multi-directional smoothing are discussed next.

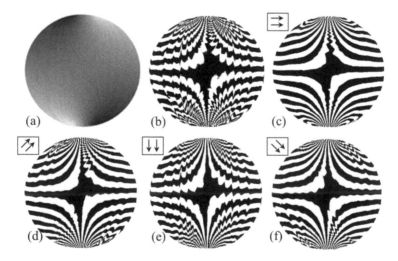

Figure 3.10. Disc under diametral compression at a load of 1578 N. (a) Unwrapped isoclinic phasemap. The feeble fringe skeletons seen here are noise. (b) Unwrapped phasemap as a binary representation showing kinks at the isochromatic locations seen in 'a'. Outlier smoothing using different scan directions (shown in top left of each image): (c) 0° scan (d) 45° scan (e) 90° scan (f) 135° scan. In each scan, the curve fitting used is RNLR with a span of 20%. Adapted from [49] with permission of SPIE.

3.9 Smoothing of isoclinics

In this section, a few methodologies for smoothing are compared and the best one is recommended for further use [49]. In fringe thinning, it is seen that the scan directions do play a role. Following that, orthogonal scans are used for developing a robust smoothing procedure. It has two steps; one is on how to remove noise in each scan direction and the next is how to combine the results suitably from different scans. Noise removal along a scan direction is done by outlier smoothing [31]. The outlier smoothing algorithm is a median absolute phase filter-based procedure in which a collection of adjacent pixels or a span is selected, and smoothing is carried out on the pixels contained within the span. The data points lying outside the trend are omitted and a local curve fitting is done by least squares analysis. Linear curve fitting is indicated with 'LR' and a second degree polynomial fit is indicated as 'NLR'. A robust version of the smoothing with outlier algorithm is indicated with prefix 'R'. In effect, 'RNLR' and 'RLR' indicate that robust outlier smoothing with polynomial and linear curve fitting, respectively. Smoothed results corresponding to individual 0°, 45°, 90° and 135° scanning directions are given as a binary plot in figures 3.10(c–f), respectively. Each scan direction is effective in removing the noise in specific zones and hence a combination of all the four scans in some form would be preferred which could result in improved noise free isoclinics. Various strategies considered for combining the individual scan results are discussed next.

3.9.1 Quality assisted smoothing

Quality guided algorithms have shown their ability in controlling the propagation of error during the execution of phase unwrapping algorithms [30–32, 44]. Among the various quality maps, the use of phase derivative variance is found to be efficient for phase shifting technique. Quality maps are generated for the results of all the four scanning directions. For a pixel location, the quality maps from all the scans are compared and smoothed data corresponding to the pixel having the highest quality is assigned. The same procedure is repeated for all the pixels in the model domain and the resulting smoothed isoclinics is shown in figure 3.11(a).

Figure 3.11. Disc under diametral compression. Binary plot in steps of 10° of smoothed isoclinic data with different algorithms: (a) quality assisted smoothing using phase derivative variance as quality measure, (b) smoothing using standard deviation approach, (c) smoothing using multi-directional progressive smoothing approach with scanning sequence of 135°, 90°, 45°, 0°. Reprinted from [49] with permission of SPIE.

3.9.2 Standard deviation (SD) assisted smoothing

In this approach, instead of quality, one makes use of standard deviation to identify the best smoothed value. Standard deviation is given by,

$$\text{SD}(i, j) = \sqrt{\frac{\sum(\theta_{i,j} - \bar{\theta}_{m,n})^2}{w^2}} \tag{3.24}$$

where for each sum, the indices (i, j) range over the $w \times w$ neighborhood of each center pixel (m, n) and $\bar{\theta}_{m,n}$ is the mean value of θ over the $w \times w$ window. The window size used for the study is 5×5. For a pixel location, the standard deviation maps of all scans are compared and smoothed data corresponding to the pixel having the lowest standard deviation is assigned to that location. The smoothed isoclinics are shown in figure 3.11(b).

3.9.3 Multi-directional progressive smoothing algorithm

Here, the smoothed isoclinic data from the first scan is smoothed by scanning in the second direction. The same procedure is progressively repeated for the other scans. The order of scanning is dictated by the problem in hand. It is found that a scanning sequence of 135°, 90°, 45°, 0° gave the best results for progressive smoothing (figure 3.11(c)) for this example. Comparing the binary representations of each of the smoothing methods, one can clearly see that the smoothing done by progressive smoothing algorithm is better than the other two algorithms. A small systematic error present in figure 3.11(c) is corrected and the comparison with theoretical results gives the least error when progressive smoothing is adopted. At lower loads where the number of fringes is very few, the mean error is within 1° and at higher loads as in the current example, it is 2.3° with a standard deviation of 1.92° [49]. The errors are high only near the load application points and in most of the model domain, the actual error remained within 1° even for the higher load case. Thus, multi-directional progressive smoothing is recommended for isoclinic data smoothing.

The multi-directional progressive smoothing approach is then applied to models, which have π-jumps and isotropic points. Adaptive smoothing procedure [49, 50] in conjunction with multi-directional progressive scanning scheme is used in each scanning direction to identify the π-jumps and isotropic points automatically and exclude them from smoothing. Earlier, this was done by artificially dividing the domain based on the knowledge of these features by the user and doing the smoothing separately and combining those results, which is tedious [31]. Table 3.4 shows the checking conditions used to take care of π-jumps and isotropic points for vertical and horizontal scanning directions. In table 3.4, '*theta_tol*' is a user-defined value used for finding the π-jumps in isoclinic data, which is problem dependent. It is used in a different context although the same variable name is used. Similar checking conditions are adopted for diagonal scans as well.

Figure 3.12 shows the dialog box for selecting the progressive smoothing parameters in the software *DigiTFP*® [47] and *Digi*Photo [48]. In progressive multi-directional smoothing, it is observed that the choice of the scanning sequence plays an important

Table 3.4. Checking conditions used by adaptive smoothing algorithm to take care of isotropic points and π-jumps.

Category	Smoothing direction	Checking condition
π-jump	Top to bottom (vertical)	$abs\left(\theta_{i,j} - \theta_{i,j+1}\right) > theta\_tol$
	Left to right (horizontal)	$abs\left(\theta_{i,j} - \theta_{i+1,j}\right) > theta\_tol$
Isotropic point	Top to bottom (vertical)	$int(\theta_{i,j}) = 0$
		$\theta_{i,j+1},\, \theta_{i,j+2} < 0 \ \&\ \theta_{i,j-1},\, \theta_{i,j-2} > 0 \,(or)$
		$\theta_{i,j+1},\, \theta_{i,j+2} > 0 \ \&\ \theta_{i,j-1},\, \theta_{i,j-2} < 0$
	Left to right (horizontal)	$int(\theta_{i,j}) = 0$
		$\theta_{i+1,j},\, \theta_{i+2,j} < 0 \ \&\ \theta_{i-1,j},\, \theta_{i-2,j} > 0 \,(or)$
		$\theta_{i+1,j},\, \theta_{i+2,j} > 0 \ \&\ \theta_{i-1,j},\, \theta_{i-2,j} < 0$

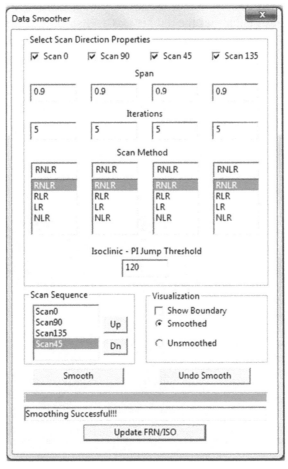

Figure 3.12. Dialog box for effecting smoothing parameters in *Digi*TFP® [47] and *Digi*Photo [48]. Courtesy: [47, 48].

role and is problem dependent. One can select the sequence of scans by appropriately ticking the boxes in the order of preference in the first row of the dialog box. The selected sequence appears in the same order in the scan sequence box at the left bottom. Even here, the sequence can be changed by using the 'Up' or 'Dn' key suitably. Different smoothing algorithms like Robust Non-Linear (RNLR), Robust Linear (RLR), Non-Linear (NLR), Linear (LR) are available for selection in the software for each scan direction individually. While selecting non-linear methods, the number of iterations can also be specified for each scan direction. The *theta_tol* can be changed in the data smoother by suitably changing the 'Isoclinic PI Jump Threshold'.

The smoothing procedure is termed local because each smoothed value is determined by the neighbouring data points defined within a span. The span defines a window of neighbouring points to be included in the smoothing calculation for each data point. The span can be expressed in terms of pixels or in terms of the percentage of the total scan width in each scan. The larger the span, the smoothed curve will follow the trend better. The selection of appropriate span length is problem specific and depends on several factors like noise level, model geometry etc. The user must select a span judiciously such that the data is not over smoothed and at the same time, noise is eliminated. However, a variety of problems can be solved with a span length ranging from 10–20 pixels or span widths of 20%–90% is possible, which will appear as 0.2–0.9 in the dialog box. The software identifies the fringe features like isotropic points and π-jumps in isoclinic data automatically by using a user specified threshold. The use of robust outlier removal methods is recommended for smoothing isoclinic data and weighted linear least squares is sufficient to smooth isochromatic data.

Problem of an angular bracket under compression (figure 3.13(a)), which has both a π-jump and an isotropic point and an internally pressurized ring (outer

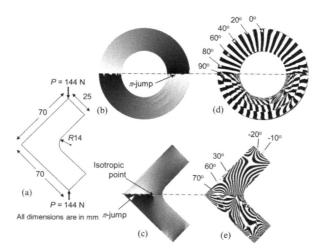

Figure 3.13. (a) Schematic of angular bracket under compression. Isoclinic phasemap with (top half) and without (bottom half) smoothing for: (b) internally pressurized ring with a smoothing sequence of 0°, 90°, 135°, 45°. The smoothing method is RNLR with a span of 90%. (c) Angular bracket under compression—the smoothing is like pressurized ring except for the zero-degree scan, 20% span is used. Binary representation of isoclinic: (d) corresponding to (b), (e) corresponding to (c). Reprinted from [49] with permission of SPIE.

diameter = 80 mm, inner diameter = 40 mm and F_σ = 12.16 N/mm/fringe, Load = 3.93 MPa) which has π-jumps along the horizontal diameter are analysed using the multi-directional progressive smoothing algorithm. The bottom half of the figures show the original data in grey scale or binary format and the top half shows the smoothed data in grey scale and binary. The grey scale images are impressive indicating a smooth variation; however, the binary representation is sensitive to even small changes and hence recommended to be plotted for accepting the results for further processing. For the angle-bracket under compression, a span width of 90% is used in all the smoothing scans. A span width of 20% is used in the horizontal and 90% is used for all the other three scans for the problem of pressurized ring. It is found that the scanning sequence 0°, 90°, 45°, 135° has worked well for these two problems.

3.10 Unwrapping of isochromatics

In isochromatic phasemap unwrapping, the integral value should be added to the fractional fringe order appropriately. It is essential that the wrapped isochromatic phasemap is free of *ambiguous* zones (equation (3.19)). A heuristic way of checking this is discussed in section 3.5, wherein the fringe features of dark field need to be seen in the isochromatic phasemap for the whole model domain if it is free of ambiguities. In the ambiguity free phasemap, how the value of fractional fringe order varies in the direction of increasing total fringe order is key in developing the algorithm for phase unwrapping. In figure 3.4, how the fractional fringe order varies in the direction of increasing total fringe order in the normal and *ambiguous* zones are presented. Depending on whether the major or minor principal stress direction is used for isoclinics unwrapping, in the whole model domain the fractional fringe order can either increase or decrease in the direction of increasing total fringe order (figure 3.3(e and f)).

A suitable seed point must be selected for which the total fringe order is known through auxiliary means. In practice, as the phasemap has high contrast, usually the integer fringe points are easily selectable, and the user must identify one of these and provide the corresponding integer fringe order determined through auxiliary means. The unwrapping process starts from the seed point and proceeds further based upon the quality map. The sign of the gradient of the adjacent pixels decides the addition or subtraction of the integral value. If it is positive, then the integer value is subtracted (k), otherwise added. Identification of the transition point to add or subtract integers is a crucial step and it is selected through a threshold value and a grey scale value of 123 is normally used. The checking condition is given as,

$$W\left(\delta_{i,j}\right) = \begin{cases} \delta_{i,j} - k & \Delta\delta_{i,j} > 123 \\ \delta_{i,j} + k & \Delta\delta_{i,j} < -123 \end{cases} \qquad (3.25)$$

where $\delta_{i,j}$ is the fractional fringe order at the corresponding pixel position (i, j), W is the unwrapping operator and $\Delta\delta_{i,j}$ is the gradient of the fractional fringe order.

The AQGPU algorithm discussed in section 3.8.2.2 is used with slight modification for isochromatic phasemap unwrapping for accommodating the condition given in equation (3.25).

3.11 Phase shifting in colour domain

Unwrapping of isochromatics is greatly simplified if *inconsistent* zones free isoclinics are used in the calculation of wrapped isochromatics while using the ten-step method so that one gets *ambiguous* zones free wrapped isochromatics [5, 6, 42, 51–53]. Further, when the model is loaded such that it has many isochromatics, isoclinics are better evaluated by processing images recorded in colour and then using the equation (3.9) for data reduction. For isochromatics, it is desirable that one uses a monochrome camera with the same image resolution as that of the colour image recorded. If not, one can even use the colour image and use only the G-channel for data processing [34] as it is found to be a good alternative from a convenience point of view.

Figure 3.14 shows the closeup of four polarisation stepped images of a plate with a hole (36 mm wide, hole diameter 10 mm, subjected to 1976 N) problem subjected to a high stress level indicated by a large density of isochromatic fringes. This example is chosen to illustrate how effective the calculation of isoclinics is, if one uses the polarisation stepped images recorded in colour. Figure 3.15 shows the six phase-shifted images recorded in colour corresponding to the optical positions 5–10 of table 3.2. Figures 3.14 and 3.15 constitute the ten phase-shifted images recorded in colour as per table 3.2. The isoclinics obtained by processing the images in figure 3.14 using equation (3.9) and unwrapping them using quality guided approach discussed in section 3.8.2.2 (using the software *Digi*TFP® [47]),

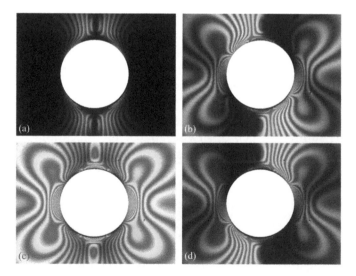

Figure 3.14. Polarisation stepped images of plate with a hole recorded in colour.

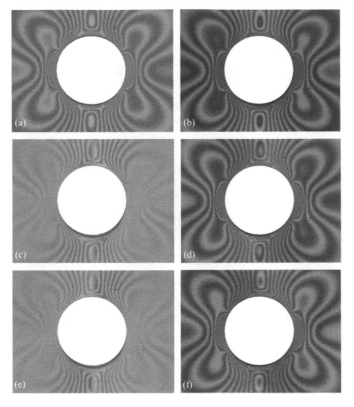

Figure 3.15. Six phase-shifted images of plate with a hole recorded in colour.

are represented as a binary image on the left half of the figure 3.16(a). In view of multi-wavelength processing, the isoclinics obtained are reasonably smooth. This result must be viewed from the perspective that this problem has an extremely high density of fringes. The results thus obtained are further smoothed by the multi-directional progressive smoothing of section 3.9.3 using RNLR in the sequence 0°, 90°, 45°, 135° with a span value of 20%. This is shown as the right half of figure 3.16(a).

The green plane images of figure 3.15 are processed using these isoclinics as input to equation (3.19). The wrapped isochromatics free of *ambiguous* zones are obtained using the software *Digi*Photo [48], and are shown in the right half of image of figure 3.16(b). One important observation is that the phasemap obtained is such that the fractional fringe order decreases in the direction of increasing total fringe order. This fact is used in conjunction with equation (3.25) for unwrapping. The result is further smoothed using progressive smoothing with NLR in the sequence 45°, 90°, 135° and 0° with the span as 10 pixels and it forms the left half of figure 3.16(b). Up to 13 fringe orders have been resolved, which demonstrates the power of the quality-guided algorithm in obtaining the results. Unlike processing analogue images, in digital domain, spatial resolution is an important issue, which affects the

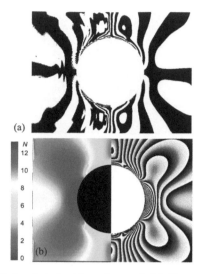

Figure 3.16. (a) Unwrapped isoclinics plotted as binary image—left half unsmoothed, and right half smoothed by progressive smoothing using RNLR in the sequence 0°, 90°, 45°, 135° with span value as 20% and π-jump threshold as 120. (b) Left half is smoothed fringe order value plotted in colour using green channel images and right half is the wrapped isochromatics obtained by using the isoclinics obtained by processing isoclinics in colour.

performance of the algorithms. The spatial resolution of the fringes in the plate with a hole problem varies from 4 pixels/fringe ($14 < N < 15$) to 161 pixels/fringe ($7 < N < 8$). Thus, the selected problem gives a wide range of fringe gradients to test the method.

3.12 Parallel unwrapping

Adaptive quality guided phase unwrapping (AQGPU) method, which is a mask-based algorithm, has helped to unwrap the photoelastic parameters throughout the model domain having complex shapes starting from a single seed point. However, when dealing with multi-body analysis, where one or more parts are joined, phase unwrapping across the segments poses problems in unwrapping. One solution is to proceed with solving each part individually, but it will be tedious if the problem has multiple parts as one has to acquire several phase-shifted sub-images for each of the parts in the model separately.

3.12.1 Multi-seeded parallel unwrapping algorithm for isoclinic evaluation

To illustrate this, the problem of two discs under diametral compression (diameter = 60 mm, load = 316 N and $F_\sigma = 12.16$ N/mm/fringe) is analysed using the ten-step PST. Phase-shifted images corresponding to the first four optical configurations given in table 3.2 for the magnified portion of the contact region (figure 3.17(a)) are given in figures 3.17(b)–(e). Isoclinic phasemap obtained from analytical solutions with respect to the major principal stress, i.e., σ_1, and minor

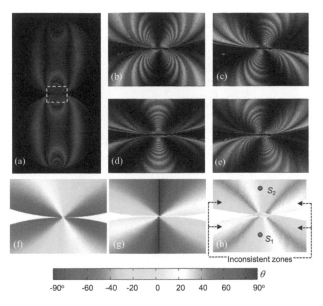

Figure 3.17. (a) Dark field isochromatics of two discs under diametral compression. Polarisation stepped images for the magnified region indicated in figure (a): (b) 0°, (c) 22.5°, (d) 45°, (e) 67.5°. Analytical isoclinic phasemap corresponding to: (f) σ_1 direction, (g) σ_2 direction, (h) experimental isoclinic phasemap obtained using equation (3.9).

principal stress, i.e., σ_2 are shown in figure 3.17(f and g), respectively. From the four phase-shifted images (figures 3.17(b)–(e)), the isoclinic parameter for the entire domain is evaluated using equation (3.9) and the phasemap is shown in figure 3.17(h). It is to be noted that, in the isoclinic values obtained from equation (3.9), one does not know the principal stress direction *a priori* and there exists an inconsistency on whether θ_c calculated using equation (3.9) corresponds to σ_1 or σ_2 direction over the domain. This is noticeably clear by comparing the phasemaps shown in figures 3.17(f)–(h) and *inconsistent* zones are marked in figure 3.17(h).

In AQGPU algorithm, a seed point needs to be selected for starting the unwrapping procedure. The basic unwrapping procedure of AQGPU [30–32] is enhanced to accommodate multiple seed points to start the unwrapping process which helps to process different regions in the model domain at the same time simultaneously. In basic AQGPU, the unwrapped pixels are sorted based on the quality values and are stored in an *adjoin list*. In the enhanced procedure, multiple *adjoin lists* are created. The number of *adjoin lists* depends upon the number of seed-point(s) selected for the problem on hand. Pixels having higher quality in the *adjoin list(s)* are taken for the next level of unwrapping and its adjacent pixels are then subsequently unwrapped and added to the *adjoin list(s)*. A check has been made on whether the adjacent pixels have already been unwrapped by any of the seed-points selected and if they are already unwrapped, these pixels are removed from the unwrapping procedure. The *adjoin list(s)* are constantly modified during the unwrapping procedure and in this way, the unwrapping route from different seed

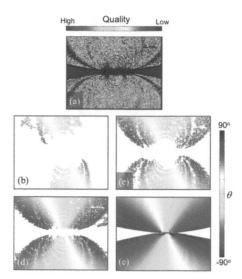

Figure 3.18. (a) Quality map of phase derivative variance. Progression of unwrapping at different stages of the process: (b) 25%, (c) 50%, (d) 75%, (e) 100%. To facilitate easy identification of the domain in intermediate stages of processing, a brown outline is drawn.

points are guided by the quality of the data. The unwrapping procedure continues till every pixel in the model domain is processed by any of the seed-points selected. The selection of the seed-points in various zones can be decided by the user such that in the case of isoclinics, it must represent one of the principal stress directions consistently in each of the model segments.

The wrapped isoclinic phasemap shown in figure 3.17(h) is unwrapped using multi-seeded parallel unwrapping algorithm. The quality map is calculated using the 'phase derivative variance' measure and is plotted in figure 3.18(a). The seed points (S_1, S_2) shown in figure 3.17(h) are used for starting the unwrapping process. They are selected in *consistent* zones, which is important. Figures 3.18(b)–(e) show the progression of the unwrapped pixels at an interval of 25%, 50%, 75% and 100% of the unwrapping process. The progression also shows how the quality has guided the unwrapping sequence. The poor-quality zones are unwrapped only towards the end. Comparing the unwrapped isoclinic phasemap obtained from parallel unwrapping algorithm with that of the theoretical isoclinic phasemap shown in figure 3.17(f), the multi-seeding has helped to get the consistent isoclinic phasemap in a single unwrapping procedure.

3.12.2 Application of multi-seeded parallel unwrapping algorithm to ball and socket joint of aero-structural component

The multi-seeded parallel unwrapping algorithm is used to solve a practical problem of a ball and socket joint used in a strap-on stage of a satellite launch vehicle (figure 3.19). The stress frozen slice of the photoelastic model is obtained from Vikram Sarabhai Space Centre, Trivandrum [54]. Figure 3.20(a) shows the dark field isochromatics of the central section of the ball and socket assembly. This problem is

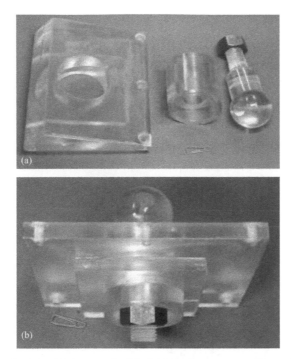

Figure 3.19. Ball and socket joint: (a) photograph of the individual components, (b) assembled joint. Courtesy: [4].

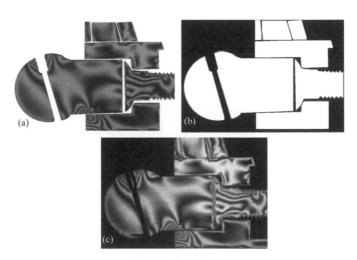

Figure 3.20. (a) Isochromatics of the ball and socket joint assembly, (b) mask used to delimit the interface with different zones. White pixels denote the area to be processed. (c) Isochromatics recorded in white light.

selected to show the usefulness of the algorithm when the model has many interfaces or joints. A boundary mask is carefully generated using the software *Digi*TFP® [47] to identify different regions in the model. The mask used for this problem for different regions of the assembly is shown in figure 3.20(b). Phasemap of wrapped

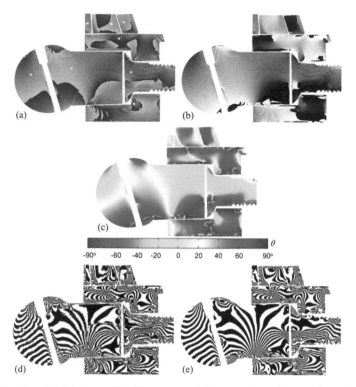

Figure 3.21. Ball and socket joint assembly. Phasemaps of: (a) wrapped isoclinics with seed points shown as yellow dots used for unwrapping each segment, (b) unwrapped isoclinics using parallel unwrapping algorithm. (c) Smoothed isoclinics using multi-directional parallel smoothing with the sequence 45°, 90°, 135°, 0°. The method of smoothing being NLR (15%), RNLR (15 pixels), NLR (15%), RNLR (15 pixels) for each scan direction, respectively. Binary representation of isoclinics: (d) unwrapped, (e) smoothed isoclinics using multi-directional parallel smoothing.

isoclinics obtained using equation (3.8) is shown in figure 3.21(a). The seed points shown as yellow dots (figure 3.21(a)) are lying uniformly in *consistent* zones of different segments and unwrapping is carried out using the multi-seeded parallel unwrapping algorithm. Phasemap of the unwrapped isoclinics is shown in figure 3.21(b). These are smoothed by progressive smoothing algorithm with the sequence 45°, 90°, 135°, 0°. The method of smoothing being NLR (15%), RNLR (15 pixels), NLR (15%), RNLR (15 pixels) for each scan direction, respectively. The colour image representation of the smoothed isoclinics is shown in figure 3.21(c). Figure 3.21(d and e) show the isoclinics as a binary plot in steps of 10° for unwrapped and smoothed isoclinics, respectively.

The unwrapped isoclinic values are used in equation (3.19) to obtain the wrapped isochromatics, which are shown in figure 3.22(a). Though the phasemap is complex, the ten-step PST ensures an *ambiguous* zones free isochromatic phasemap. It can be verified by comparing figure 3.22(a) with the dark field isochromatics of figure 3.20(a), exhibiting similar features. This is subsequently unwrapped using the software *Digi*Photo [48] to obtain the total fringe order using multiple seed points shown as

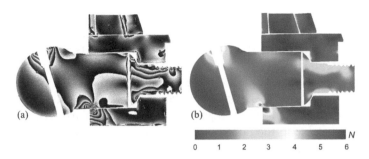

Figure 3.22. Ball and socket joint assembly. (a) Phasemaps of wrapped isochromatics with seed points used for unwrapping, (b) smoothed isochromatics using multi-directional progressive smoothing with the sequence 45°, 90°, 135°, 0°. The method of smoothing being NLR (10 pixels), NLR (15 pixels), NLR (10 pixels), NLR (10 pixels) for each scan direction, respectively.

red dots in figure 3.22(a). The results are smoothed by the progressive smoothing algorithm with the sequence 45°, 90°, 135°, 0°. The method of smoothing being NLR (10 pixels), NLR (15 pixels), NLR (10 pixels), NLR (10 pixels) for each scan direction, respectively. The smoothed isochromatics are shown as a colour plot in figure 3.22(b). It is to be noted that this problem has eight segments and solving it individually for each domain separately can be tedious. However, with the multi-seeded parallel unwrapping algorithm, the entire model is unwrapped in one go.

3.13 Developments in digital photoelastic hardware and software

One of the early approaches in the development of digital polariscopes was to control the optical elements by a stepper motor so that manual rotation is avoided [3]. Aben *et al* [3] have developed a compact digital polariscope for glass stress analysis that intelligently uses the reversibility of optical systems and alters the polarisation state of the input light. They have used a six-step PST involving a circular polariscope, a subset of the ten-step method. Further, as the retardation levels are quite low in glass samples, hence the modifications introduced by Ajovalasit *et al* [18] for using both the left and right circularly polarised lights are not needed for such applications. This implies that only the optical elements after the model are to be changed for getting the phase-shifted images. In reversibility of optical systems, the light source and camera can be interchanged. Using this, Aben *et al* [55] kept the optical elements attached to the camera as constant and varied only the optical elements after the light source. To minimise the error due to rotation and misalignment between optical elements they have used pre-aligned polariser and achromatic quarter-wave plate combination that are brought in front of the light source and controlled by a stepper motor. This has dramatically improved the compactness of the equipment, and accuracy of results, but the drawback is that the field of view is only about 25 mm.

The other challenge is extending the use of PST to solve time varying problems. Although it is well established that for accurate photoelastic parameter evaluation, a ten-step method is desirable, at the cost of accuracy, simpler methods involving four steps have been attempted for recording time varying phenomena [56–58].

The approach has been to use multiple cameras [56] or optically split the image [57] or use a split lens [58], each recording different polarisation stepped image on the same camera image plane. In all these approaches, optical fidelity is well maintained but the optical arrangements are not the optimal ones to use. Extensive research on digital photoelasticity shows that the individual choice of optical elements does play a role in the quality of experimentally recorded images. So, a design solution is achieved by making suitable compromises in devising the digital polariscopes. Further, little information is available in the public domain and they are sold as digital polariscopes with appropriate software bundled into them, thus costing a fortune to buy them.

With developments in microfabrication technology, the use of sub-wavelength grating as a retarder has been developed [59, 60]. These exhibit birefringence due to their form and the phase retardation is controlled by the groove depth and the optic axis coincides with the grating direction. This is integrated with the camera as a single hardware for digital photoelastic analysis [61]. The greatest advantage is that the system is very compact and there are no moving parts—thus it is rugged, even for highly vibrating environments. In such systems, micro retarders of pixel size in an array of 2×2 is replicated over the CCD sensor as shown in figure 3.23(a) [62]. Four light intensity distributions corresponding to four retarders are obtained as a single image from which each phase-stepped image needs to be culled out. It is important to note that the spatial positions of the four light intensity distributions do not mutually correspond to the same point and these are determined by suitable interpolation of the light intensity of neighbouring pixels. This is an approximation and can greatly affect the quality of the results. The phase shifts possible in this approach are based on the optimality of the grating formation and not based on the best arrangements

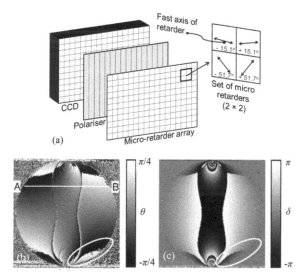

Figure 3.23. (a) Micro retarder array showing the overall arrangement and the details of the close-up of retarder array of 2×2. Reprinted from [62] with permission from Elsevier. Phasemaps obtained using this hardware showing inconsistent and ambiguous zones and only a sample zone is encircled: (b) isoclinic phasemap, (c) isochromatic phasemap. Reprinted from [61] with permission of SPIE.

for extracting photoelastic parameters. The processed images [61] contain *inconsistent* and *ambiguous* zones (figure 3.23(b) and (c)) unresolved; thus, affecting the evaluation of even isochromatic fringe orders. This technology is in its early stages of development and further improvements are essential for its success.

Standalone digital photoelastic software have also been developed by various groups and only a few of them are available commercially [47, 48]. The advantage of such an approach is that any available polariscope can be used as a digital polariscope provided a suitable digital camera is used for image recording. While using a polariscope, it needs to be calibrated to match the theoretical development of equations, which has been discussed in section 3.4.1.

Initially, true colour cameras that record the *R*, *G*, and *B* were the only ones available for colour image recording with independent filters for each image plane and the individual colour image planes are fully recorded for the complete image. Such cameras ensure high image fidelity, and the individual colour image plane information can be extracted fully over the image, but these cameras are bulky as per modern standards. With advancements on miniaturising hardware, compact cameras that use colour filter arrays (CFA) have been developed [63]. The GRGB Bayer configuration of 2×2 replicated over the CCD sensor (figure 3.24) is the most common arrangement used [64]. Here again, like the micro retarders integrated with the CCD sensor, information of each colour plane signal is not available for the same point and interpolations are needed to complete the information. The process of interpolation is termed as demosaicking [63]. It is a digital image process used to reconstruct a full colour image from the incomplete colour samples output from an image sensor overlaid with a colour filter array. Such cameras are good enough for visual rendering of general photography. However, they can lead to errors if used for PSTs, as one would require the intensity data at each point as provided by the experiment. This is particularly so at high fringe gradient zones as intensity interpolations can lead to unpredictable results. A computational study on the influence of demosaicking algorithms in digital photoelasticity has been recently reported [65]. They have analysed how various demosaicking methods influence digital photoelastic techniques (PST, TFP) by considering multiple polariscope configurations, image resolution and light source. Their study revealed that the demosaicking algorithm has minimal influence if the image resolution is higher than 512×512 pixels and recommends the usage of a fluorescent light source for better fringe demodulation.

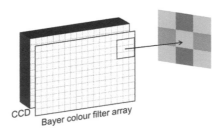

CCD
Bayer colour filter array

Figure 3.24. Compact colour camera with Bayer colour filter array.

3.14 Closure

The development of phase shifting techniques has revolutionised the data acquisition and processing of phase information of various optical techniques including photo-elasticity. Although one may not require an external unit to introduce the requisite phase shifts, the existence of two phase information—one relating to the isoclinics and the other relating to isochromatics has posed a unique challenge in photo-elasticity. The specific zones such as *inconsistent* and *ambiguous* zones are unique to photoelasticity and the very recognition of the existence of such zones has helped in solving the interaction of isoclinics in isochromatic evaluation. The evaluation of isoclinics uniquely representing one of the principal stress directions over the model domain is important for the correct evaluation of isochromatics as these are interdependent.

The ten-step method has intelligently combined the use of conventional plane polariscope and generic circular polariscope arrangements for accurate evaluation of both isoclinics and isochromatics. High quality isoclinics evaluation is possible using a plane polariscope and the error propagation is minimised by the use of adaptive quality guided phase unwrapping (AQGPU). The use of multi-directional progressive smoothing of isoclinic data particularly in the presence of many isochromatics has helped in minimising the noise. With multi-seed parallel unwrapping, even problems with multiple domains can be conveniently handled.

The ten-step PST with colour images for isoclinics and monochrome images for isochromatics evaluation can provide the highest quality results for photoelastic parameters. These results can be used as a benchmark to assess the results of other simpler digital photoelastic methods that are more suited to the industrial environment (chapter 4). The role of carrier fringe method in glass stress analysis, which is simple to implement has been compared with PST in chapter 2 for its efficacy. Similarly, the use of colour information for the determination of isochromatic fringe orders is compared with PST in the next chapter. Such methods require only one image to be processed and suitable for even studying time varying phenomenon conveniently.

While using PST in colour domain, it is recommended to use true colour cameras for the first four steps (isoclinic recording) and if possible, use a monochrome camera of the same image resolution to record the next six images of the ten-step method. Such an approach would give the best results even with a high isochromatic fringe density of 0.1 fringe/pixel (10 pixels/fringe). If a suitable monochrome camera is not available, then one can use the green channel of the true colour camera for the best results. The next chapter deals with the use of colour information in evaluating the photoelastic parameters.

Exercises

1. What has influenced the paradigm shift in data acquisition and processing of data in optical methods? Explain phase shifting in photoelasticity by comparing it with Tardy's method of compensation.

2. A circular polariscope is constructed using various optical arrangements of the elements. Establish using Jones Calculus, which of the arrangements correspond to bright and dark fields?

α	ξ	η	β
$\pi/2$	$3\pi/4$	0	$\pi/4$
$\pi/2$	$3\pi/4$	0	$3\pi/4$

3. Give a brief account on the genesis of the ten-step method in photoelasticity.
4. How are fractional retardation and isoclinic angle calculated in the ten-step method? Discuss the range of these parameters obtained directly from the trigonometric equation
5. How do you calibrate a digital polariscope?
6. What is the basic whole field representation of fractional retardation and how is it plotted?
7. What do you understand by *ambiguous* zones in isochromatic phasemaps and *inconsistent* zones in isoclinic phasemaps? Are they interrelated? Elaborate.
8. What is the criterion to determine the *ambiguous* zones? Identify the *ambiguous* zones and *inconsistent* zones for the following problem.

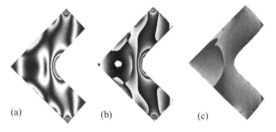

(a) (b) (c)

9. How is an isochromatic phasemap that is free of ambiguous zones arrived at in the ten-step method?
10. One can observe in the grey scale representation of the unwrapped isoclinic phasemap, a faint appearance of isochromatics. What are these? Is it desirable to have them?
11. In digital photoelasticity, inverse trigonometric functions are employed that lead to multiple solutions. A similar problem exists in evaluating the principal stress directions using the expression of θ in terms of in-plane stress components in 2-D problems. Have you ever recognised that? List your observations on this.
   a. Although for employing failure theories one is accustomed to evaluating the principal stresses and labelling them as σ_1 and σ_2, have you ever attempted to get the associated principal stress directions? Suppose you have knowledge of Mohr's circle; can you use it effectively to achieve this? Explain in detail.

b. In a 2-D problem, for a given stress tensor, without resorting to the Mohr's circle approach, how would you find the principal stress and their corresponding directions uniquely?

Note: For questions 12–20, I_1 to I_{10} correspond to the ten-step phase shifted intensities elaborated in table 3.2.

12. Using the intensity values given, determine the isoclinic angle at the point of interest. Comment on the use of 'atan' and 'atan2' functions for isoclinic evaluation.

Arrangement	I_1	I_2	I_3	I_4
Intensity	49	150	35	200

13. Intensities I_1 to I_4 recorded at a point (23.26, 10.43 mm) in a thick ring under internal pressure (outer dia. = 60 mm, inner dia. = 30 mm, thickness = 6 mm, internal pressure = 6 MPa, F_σ = 12 N/mm/ fringe) are tabulated with centre as the origin. Find the isoclinic angle at the point. From Lame's solution, the stress components at this point are: $\sigma_r = -0.769$ MPa, $\sigma_\theta = 4.770$ MPa, $\tau_{r\theta} = 0$ MPa.

Arrangement	I_1	I_2	I_3	I_4
Intensity	162	25	149	140

14. For a circular disc made of epoxy (F_σ = 12 N/mm/fringe) under diametral compression of 800 N, the R, G, B values recorded at a point (6.46, 25.85 mm) for the first four arrangements in plane polariscope are tabulated. The disc has a thickness of 6 mm and a diameter of 60 mm. Evaluate the principal stress direction at the point and compare the answer with the one given by the theoretical stress field equations from chapter 1.

Arrangement		I_1	I_2	I_3	I_4
Intensity	R	201	146	8	64
	G	222	160	9	70
	B	176	128	7	56

15. The following table summarizes the intensities observed in plane and circular polariscope arrangements corresponding to a point (24.09, 4.25 mm) in a thick ring under internal pressure (outer dia. = 60 mm, inner dia. = 30 mm, thickness = 6 mm, internal pressure = 6 MPa, F_σ = 12 N/mm/ fringe). Determine the isoclinic value using both the arrangements. Which arrangement is preferred for isoclinics evaluation? Base your assessments by comparing the result with the theoretical solution. The theoretical values of

stress components at this point are: $\sigma_r = -1.007$ MPa, $\sigma_\theta = 5.008$ MPa, $\tau_{r\theta} = 0$ MPa.

Arrangement	I_1	I_2	I_3	I_4	I_5	I_6	I_7	I_8	I_9	I_{10}
Intensity	2	2	5	5	170	254	170	171	189	188

16. An experimentalist uses the ten-step phase shifting technique to capture the intensities at a point in a ring under diametral compression. He makes an error in capturing intensity by keeping the slow axis of the first quarter wave plate at 45° in the 8th arrangement (4th arrangement of circular polariscope) and hence discards the value. Neglecting the mismatch of the quarter-wave plates, what would be the possible integer value of intensity captured in the 8th step. (Hint: being an ideal polariscope, consider that the value of principal stress direction by both plane and circular polariscopes be the same).

Arrangement	I_1	I_2	I_3	I_4	I_5	I_6	I_7	I_8	I_9	I_{10}
Intensity	126	9	112	5	191	168	91		237	168

17. For a point (7.85, 10.15 mm) in a circular disc under diametral compression (dia. = 60 mm, thickness = 6 mm, load = 800 N, $F_\sigma = 12$ N/mm/fringe), the intensities recorded using ten-step phase shifting method are tabulated. Using the data, find the unwrapped isoclinic angle corresponding to the major principal stress direction and compute the fractional retardation. The theoretical values of stresses at this point are: $\sigma_x = 0.840$ MPa, $\sigma_y = -3.745$ MPa, $\tau_{xy} = 0.882$ MPa.

Arrangement	I_1	I_2	I_3	I_4	I_5	I_6	I_7	I_8	I_9	I_{10}
Intensity		90	102	235	230	34	252	171	201	188	155

18. The unwrapped isoclinic phase map with respect to the major principal stress direction is to be used in evaluating fractional retardation at a point (6.46, 25.85 mm) in a circular disc. The disc is loaded to 800 N in diametral compression having a radius of 30 mm and a thickness of 6 mm. The F_σ value for the material of the disc is 12 N/mm/fringe. The theoretical values of stress components at this point are: $\sigma_x = -2.833$ MPa, $\sigma_y = -1.812$ MPa, $\tau_{xy} = 2.547$ MPa. Determine the fractional retardation from the intensities recorded in the table. If the value of θ is not unwrapped, what would be the value of the fractional retardation? Verify your answer by comparing it with theoretical values.

Arrangement	I_1	I_2	I_3	I_4	I_5	I_6	I_7	I_8	I_9	I_{10}
Intensity	238	202	47	134	76	243	225	190	119	169

19. For the problem of a plate with a hole, the following table summarizes the intensities recorded from monochrome and colour images at a point. The isoclinic angle theoretically evaluated for this point is 17°. Compare the use of the data from colour and monochrome images to evaluate the isoclinic angle and retardation. Comment on how one can use the given data to obtain the highest quality results for photoelastic parameters.

Arrangement		I_1	I_2	I_3	I_4	I_5	I_6	I_7	I_8	I_9	I_{10}
Intensity	R	42	25	85	157	119	76	152	86	82	141
	G	84	31	173	223	87	135	126	117	132	122
	B	28	25	81	103	127	44	117	91	100	84
	Monochrome	50	29	140	190	101	107	133	105	113	123

20. A 36 mm wide plate of thickness 6 mm and length 250 mm with a hole of 10 mm diameter is subjected to a tensile load of 1976 N. The R, G, B values at a point corresponding to plane and circular polariscope arrangements are summarized in the table. Find the isoclinic angle at the point. How will you use the intensities recorded to get the fractional retardation at the point of interest? Justify your approach by suitable arguments. Estimate the fractional retardation. By Tardy's method of compensation the fringe order at the point is evaluated to be 1.53. Calculate the % deviation.

Arrangement		I_1	I_2	I_3	I_4	I_5	I_6	I_7	I_8	I_9	I_{10}
Intensity	R	42	25	85	157	119	76	152	86	82	141
	G	84	31	173	223	87	135	126	117	132	122
	B	28	25	81	103	127	44	117	91	100	84

21. Can you employ digital photoelastic techniques using the commercially available conventional polariscopes? What restrictions do you foresee?
22. Consider a case where a quarter-wave plate is unmatched to the light source is used and let the error be ε. How will you model this unmatched quarter-wave plate and accommodate this error in Jones calculus?

References

[1] William N and Sharpe J 2008 *Springer Handbook of Experimental Solid Mechanics* ed W N Sharpe (Boston, MA: Springer)
[2] Ramesh K 2000 *Digital Photoelasticity Advanced Techniques and Applications* (Berlin: Springer) https://link.springer.com/book/10.1007%2F978-3-642-59723-7

[3] Ramesh K 2008 Photoelasticity *Springer Handbook of Experimental Solid Mechanics* ed J William and N Sharpe (Boston, MA: Springer) pp 701–42 https://home.iitm.ac.in/kramesh/ESM.html

[4] Ramesh K 2009 *e-book on Experimental Stress Analysis* 1st edn (Chennai, India: IIT Madras) https://home.iitm.ac.in/kramesh/ESA.html

[5] Ramesh K 2015 Digital photoelasticity *Digital Optical Measurement Techniques and Applications* ed P Rastogi (Boston, MA: Artech House) pp 289–344

[6] Ramesh K, Kasimayan T and Neethi Simon B 2011 Digital photoelasticity—a comprehensive review *J. Strain Anal. Eng. Des.* **46** 245–66

[7] Ramesh K and Mangal S K 1998 Data acquisition techniques in digital photoelasticity: a review *Opt. Lasers Eng.* **30** 53–75

[8] Ajovalasit A, Barone S and Petrucci G 1998 A review of automated methods for the collection and analysis of photoelastic data *J. Strain Anal. Eng. Des.* **33** 75–91

[9] Asundi A 1993 Phase shifting in photoelasticity *Exp. Tech.* **17** 19–23

[10] Hecker F W and Morche B 1986 Computer-aided measurement of relative retardations in plane photoelasticity *Experimental Stress Analysis* ed H Wieringa (Dordrecht: Springer) pp 535–42 https://doi.org/10.1007/978-94-009-4416-9_58

[11] Kihara T 1990 Automatic whole-field measurement of photoelasticity using linear polarized incident light *IX Int. Conf. of Experimental Mechanics (Copenhagen)* 821–7

[12] Patterson E A and Wang Z F 1991 Towards full field automated photoelastic analysis of complex components *Strain* **27** 49–53

[13] Sarma A V S S S R, Pillai S A, Subramanian G and Varadan T K 1992 Computerized image processing for whole-field determination of isoclinics and isochromatics *Exp. Mech.* **32** 24–9

[14] Ramesh K and Ganapathy V 1996 Phase-shifting methodologies in photoelastic analysis—the application of Jones calculus *J. Strain Anal. Eng. Des.* **31** 423–32

[15] Mangal S K and Ramesh K 1998 Factors affecting the intensity distribution in high stress gradient zones in digital photoelasticity *Strain* **34** 133–4

[16] Ramesh K and Sreedhar D 1998 Optically enhanced tiling (OET) in digital fringe pattern analysis *Strain* **34** 127–30

[17] Yao X-F, Jian L-H, Xu W, Jin G-C and Yeh H-Y 2005 Digital shifting photoelasticity with optical enlarged unwrapping technology for local stress measurement *Opt. Laser Technol.* **37** 582–9

[18] Ajovalasit A, Barone S and Petrucci G 1998 A method for reducing the influence of quarter-wave plate errors in phase stepping photoelasticity *J. Strain Anal. Eng. Des.* **33** 207–16

[19] Prashant P L and Ramesh K 2006 Genesis of various optical arrangements of circular polariscope in digital photoelasticity *J. Aerosp. Sci. Technol.* **58** 117–32

[20] Sai Prasad V R K 2003 Background intensity effects on various phase shifting techniques in photoelasticity *Conf. on Optics and Photonics in Engineering (New Delhi)* (ABC Enterprises) pp 183–6

[21] Brown G M and Sullivan J L 1990 The computer-aided holophotoelastic method *Exp. Mech.* **30** 135–44

[22] Chen T Y and Lin C H 1998 Whole-field digital measurement of principal stress directions in photoelasticity *Opt. Lasers Eng.* **30** 527–37

[23] Mangal S K and Ramesh K 1999 Use of multiple loads to extract continuous isoclinic fringes by phase shifting technique *Strain* **35** 15–7

[24] Petrucci G 1997 Full-field automatic evaluation of an isoclinic parameter in white light *Exp. Mech.* **37** 420–6

[25] Nurse A D 1997 Full-field automated photoelasticity by use of a three-wavelength approach to phase stepping *Appl. Opt.* **36** 5781–6

[26] Barone S, Burriesci G and Petrucci G 2002 Computer aided photoelasticity by an optimum phase stepping method *Exp. Mech.* **42** 132–9

[27] Ramji M and Ramesh K 2006 A new six-step phase shifting technique using mixed-polariscope in digital photoelasticity *Key Eng. Mater.* **326–328** 35–8

[28] Ramji M, Gadre V Y and Ramesh K 2006 Comparative study of evaluation of primary isoclinic data by various spatial domain methods in digital photoelasticity *J. Strain Anal. Eng. Des.* **41** 333–48

[29] Zhenkun L, Dazhen Y and Wanming Y 2003 Whole-field determination of isoclinic parameter by five-step color phase shifting and its error analysis *Opt. Lasers Eng.* **40** 189–200

[30] Ramji M and Ramesh K 2008 Stress separation in digital photoelasticity part a—photoelastic data unwrapping and smoothing *J. Aerosp. Sci. Technol.* **60** 5–15

[31] Ramji M and Ramesh K 2008 Whole field evaluation of stress components in digital photoelasticity—issues, implementation and application *Opt. Lasers Eng.* **46** 257–71

[32] Ramji M and Ramesh K 2010 Adaptive quality guided phase unwrapping algorithm for whole-field digital photoelastic parameter estimation of complex models *Strain* **46** 184–94

[33] Ramji M and Prasath R G R 2011 Sensitivity of isoclinic data using various phase shifting techniques in digital photoelasticity towards generalized error sources *Opt. Lasers Eng.* **49** 1153–67

[34] Ramesh K and Deshmukh S 1997 Automation of white light photoelasticity by phase-shifting technique using colour image processing hardware *Opt. Lasers Eng.* **28** 47–60

[35] Ji W and Patterson E A 1998 Simulation of errors in automated photoelasticity *Exp. Mech.* **38** 132–9

[36] Tamrakar D K and Ramesh K 2001 Simulation of error in digital photoelasticity by Jones calculus *Strain* **37** 105–12

[37] Ramji M, Prasath R G R and Ramesh K 2009 A generic error simulation in digital photoelasticity by Jones calculus *J. Aerosp. Sci. Technol.* **61** 475–81

[38] Ajovalasit A, Petrucci G and Scafidi M 2007 Phase shifting photoelasticity in white light *Opt. Lasers Eng.* **45** 596–611

[39] Ramesh K, Ramakrishnan V and Ramya C 2015 New initiatives in single-colour image-based fringe order estimation in digital photoelasticity *J. Strain Anal. Eng. Des.* **50** 488–504

[40] Ramesh K 2017 P_Scope®—a virtual polariscope *Photomechanics Lab.* (IIT Madras) https://home.iitm.ac.in/kramesh/p_scope.html (accessed Apr. 06, 2021)

[41] Quiroga J A and González-Cano A 1997 Phase measuring algorithm for extraction of isochromatics of photoelastic fringe patterns *Appl. Opt.* **36** 8397–402

[42] Ashokan K and Ramesh K 2006 A novel approach for ambiguity removal in isochromatic phasemap in digital photoelasticity *Meas. Sci. Technol.* **17** 2891–6

[43] Sai Prasad V, Madhu K and Ramesh K 2004 Towards effective phase unwrapping in digital photoelasticity *Opt. Lasers Eng.* **42** 421–36

[44] Ramji M, Nithila E, Devvrath K and Ramesh K 2008 Assessment of autonomous phase unwrapping of isochromatic phase maps in digital photoelasticity *Sadhana* **33** 27–44

[45] Siegmann P, Backman D and Patterson E A 2005 A robust approach to demodulating and unwrapping phase-stepped photoelastic data *Exp. Mech.* **45** 278–89

[46] Madhu K R and Ramesh K 2007 New boundary information encoding and autoseeding for effective phase unwrapping of specimens with cutouts *Strain* **43** 54–7

[47] Ramesh K 2017 *DigiTFP®*-software for digital twelve fringe photoelasticity *Photomechanics Lab* (IIT Madras) https://home.iitm.ac.in/kramesh/dtfp.html (accessed Apr. 06, 2021)

[48] Ramesh K 2016 *Digi*Photo digital photoelasticity software for employing phase shifting technique *Photomechanics Lab.* (IIT Madras) https://home.iitm.ac.in/kramesh/dphoto.html (accessed Apr. 06, 2021)

[49] Ramesh K, Hariprasad M P and Ramakrishnan V 2015 Robust multidirectional smoothing of isoclinic parameter in digital photoelasticity *Opt. Eng.* **54** 081205.1–9

[50] Kasimayan T and Ramesh K 2011 Adaptive smoothing for isoclinic parameter evaluation in digital photoelasticity *Strain* **47** e371–5

[51] Vinayak G, Mangal S K and Ramesh K 2001 Reasons for ambiguity in retardation calculations in phase shifting technique *Strain* **37** 53–4

[52] Prashant P L, Madhu K R and Ramesh K 2005 *New Initiatives in Phase Unwrapping in Digital Photoelasticity* ed J F Lopez, C Quan, F S Chau, F V Fernandez, J M Lopez-Villegas, A Asundi, B S Wong, J M de la Rosa and C T Lim (Bellingham, WA: SPIE) pp 192–7 https://doi.org/10.1117/12.621501

[53] Neethi Simon B and Ramesh K 2011 Identification of inconsistent zones in digital photo-elasticity *Strain* **47** 382–5

[54] Ashokan K, Ramesh K, Pillai S A and Philip J 2007 Digital photoelastic evaluation of isochromatic fringe order in a 3D model of complex rocket motor strap-on joint *Experimental Analysis of Nano and Engineering Materials and Structures* (Dordrecht: Springer) pp 85–6 https://doi.org/10.1007/978-1-4020-6239-1_41

[55] Aben H K, Errapart A, Ainola L and Anton J 2005 Photoelastic tomography for residual stress measurement in glass *Opt. Eng.* **44** 1–8

[56] Patterson E A and Wang Z F 1998 Simultaneous observation of phase-stepped images for automated photoelasticity *J. Strain Anal. Eng. Des.* **33** 1–15

[57] Hobbs J W, Greene R J and Patterson E A 2003 A novel instrument for transient photoelasticity *Exp. Mech.* **43** 403–9

[58] Lesniak J, Zhang S J and Patterson E A 2004 Design and evaluation of the poleidoscope: a novel digital polariscope *Exp. Mech.* **44** 128–35

[59] Kikuta H, Haccho H, Iwata K, Hamamoto T, Toyota H and Yotsuya T 2001 Real-time polarimeter with a form-birefringent micro retarder array *Proc. SPIE* **4416** 19–22

[60] Kikuta H, Toyota H and Yu W 2003 Optical elements with subwavelength structured surfaces *Opt. Rev.* **10** 63–73

[61] Yoneyama S, Kikuta H and Moriwaki K 2006 Simultaneous observation of phase-stepped photoelastic fringes using a pixelated microretarder array *Opt. Eng.* **45** 1–7

[62] Ramesh K and Sasikumar S 2020 Digital photoelasticity: recent developments and diverse applications *Opt. Lasers Eng.* **135** 106186

[63] Ramanath R, Snyder W E, Bilbro G L and Sander W A 2002 Demosaicking methods for Bayer color arrays *J. Electron. Imaging* **11** 306

[64] Duran J and Buades A 2015 A demosaicking algorithm with adaptive inter-channel correlation *Image Process. Line* **5** 311–27

[65] Briñez-de León J C, Restrepo-Martínez A and Branch-Bedoya J W 2019 Computational analysis of Bayer colour filter arrays and demosaicking algorithms in digital photoelasticity *Opt. Lasers Eng.* **122** 195–208

IOP Publishing

Developments in Photoelasticity
A renaissance
K Ramesh

Chapter 4

Total fringe order photoelasticity

The use of colour information that has been in vogue in conventional photoelasticity for qualitative appreciation of fringe ordering has been extended into a systematic quantitative analysis with the use of digital colour cameras and the development of appropriate colour processing methodologies and software. The range of fringe orders that could be resolved with RGB colour model can be expanded to process up to twelve fringes with the combination of both RGB and hue-saturation-value (HSV) colour models is established through different example problems. The choice of calibration specimen, and the need for using a calibration table that incorporates the fringe gradient information are discussed. Methods to ensure fringe order continuity and the associated need for appropriate scanning to achieve this are illustrated with the use of both simply and multiply connected domains. A new measure to identify the fringe gradient direction digitally, thereby accommodating this in the scanning to minimise error propagation—christened as fringe resolution-guided scanning (FRSTFP) is discussed. The various colour adaptation techniques to suitably modify the existing calibration table for solving a range of problems to simplify the analysis are also discussed. The applicability of the methodology developed is used for solving the crack-tip stress fields in an edge heated plate and for resolving the fringes in situations that have excessive noise such as those in stereo-lithographic models, or domains that are enlarged as in the case of microcapsules that highlight even small markings. The use of the five-step/four-step method to evaluate both isochromatics and isoclinics by solving a problem in orthodontics has demonstrated that a conventional polariscope with appropriate software for processing is sufficient to do digital photoelastic analysis[1].

---

[1] Parts of this chapter have been adapted with permission of Elsevier from the author's own work in [31], [32], [40], [8], [53], with the permission of SAGE from the author's own work in [4] and with the permission of Springer from the author's own work in [9], [34].

doi:10.1088/978-0-7503-2472-4ch4

4.1 Introduction

Among the various experimental techniques, photoelasticity has the unique distinction of providing fringe data in colour. The use of a colour code to identify fringe gradient direction and to assign the total fringe order approximately up to three fringes has been in use in conventional photoelasticity. Methodologies such as three-fringe photoelasticity (TFP) [1] and RGB photoelasticity [2, 3] have been developed to process the colour information to extract fringe order data over the complete model domain but are restricted to only three fringe orders initially. With further understanding, the methodologies have been extended to process up to twelve fringe orders [4]. The quest for increasing this limit is still there but most problems of practical interest could be solved if twelve fringes could be resolved successfully in the domain. As one gets the total fringe order by processing just one image, TFP can also be re-christened as total fringe order photoelasticity and it covers all its variants since its first use. The very first work of using digital information in photoelasticity for monochrome images was named half-fringe photoelasticity in 1983 [5]. Following that, in the colour domain, as the colours are distinct up to three fringes, the term three-fringe photoelasticity was originally coined. Half-fringe photoelasticity also relied on comparing the grey scale value to the fractional fringe order. However, this is completely replaced by the phase shifting techniques (PSTs) in the mono-chrome domain. In the colour domain, there have been sustained developments and TFP is now a reliable tool to evaluate the isochromatic fringe orders.

In essence, in TFP, the colour information is used to label the total fringe order at a point in each application image. It is easier said than done as several aspects influence the process. Generation of calibration tables, ensuring fringe order continuity, application to models of arbitrary geometry, multiply connected domains, external marks influencing the generation of noise, circumventing or minimising noise propagation due to low spatial resolution, easy adaptability to industrial environments are the issues to be considered.

Ideally, for each experiment, one must develop a calibration table for colour interpretation of the total fringe order. With developments such as colour adaptation [6], the adaptability of TFP to industrial environments has phenomenally increased. The need to check for fringe continuity necessitated the development of suitable scanning schemes to cover the entire model domain and reduce the propagation of noise through suitable criteria. Complex industrial models could now be handled effectively with the development of sophisticated scanning schemes [7, 8] to employ the methodology with varying spatial resolutions of the fringe field.

In this chapter, the intensity of light transmitted by the polariscope when white light is used is presented to establish the ease in adopting a calibration table approach for fringe ordering. Subsequently, the basics of three-fringe photoelasticity using R, G, and B values, refining the fringe order evaluation by using fringe order continuity, the generation of a calibration table and the use of a merged calibration table using different loads are discussed. Following this, how the fringe order evaluation can be expanded to higher fringe orders with various colour models is discussed. The problem of a plate with a hole is chosen to evaluate the fringe orders up to twelve using the

RGB–HSV colour model and the results are compared with those obtained by phase shifting technique. The various colour adaptation methods are discussed next, and a comparative study has also been done to bring out their relative merits.

The inadequacy of the commonly available scanning schemes to completely cover the entire model domain is discussed with the help of simply and multiply connected domains. The example of a crack-tip stress field is chosen to illustrate the influence of low spatial resolution of the fringes near the crack-tip in precipitating noise propagation. The details of a novel fringe resolution-guided scanning christened as FRSTFP is then presented. The applicability of FRSTFP to handle multiple pores in a stereolithographic model is demonstrated. In poorly recorded images, normalization schemes can help to extract data. The use of normalization schemes in the context of TFP is then discussed.

The use of a conventional polariscope for digital photoelastic analysis is fully demonstrated in TFP as it just requires a single isochromatic image. In view of this, it is also easily suitable for studying time varying phenomena. Applicability of TFP to solve crack-tip stress field for an edge heated plate is presented. With four polarisation stepped images that require only a conventional plane polariscope, one can also determine the isoclinic parameter over the model domain. By combining TFP and polarisation stepping, a five-step method in evaluating all the photoelastic parameters is possible. As one can construct dark field isochromatics from four polarisation stepped images recorded in white light, complete digital photoelastic analysis is possible with just a conventional plane polariscope! This is demonstrated by solving a problem in orthodontics.

4.2 Intensity of light transmitted in white light for various polariscope arrangements

In chapter 3, intensity equations have been elegantly developed while using a monochromatic light source and the photoelastic parameters have also been obtained by directly processing the intensity data. While using a white light source, there are difficulties in quantifying the intensity of light transmitted in a simplistic sense. The colour camera does not have a flat response for all the wavelengths and its spectral response $F(\lambda)$ is a function of the wavelength. Further, the amplitudes of light corresponding to each of the wavelengths of the light source may not be the same and it will be different for different sources. In addition to these effects, one may have to include a function $T(\lambda)$ which is the sum total of the transmission behaviour of all the optical elements as a function of wavelength forming a particular polariscope arrangement. Finally, in view of a spectrum of colours, the perceived intensity will be an integration of the individual light intensity corresponding to each of the wavelengths in the spectrum. For a plane polariscope, in dark-field arrangement, with the analyser at an angle β, equation (3.3) could be modified to find the intensity I_p in a dark-field plane polariscope as [9, 10],

$$I_p = I_b + \left[\frac{1}{(\lambda_2 - \lambda_1)} \int_{\lambda_1}^{\lambda_2} T(\lambda)F(\lambda)I_a(\lambda) \sin^2 \frac{\delta(\lambda)}{2} d\lambda \right] \sin^2 2(\theta - \beta) \qquad (4.1)$$

where I_b is the total background light intensity for the entire spectrum of colours, the factor $(\lambda_2 - \lambda_1)$ in the denominator is the normalizing factor [9] such that the intensity varies in the range 0–1 from dark- to bright-field arrangement. Normally, one would like to express the retardation $\delta(\lambda)$ with respect to a reference wavelength λ_{ref}.

$$\delta(\lambda) = \frac{\delta_{ref} F_{\sigma_{ref}}}{F_\sigma(\lambda)} = \frac{2\pi N_{ref} F_{\sigma_{ref}}}{F_\sigma(\lambda)} \tag{4.2}$$

In most studies, $T(\lambda)$ is assumed to be unity. Substituting equation (4.2) in equation (4.1), one gets

$$I_p = I_b + \left[\frac{1}{(\lambda_2 - \lambda_1)} \int_{\lambda_1}^{\lambda_2} F(\lambda) I_a(\lambda) \sin^2 \frac{\pi N_{ref} F_{\sigma_{ref}}}{F_\sigma(\lambda)} d\lambda \right] \sin^2 2(\theta - \beta) \tag{4.3}$$

where N_{ref} is the total fringe order corresponding to the reference wavelength.

A circular polariscope is normally constructed with quarter-wave plates meant for a specific wavelength. While using white light, the quarter-wave plate no longer provides a phase shift of $\pi/2$ for all the wavelengths and in general, it behaves like a retarder. If the reference wavelength for which the quarter-wave plate is matched is say λ_{ref}, the error introduced for all other wavelengths (λ) is,

$$\varepsilon = \frac{\pi}{2} \left(\frac{\lambda}{\lambda_{ref}} - 1 \right) \tag{4.4}$$

Although the error due to quarter-wave plate can be minimised using achromatic quarter-wave plates, it cannot be ruled out completely. Considering the quarter-wave plate as a retarder with a retardation of $(\pi/2 + \varepsilon)$, the intensity of light transmitted for a dark-field circular polariscope (I_d), with crossed quarter-wave plates, for a particular wavelength is,

$$I_d(\lambda) = I_a(\lambda) \sin^2 \left(\frac{\pi N_{ref} F_{\sigma_{ref}}}{F_\sigma(\lambda)} \right) [1 - \cos^2 2\theta \sin^2 \varepsilon] \tag{4.5}$$

For the complete spectrum of white light, including the background light intensity, equation (4.5) changes to

$$I_d = I_b + \frac{1}{(\lambda_2 - \lambda_1)} \int_{\lambda_1}^{\lambda_2} F(\lambda) I_a(\lambda) \sin^2 \left(\frac{\pi N_{ref} F_{\sigma_{ref}}}{F_\sigma(\lambda)} \right) [1 - \cos^2 2\theta \sin^2 \varepsilon] \, d\lambda \tag{4.6}$$

Though the intensity equations presented in this section appear quite complex, judicious approximations are invoked while using them for practical applications. One of the simplest approaches is to approximate each image plane as a mono-chrome image for image processing. Ramesh and Deshmukh [11] proposed the use of green plane of the colour image recorded by a colour CCD camera for phase shifting. Although approximate, the methodology can give both isoclinics and isochromatics for a given problem.

4.3 Basics of three-fringe photoelasticity

Equations (4.3) and (4.6) show that the mathematical modelling of the RGB intensity variation of the output image requires proper consideration of various factors such as quarter-wave plate error, dispersion of the stress-optic coefficient, spectral response of the camera, spectral composition of the light source, and the transmission response of the polariscope components [9, 10]. It is exceedingly difficult to accurately estimate all these parameters. Nevertheless, for each data point, one can digitally extract the relevant grey scale value of the three image planes R, G, and B. From a model of known fringe order variation, such as a beam under four-point bending, it is possible to associate a total fringe order for each combination of R, G, and B values. The basic calibration table will have a collection of R, G, and B values for the fringe orders varying from 0–3, usually of an equal number of data points between integer fringe orders. In practice, an approach based on a calibration table is generally followed for finding the total fringe order [9, 10]. Although, isoclinic values cannot be obtained through this method, a large class of problems exist wherein isochromatics themselves yield rich data for further analysis.

The most direct approach is to calculate the least squares error term (e_i) for each row i in the calibration table and find the fringe order for which e_i is the minimum using the following equation [1].

$$e_i = \sqrt{(R - R_i)^2 + (B - B_i)^2 + (G - G_i)^2} \qquad (4.7)$$

where, R, G, and B values correspond to the grey scale intensities of the application image and those with the subscript 'i' are from the calibration table. A visual representation of the data processed is always desirable to get an overall idea of the success of the method. The fringe order data thus obtained can be converted into a set of grey level values using the following equation.

$$g(x, y) = \mathrm{INT}\left[\frac{255}{N_{max}} \times f(x, y)\right] = \mathrm{INT}[R] \qquad (4.8)$$

where $f(x, y)$ is the fringe order at point (x, y), N_{max} is the maximum fringe order of the calibration table, $g(x, y)$ is the grey level value at the point (x, y) and INT $[R]$ is the nearest integer of R. The symbol 'R' is used here locally to represent a positive real number.

Although, the least squares error method using the calibration table is very simple in implementation, it is prone to error at several locations in the model domain due to repetition of colours [12]. It can lead to false estimation of fringe order (N) at some locations, and spatial continuity of N is lost. Figure 4.1(a) shows the colour image of the isochromatics recorded for a disc under diametral compression, figure 4.1(b) shows the dark field isochromatics of the calibration specimen (beam under four-point bending) and figure 4.1(c) shows the grey scale representation of the total fringe order evaluated by equation (4.7) and plotted using equation (4.8). In view of the repetition of colours, one observes several patches in figure 4.1(c) rather than a smooth variation of the grey scale values.

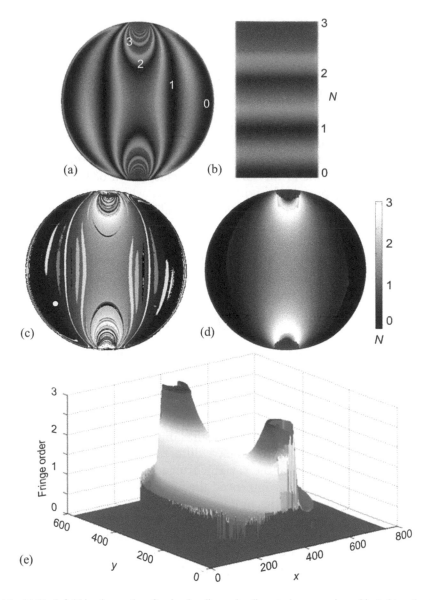

Figure 4.1. (a) Dark field isochromatics of a circular disc under diametral compression subjected to a load of 492 N recorded using incandescent lamp. (b) Dark field isochromatics of the calibration specimen. (c) Grey scale representation of the fringe order values obtained using the colour difference formula. Abrupt jumps show inaccurate fringe order estimation due to colour repetition. (d) Grey scale representation of the fringe order results obtained after refining. Seed point selected for refining is shown as a yellow dot in (c). (e) 3-D plot of values in (d).

Madhu and Ramesh [12] brought out that the patches seen in the grey scale image (figure 4.1(c)) are jumps in fringe order values due to repetition of colours in the calibration table. They proposed refined three-fringe photoelasticity (RTFP) to remove these jumps based on fringe order continuity. In RTFP, equation (4.7) is modified as [12],

$$e_i = \sqrt{(R - R_i)^2 + (B - B_i)^2 + (G - G_i)^2 + (N_p - N)^2 K^2} \qquad (4.9)$$

The additional term N_p in equation (4.9) is the fringe order of the neighbouring resolved pixel and N is the fringe order at the current checking point of the calibration table. The parameter K is determined interactively for refining the results and lies between 20 and 200 [10, 13, 14]. Figure 4.1(d) shows a smooth variation of the grey scale values obtained using RTFP. A 3-D plot of the fringe order variation is shown in figure 4.1(e). The averaged value of the adjacent neighbouring resolved pixels as N_p in equation (4.9) would enhance the accuracy further and such a provision is routinely done in the software implementation of the scheme.

Quiroga et al [15] proposed another method in which only a window of the calibration table is searched instead of the entire calibration table to minimise colour repetition. Ajovalasit et al [16] extended this idea and proposed a criterion for selecting the window size based on the index of the calibration table. The window search method assures identification of the local minima, which in turn ensures the spatial continuity of the fringe orders. For any data point, equation (4.7) is evaluated only for a limited window, which is expressed as [16].

$$e_i = \sqrt{(R - R_i)^2 + (G - G_i)^2 + (B - B_i)^2} \, ; \, i \in [i_c - \Delta i, \, i_c + \Delta i] \qquad (4.10)$$

where, i_c is the index of the fringe order value of the neighbouring pixel in the calibration table and the span Δi of the window is given by,

$$\Delta i = 0.4 \times \text{Total number of pixels in the calibration table}/N_{\max} \qquad (4.11)$$

Equation (4.11) implicitly assumes an equal number of pixels (data points) between the integer fringe orders in the calibration table, which need not be the case in general. It may be desirable to have different numbers of pixels between the integer fringe orders to capture the variable fringe gradients in an image.

Kale and Ramesh [7] further improved the window search method by using the fringe order value instead of the index of the calibration table for selecting the window that enabled the use of a merged calibration table. Merged calibration table is obtained by merging several calibration tables obtained at different loads, which is discussed in section 4.4. In this, as the loads are increased, higher fringe orders are obtained resulting in each of them having a different number of entries between consecutive integer fringes and hence, can accommodate problems having a variety of fringe gradients by virtue of having finer data. In the merged table, the number of pixels between the adjacent integer fringe orders will be larger to conveniently resolve even exceptionally low fringe gradient zones. With colour adaptation techniques (section 4.6), it is worthwhile to generate a merged calibration table as they can be profitably used for a variety of situations by suitably modifying them.

In the modified window search method, the window is determined based on fringe order increments and equation (4.10) is modified as [7],

$$e_i = \sqrt{(R - R_i)^2 + (G - G_i)^2 + (B - B_i)^2} \, ; \, N \in [N_p - \Delta N, \, N_p + \Delta N] \qquad (4.12)$$

where, ΔN is determined to be 0.4 following equation (4.11). The average value of the fringe orders of all the neighbouring resolved pixels is used as N_p in equation (4.12).

4.4 Calibration specimens and generation of a merged calibration table

Calibration is an essential step in TFP wherein one generates a table containing R, G, and B values and the corresponding fringe order values obtained from a calibration specimen. Usually used calibration specimens are specimen under four-point bending [1], eccentrically loaded tensile specimen [17] and eccentrically loaded channel shaped specimen [16]. Figure 4.2 illustrates the geometry and loading arrangements for various calibration specimens used for TFP. In all these cases, the zone in the model sufficiently away from the loading points is considered for the generation of the calibration table. Among these, the four-point bent specimen is the simplest model for generating the calibration table up to fringe order 3.

For calibration involving higher fringe orders, the selection of an appropriate calibration specimen is important. For the three specimens (figure 4.2) having a height (h) of 50 mm and thickness of 6 mm, the maximum fringe order obtainable is calculated theoretically for a load of 700 N, by taking the material stress fringe value

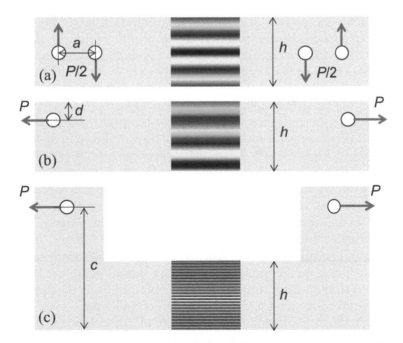

Figure 4.2. Calibration specimens used in TFP: (a) four-point-bent specimen, (b) eccentrically loaded tensile specimen, (c) eccentrically loaded channel shaped specimen. Simulated fringes are embedded at the region of interest to illustrate the maximum fringes obtainable for the same given load. The relevant loading and geometric parameters used are ($P = 700$ N, $F_\sigma = 12$ N/mm/fringe, thickness = 6 mm, $a = 25$ mm, $c = 100$ mm, $d = 10$ mm, and $h = 50$ mm).

as 12 N/mm/fringe for a reference wavelength of 546.1 nm. They are found to be 1.8, 3, and 12 for four-point-bent specimen, eccentrically loaded tensile specimen, and eccentrically loaded channel shaped specimen, respectively. The simulated fringe patterns obtained for the specimens are also shown in figure 4.2. It is found that the channel shaped specimen gives the maximum fringe orders for a given load. Unlike the four-point bent specimen which is just a rectangular strip, the channel shaped specimen is to be made especially for calibration. With developments in colour adaptation techniques, the need for calibration for each experiment is excluded and hence it is worthwhile to make the channel shaped calibration specimen to extend the technique beyond three fringe orders.

The calibration table generated is linked to the light source, polariscope, and the camera. It is desirable that the number of pixels per fringe order in the calibration table (fringe order resolution) is as high as possible to obtain good results using TFP. The fringe order resolution may be increased either by acquiring a magnified image of the calibration zone of the specimen or by merging the calibration tables obtained at different loads for the same zone. Acquiring a magnified image would change the zoom settings of the camera, resulting in some minor variations in the colour information.

It is recommended that the calibration specimen is loaded incrementally and for each of these load steps, colour data is extracted from the central region of the specimen. The median of the R, G, and B values extracted along 25 pixels on either side of the central line is used to generate the calibration table. Figures 4.3(a) to 4.3(c) show the dark-field isochromatics of the central region of the calibration specimen for three load steps (represented horizontally) to generate individual calibration tables for 0–3, 0–6, and 0–12 fringes.

Assigning appropriate fringe orders for the colour extracted is an equally important aspect of the calibration. Also, one would like to know which wavelength of light be taken as the reference wavelength for the fringe orders assigned. Grey scale image and the green plane image of figure 4.3(c) are shown in figures 4.3(d) and 4.3(e), respectively. Figure 4.3(f) shows the theoretical plot of fringe orders for a wavelength of 546.1 nm corresponding to the green wavelength. On scrutiny, figures 4.3(d) to 4.3(f) are similar, which indicates that the green plane image can be considered as the monochrome image corresponding to a wavelength of 546.1 nm. To obtain the theoretical plot (figure 4.3(f)), the value of the material stress fringe value of the calibration specimen determined using a sodium vapour lamp (589.3 nm) is normalized to the green wavelength of 546.1 nm and is found to be 10.29 N/mm/fringe. As the fringe order for the calibration table is assigned based on the green wavelength, the fringe order results obtained using the processed colour information would correspond to the wavelength of 546.1 nm.

Figure 4.3(g) shows a pictorial representation of the colour table obtained by merging five different tables of various pixel resolutions (0–1, 0–3, 0–6, 0–9, 0–12 fringe orders) into one. This representation in a sense illustrates the capability of the calibration table to accommodate fringe gradients in actual specimens. The variations of pixel resolution for the selected load steps and the merged calibration table are given in table 4.1. In the merged calibration table, the pixel resolution is relatively high for

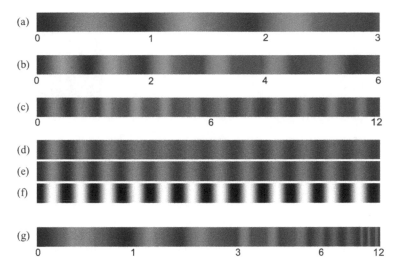

Figure 4.3. Calibration images represented horizontally for maximum fringe orders of (a) 3 (b) 6 (c) 12. (d) Grey scale image of the colour image, (e) green channel of the colour image, (f) theoretically obtained image for a wavelength of 546.1 nm. (g) Image representation showing variation of fringe gradient obtained by merging different calibration tables. Reprinted from [4] with permission of SAGE Publishing.

Table 4.1. Fringe order resolutions obtained from selected load steps and the merged calibration table. Reprinted from [4] with permission of SAGE Publishing.

Fringe order range	Selected load steps		Merged calibration table (no. of pixels)
	Load applied (N)	No. of pixels	
0–1	80	160	370
1–2, 2–3	158	94	210
3–4, 4–5, 5–6	317	50	114
6–7, 7–8, 8–9	475	34	64
9–10, 10–11, 11–12	633	25	25

lower fringe orders (370 pixels) and it reduces as one goes to higher fringe orders (25 pixels). It is reported that it is desirable to have at least 10 pixels per fringe order [18] while recording both the application and calibration specimens.

To illustrate the role of fringe order resolutions, the benchmark problem of a circular disc under diametral compression (diameter = 60 mm, load = 492 N) shown in figure 4.1 is solved using different calibration tables. The isochromatics are demodulated using equation (4.7) followed by refining (equation (4.12)) using calibration tables having pixel resolutions of 15 and 30 pixels per fringe order. The problem is also solved using the merged calibration table. Figure 4.4 shows the variation of fringe orders along the diameter of the disc. The variation of the fringe orders becomes more and more uniform with increasing fringe order resolution of the calibration table used.

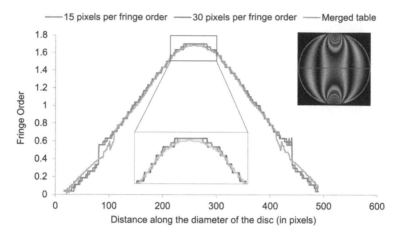

Figure 4.4. Fringe order variation along the diametral line of a disc under diametral compression obtained using calibration tables of different fringe order resolutions. Use of the merged calibration table provides a smooth variation of fringe order. Reprinted from [4] with permission of SAGE Publishing.

4.5 Twelve-fringe photoelasticity/ Total fringe order photoelasticity

4.5.1 Exploration of various colour models

Figure 4.5(a) shows the modulation of red, green, and blue signals as a function of fringe order while using a fluorescent light source. It can be observed that the modulation of the blue light is much lower after fringe order 3 as compared to red and green. It is found that the modulation of signals depends on the tint of the photoelastic specimens as well as the light source used. For strongly tinted specimens, the modulation of the blue channel will be significantly less compared to red and green channels.

The hue-saturation-value (HSV) colour representation has also been used in image processing and machine vision applications, since it describes colours better [19]. To explore the use of the HSV colour model for photoelastic specimens, the RGB values in figure 4.5(a) are converted to HSV using the following relations,

$$H = \begin{cases} \dfrac{1}{6} \times rem\left(\dfrac{G - B}{I_{\text{diff}}}, 6\right), & \text{if } I_{\max} = R \\[2mm] \dfrac{1}{6} \times \left(\dfrac{B - R}{I_{\text{diff}}} + 2\right), & \text{if } I_{\max} = G; \quad S = \dfrac{I_{\text{diff}}}{I_{\max}}; \quad V = \dfrac{I_{\max}}{255} \\[2mm] \dfrac{1}{6} \times \left(\dfrac{R - G}{I_{\text{diff}}} + 4\right), & \text{if } I_{\max} = B \end{cases} \quad (4.13)$$

where, $I_{\max}$ = maximum (R, G, B), $I_{\min}$ = minimum (R, G, B), $I_{\text{diff}} = (I_{\max} - I_{\min})$ and $rem\,(a, b)$ returns the remainder after division of a by b. Using equation (4.13) one gets the H, S and V in the range 0–1. These are scaled to 0–255 for our use. Figure 4.5(b) shows a uniform modulation of HSV for the entire range of fringe orders. The colour difference formula for the HSV model is written as,

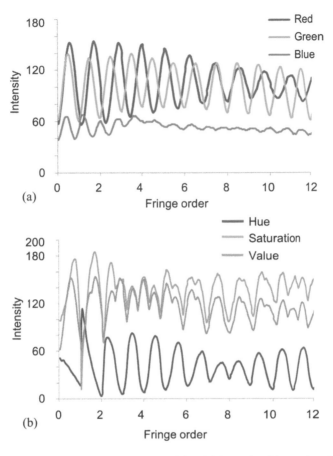

Figure 4.5. Variation of the (a) RGB and (b) HSV intensities with increasing fringe order along the central line of the calibration specimen. Reprinted from [4] with permission of SAGE Publishing.

$$e_i = \sqrt{(H - H_i)^2 + (S - S_i)^2 + (V - V_i)^2} \qquad (4.14)$$

The performance of various colour difference formulae incorporating **RGB** and/ or **HSV** colour spaces is evaluated using the problem of a thick ring subjected to internal pressure (load = 3.93 MPa, thickness = 6 mm). Figure 4.6(a) shows the dark field isochromatics in the ring and it can be observed that the maximum fringe order is close to 6 and the minimum fringe order is in between 1 and 2. It is to be noted that in the fringe field selected, there is no zeroth fringe order. In view of the lack of zeroth fringe order, determination of fringe order for a pressure vessel is quite tricky if only a monochromatic light source is used. However, the use of white light provides a colour variation for fringe order determination.

A quantitative assessment of the performance of various colour difference equations involving **RGB** and/or **HSV** is made by comparing the whole field fringe order results with the one obtained using the ten-step phase shifting technique [20–22]. Figure 4.6(b) shows the fringe order result obtained using ten-step phase shifting represented (section 3.11) as a grey scale image.

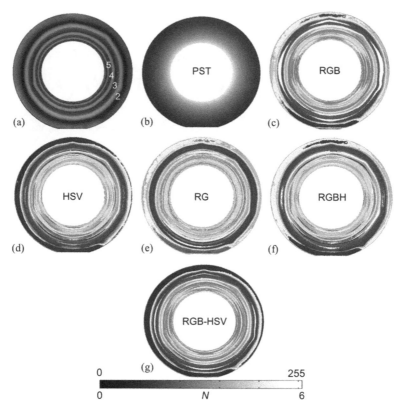

Figure 4.6. (a) Dark-field isochromatics in an epoxy ring subjected to internal pressure. (b) Smooth variation of fringe order obtained by phase shifting represented in grey scale. Fringe orders obtained by various colour difference formulae (c) RGB (d) HSV (e) RG (f) RGBH (g) RGB-HSV. Fringe order variation is not smooth and the percentage of resolved zone is the highest for (g) and the least for (e). Adapted from [4] with permission of SAGE Publishing.

Figure 4.6(c) shows the fringe order data obtained using equation (4.7). Comparing this with figure 4.6(b), only a few areas are resolved correctly. Next, the use of HSV colour model instead of RGB model is explored. Figure 4.6(d) shows the fringe orders obtained using equation (4.14). The results are better compared to the one shown in figure 4.6(c). Since figure 4.5(a) indicates that the modulations of R and G are good up to 12 fringe orders, the use of only red and green terms in equation (4.7) is then explored. The result obtained by considering only RG is shown in figure 4.6(e) and it is found that the resolved area is reduced. Next, to compensate for the loss of modulation in B, as the modulation of hue is good, following Grewal *et al* [23], use of RGB and H is studied, which is shown in figure 4.6(f). It is found to be only a shade better. Finally, the combined use of RGB as well as the HSV colour model is studied. The colour difference formula now becomes,

$$e_i = \sqrt{(R - R_i)^2 + (G - G_i)^2 + (B - B_i)^2 + (H - H_i)^2 + (S - S_i)^2 + (V - V_i)^2} \quad (4.15)$$

Figure 4.6(g) shows the fringe order estimated using the RGB–HSV colour difference formula (equation (4.15)). Qualitatively, the area of correctly resolved pixels is maximum using the combined RGB–HSV formula. Table 4.2 quantifies the performance of the five colour difference formulae in terms of the percentage of resolved area over the domain. It can be observed that the percentage of resolved zones increases with the increase in the number of parameters used in the colour difference formula. The combined use of RGB–HSV (equation (4.15)) is found to be the best by resolving 47% of the area of the model. The mean errors in the least squares results are found to decrease as more area is resolved correctly. To select appropriate seed points for further refining, it is better to have a larger area correctly resolved in the initial least squares stage itself. In view of this observation, in subsequent discussions, only the RGB–HSV and the conventionally used RGB colour models are explored further. Window search method using RGB–HSV colour spaces can be written as,

$$e_i = \sqrt{(R - R_i)^2 + (G - G_i)^2 + (B - B_i)^2 + (H - H_i)^2 + (S - S_i)^2 + (V - V_i)^2}$$
$$N \in [N_P - \Delta N, \ N_P + \Delta N] \tag{4.16}$$

For refining, the model domain is identified by a boundary mask. In section 3.7, the dialog box available in *DigiTFP*® [24] to obtain the boundary mask using geometric primitives has been discussed. Figure 4.7(a and b) shows the refined results by taking four seed points using RGB and RGB–HSV models. It is noteworthy to mention that the model is completely resolved only using RGB–HSV. It closely follows the phase shifted results shown in figure 4.6(b), whereas the RGB model is unable to resolve some zones having higher fringe order in the range of 4–6. The absolute mean error in the results is 0.21 and 0.15 fringe orders for the RGB and RGB–HSV models, respectively. It can be observed from figure 4.7 (a and b) that the results obtained even after incorporating the fringe order continuity have some minor jumps. These small jumps seen in the fringe order transition zones, especially when the fringe density is higher, are due to the sudden transition of colours and its localised variations. Hence, one must smooth the results obtained by an appropriate smoothing scheme.

Unlike the smoothing of isoclinic data, where the identification and subsequent elimination of outliers is the focus, the objective of smoothing isochromatics data is to remove minor jumps that may arise due to localised colour variations, noise in the

Table 4.2. Performance of various colour difference formulae in fringe order estimation. Reprinted from [4] with permission of SAGE Publishing.

Colour parameters	% Resolved area	Mean error
RGB	32.36	1.51
HSV	39.07	0.98
RG	27.40	1.52
RGBH	30.65	1.39
RGB-HSV	47.03	0.94

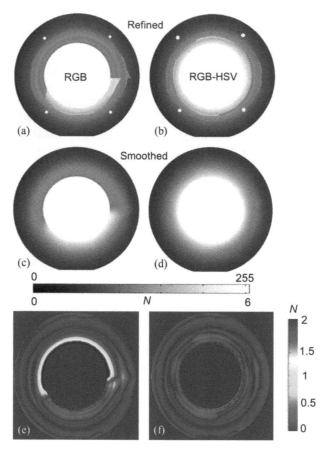

Figure 4.7. Grey scale representation of fringe orders obtained after refining by different colour difference formulae using four seed points (shown as yellow dots): (a) RGB model, (b) RGB–HSV model. Fringe order results obtained after multi-directional smoothing: (c) RGB model, (d) RGB–HSV model. Whole field error plot expressed in fringe orders with respect to phase shifting results: (e) RGB model, (f) RGB–HSV model. Adapted from [4] with permission of SAGE Publishing.

models, or insufficient resolution of the calibration table. The smoothing technique involves local regression using weighted linear least squares [25, 26]. Smoothing is performed on a one-dimensional array of fringe order values and hence initially, the boundary points of the model are identified from the boundary mask. The scanning sequence is 135°, 90°, 45°, and 0° (section 3.9.3). Figure 4.7(c and d) shows the results obtained after smoothing obtained by RGB and RGB–HSV colour difference formulae. The whole field error plots in comparison to figure 4.6(b) are shown in figure 4.7(e and f), respectively. It can be clearly observed from figure 4.7(e) that the RGB difference formula is unable to correctly resolve some zones of the high fringe order areas and the maximum error is found to be greater than 2 fringe orders near the inner diameter of the ring. The absolute mean error in the fringe order obtained finally after smoothing is 0.19 fringe orders and 0.1 fringe orders for the RGB and RGB–HSV models, respectively. The corresponding standard deviations are 0.19 and 0.036, respectively. Hence, the RGB–HSV model is better compared to the conventional RGB colour model.

4.5.2 Resolving fringe orders up to 12

The problem of a finite plate with a hole (36 mm wide, hole diameter 10 mm, subjected to 1976 N) is chosen since the fringe shape is distinctly different from that of the calibration specimen. Further, it has two sinks in the fringe field, which pose a challenge in resolving the fringe orders. Unlike processing analogue images, spatial resolution is an important issue in the digital domain, which affects the performance of the algorithms. The spatial resolution of the fringes in the plate with hole problem varies from 4 pixels/fringe ($14 < N < 15$) to 161 pixels/fringe ($7 < N < 8$). Thus, the selected problem gives a wide range of fringe gradients to test the method.

Figure 4.8(a) shows the dark field isochromatics and it can be observed that the fringe field is quite dense and the percentage of the resolved area after the initial least squares step is 16.51% and 17.52% using RGB and RGB–HSV models, respectively. In view of the high-density fringes and the percentage of the resolved area being comparatively less in relation to earlier problems discussed, only the RGB–HSV model is considered for further discussions. Figure 4.8(b) shows the least squares result obtained using the RGB–HSV model. Although it looks very noisy, refining and the subsequent smoothing process improves it considerably. Figure 4.8(c) shows the unwrapped fringe order for this problem obtained by ten-step phase shifting method (section 3.11). These results are used as a benchmark to compare the quality of results by TFP.

Figure 4.9(a) shows the refined and smoothed results of fringe orders for the isochromatics of figure 4.8(a) obtained by RGB–HSV model. Refining is achieved by selecting six seed points at different locations dictated by the fringe field that are marked in figure 4.9(a). The refining process is done by the most sophisticated scanning known as fringe order resolution-guided scanning known as FRSTFP. The details of the scanning scheme are discussed in section 4.9. The error plot in comparison to the PST results for the whole domain is shown in figure 4.9(b). It can be observed that the error zones are at the areas of high fringe concentration where the fringe resolution is less than 10 pixels. Figure 4.9(c and d) shows the variation of fringe order along two horizontal lines *EF* and *GH* shown in figure 4.9(a). The lines are chosen such that they cover a range of fringe orders and capture their gradients. It can be seen from figure 4.9(c) that the results closely follow the phase shifting results. The spatial resolution of the fringe patterns decreases towards the hole, becoming as low as 4 pixels per fringe order at $14 < N < 15$. It can be observed from figure 4.9(d) that the PST algorithm has captured up to 11.5 fringe orders and TFP using the RGB–HSV model is able to capture up to 10.75 fringe orders. This limitation is essentially due to the lack of spatial resolution and the method has worked well up to a fringe resolution of 10 pixels per fringe order, which tallies with the observation made by Ajovalasit *et al* [18].

Figure 4.10 shows the dialog box in *Digi*TFP® that summarises the sequence of steps from top to bottom, needed to perform the TFP analysis. The first step is to select whether one has to perform the fringe order evaluation within a tile or a boundary

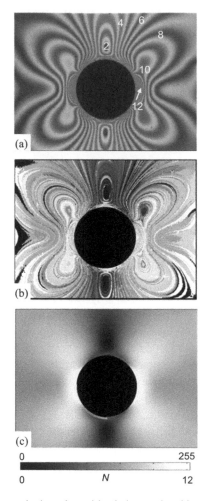

Figure 4.8. (a) Dark-field isochromatics in a plate with a hole sample subjected to a load of 1976 N. (b) Fringe order obtained by RGB–HSV model (equation (4.15)). (c) Smooth variation of fringe order obtained by phase shifting technique. Adapted from [4] with permission of SAGE Publishing.

mask. A tile is usually a rectangular window, whereas the boundary mask can be of any arbitrary shape to cover any model domain. Geometric primitives are used to generate the boundary mask and the relevant dialog box has been discussed in section 3.7.

The next step is whether one would like to work on the actual image or the normalized image. The details of the normalization scheme are discussed in section 4.10. For poorly recorded images, one may need to use the normalization scheme, otherwise, in general, one may use the actual image for processing. Next, one has to select the relevant calibration table and also specify the minimum and the maximum fringe orders that need to be determined for the given problem. This is optional information to minimise the influence of colour repetition affecting fringe order continuity. To minimise the need for generating the calibration table afresh for each problem situation, colour adaptation techniques are developed, and these are

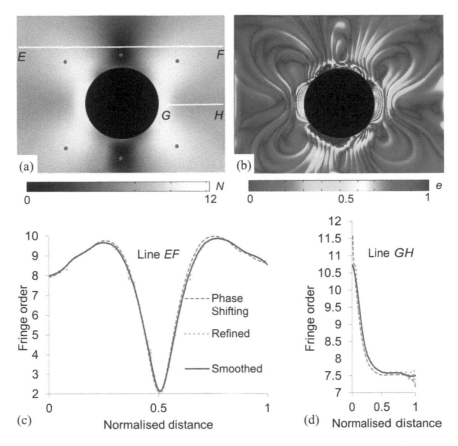

Figure 4.9. (a) Refined and smoothed fringe order results obtained using six seed points for isochromatics shown in figure 4.8(a) by RGB–HSV model. (b) Whole field error plot expressed in fringe orders obtained with respect to phase shifting. Variation of fringe order along lines (c) *EF* and (d) *GH* shown in (a). Adapted from [4] with permission of SAGE Publishing.

discussed in section 4.6. Normally, one may have to choose a suitable colour adapted calibration table. Then one may choose the relevant colour model for data reduction. It can be RGB, HSV, RG or a combination of RGB–HSV. The 'Generate FRN' button (this will provide the whole field fringe order data as a .frn file) will do the basic least squares search based on the chosen colour model. This data needs to be refined for fringe order continuity and then smoothed for further processing. The refining process requires the selection of appropriate scanning schemes that are discussed in sections 4.7–4.9.

4.6 Colour adaptation techniques

The accurate fringe demodulation in TFP depends on the similarity in colour characteristics between the application and the calibration images. In other words, it depends on the similarity in pixel distribution of the application and the calibration images in RGB colour space. The colour mismatch or tint variation can be

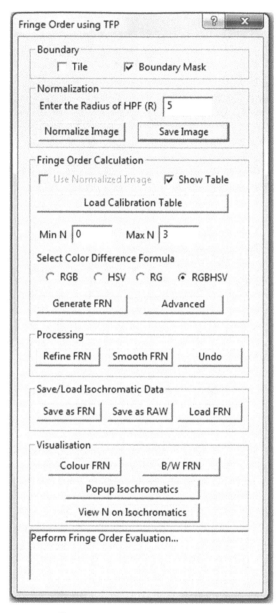

Figure 4.10. Dialog box in *Digi*TFP® [24] that summarises the sequence of steps needed to perform the TFP analysis. Courtesy: [24].

emanating from (i) *light source*—its emission spectra, colour rendering index (CRI), non-uniformity in illumination (ii) *optical elements*—transmission spectra of polarising elements, quarter-wave plate error (iii) *camera*—relative spectral response, calorimetric characterization, dynamic range, spectral exposure level, and white balance (iv) *specimen material*—material response, transmission spectra, dispersion of birefringence (wavelength dependence), composition, ageing effects, annealing and stress freezing [27]. If the polariscope, light source, camera, and ambient conditions remain the same, the calibration table obtained for one of the specimens can be used for repeated experiments. However, there could be a small tint variation of the specimens due to various aspects mentioned in (iv). In such a case, in the normal course, one must develop a new calibration table meant for that material. This can be tedious and here, the colour adaptation schemes offer hope for a simplification to use the already available calibration table by suitably modifying the same. Many colour adaptations methods have been developed and one should know the appropriate method for a given situation. The modifications to be done to achieve this on the original calibration table for one-point, two-point, three-point, and colour transfer methods are discussed next.

Colour adaptation techniques [6, 14, 28, 29] have been originally introduced to tackle the issue of small tint variation between the application and the calibration images. The first effort towards this was made by Madhu *et al* [6] in 2007. In one-point colour adaptation technique, Madhu *et al* [6] made use of the bright field images corresponding to zero load on the application and calibration specimens to modify the calibration table. To avoid slight variations that may be present in the *R, G, B* values, they resorted to using the average of individual *R, G, B* intensities inside a small tile within the image. Let R_z^a, G_z^a, B_z^a denote the averaged RGB values for the application specimen and R_z^c, G_z^c, B_z^c denote the same for the calibration specimen and the subscript 'z' denotes that these are obtained for zero load. Then the modified intensity values of the calibration table are given by,

$$R_{mi}^c = \frac{R_z^a}{R_z^c} R_i^c; \quad G_{mi}^c = \frac{G_z^a}{G_z^c} G_i^c; \quad B_{mi}^c = \frac{B_z^a}{B_z^c} B_i^c \tag{4.17}$$

where R_{mi}^c, G_{mi}^c, B_{mi}^c indicate the RGB values at the ith row in the modified calibration table and R_i^c, G_i^c, B_i^c are the RGB values at the ith row in the original calibration table. In the subsequent discussions, the superscripts *a* and *c* denote the application and the calibration specimens, respectively. The subscript *m* denotes the entry in the modified calibration table.

In two-point colour adaptation proposed by Neethi Simon and Ramesh [14], let the maximum and minimum intensity in the application image be represented as R_{Max}^a, G_{Max}^a, B_{Max}^a, R_{Min}^a, G_{Min}^a, B_{Min}^a and that in the calibration table be R_{Max}^c, G_{Max}^c, B_{Max}^c, R_{Min}^c, G_{Min}^c, B_{Min}^c. The ith row entries in the original calibration table R_i^c, G_i^c, B_i^c gets modified as R_{mi}^c, G_{mi}^c, B_{mi}^c given by,

$$R_{mi}^c = \left[\left(\frac{R_{Max}^a - R_{Min}^a}{R_{Max}^c - R_{Min}^c} \right) \times (R_i^c - R_{Min}^c) \right] + R_{Min}^a$$

$$G_{mi}^c = \left[\left(\frac{G_{Max}^a - G_{Min}^a}{G_{Max}^c - G_{Min}^c} \right) \times (G_i^c - G_{Min}^c) \right] + G_{Min}^a \tag{4.18}$$

$$B_{mi}^c = \left[\left(\frac{B_{Max}^a - B_{Min}^a}{B_{Max}^c - B_{Min}^c} \right) \times (B_i^c - B_{Min}^c) \right] + B_{Min}^a$$

Swain *et al* [29] reported that the mean intensities of colours in the application image and the calibration table have a nonlinear dependency and proposed a three-point colour adaptation to address this nonlinearity between the dependent and independent variables. Let I_{Min}^c, I_{Mean}^c, I_{Max}^c be the independent variables which denotes the minimum, mean, and maximum intensities of the calibration table and I_{Min}^a, I_{Mean}^a, I_{Max}^a be the dependent variables that corresponds to the minimum, mean, and maximum intensities of the application image, where I indicates any of the R, G, B intensities. For an entry at the ith row of the calibration table I_i^c, the modified value I_{mi}^c is given by,

$$I_{mi}^c = I_{\text{Min}}^a \frac{(I_i^c - I_{\text{Mean}}^c)(I_i^c - I_{\text{Max}}^c)}{(I_{\text{Min}}^c - I_{\text{Mean}}^c)(I_{\text{Min}}^c - I_{\text{Max}}^c)} + I_{\text{Mean}}^a \frac{(I_i^c - I_{\text{Min}}^c)(I_i^c - I_{\text{Max}}^c)}{(I_{\text{Mean}}^c - I_{\text{Min}}^c)(I_{\text{Mean}}^c - I_{\text{Max}}^c)}$$
$$+ I_{\text{Max}}^a \frac{(I_i^c - I_{\text{Min}}^c)(I_i^c - I_{\text{Mean}}^c)}{(I_{\text{Max}}^c - I_{\text{Min}}^c)(I_{\text{Max}}^c - I_{\text{Mean}}^c)} \tag{4.19}$$

In specific situations, the colour adaptation techniques [6, 14, 28, 29] can fail. Consider figure 4.11, which contains a bright red spot that may be 'hot pixels' that develop at long exposure or high ISO photography. These kinds of noise can develop due to lighting or photographic artefacts. The positions of maximum and minimum R, G, B intensity values are marked in figure 4.11. On scrutiny, one can see that the maximum red intensity (R_{Max}^a) and minimum blue intensity (B_{Min}^a) are wrongly identified at the red spot. Moreover, commonly developed external marks on the specimen in an industrial scenario, and the inclusion of background pixels due to erroneous image domain mask selection can alter the minimum intensity value. In these cases, the existing colour adaptation schemes would lead to error since these use the maximum and minimum intensity values as the interpolation parameters and are sensitive to changes in these parameters. This calls for the exploration of new colour matching strategies. Colour transfer [30] is the counterpart to colour

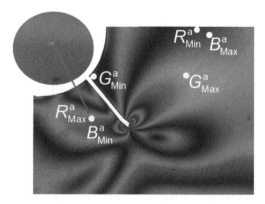

Figure 4.11. The presence of photographic artefact (hot pixels) as red spot in biaxially loaded cruciform specimen with an inclined crack. The positions of maximum and minimum colour channel intensities are marked on the image along with the magnified view of the red spot. Reprinted from [31] with permission from Elsevier.

adaptation used in the field of image processing to match colour characteristics between natural scenes.

Sachin and Ramesh [31] developed in 2020, a new colour transfer method in which the mean and standard deviation (SD) of intensities of the application specimen image (R^a_{Mean}, G^a_{Mean}, B^a_{Mean}, R^a_{SD}, G^a_{SD}, B^a_{SD}) and calibration table (or image) (R^c_{Mean}, G^c_{Mean}, B^c_{Mean}, R^c_{SD}, G^c_{SD}, B^c_{SD}) are used to modify the entries in the calibration table (R^c_i, G^c_i, B^c_i) to generate a modified calibration table (R^c_{mi}, G^c_{mi}, B^c_{mi}) given by,

$$R^c_{mi} = \frac{R^a_{\text{SD}}}{R^c_{\text{SD}}}(R^c_i - R^c_{\text{Mean}}) + R^a_{\text{Mean}} \tag{4.20}$$

$$G^c_{mi} = \frac{G^a_{\text{SD}}}{G^c_{\text{SD}}}(G^c_i - G^c_{\text{Mean}}) + G^a_{\text{Mean}} \tag{4.21}$$

$$B^c_{mi} = \frac{B^a_{\text{SD}}}{B^c_{\text{SD}}}(B^c_i - B^c_{\text{Mean}}) + B^a_{\text{Mean}} \tag{4.22}$$

The choice of mean and standard deviation, which are statistical quantities, help to capture the overall pixel distribution of the images better.

From the discussions, the one-point colour adaptation technique has kickstarted many methods for colour adaptations. The later ones like two-point, three-point and the recent one of colour transfer need to be compared for their relative merits.

4.6.1 Comparative study of the colour transfer method with two-point and three-point colour adaptation schemes

The fringe pattern shown in figure 4.11 is taken up to illustrate the comparison between the methods. Only the essence of the image domain mask that eliminates crack faces and the high stress gradient zone (low fringe resolution zone) near the crack tip are shown in figure 4.12(a). The outer boundaries of the mask covering the image are not shown fully here due to space constraints. Figure 4.12(b) shows the application image within the domain mask. As the maximum fringe order in the domain is below 4, a 0–4 fringe order calibration table of epoxy, which maintains the same image grabbing condition as that of the application specimen is used for effecting different colour adaptation schemes. The standard calibration table is separately modified by two-point adaptation, three-point adaptation, and the colour transfer method.

To compare the total fringe orders evaluated using modified calibration tables by various methods, the analytical solution of the crack-tip fringe field is evaluated based on the methods discussed in chapter 2. Using data collected from the fringe field, by performing an over-deterministic non-linear least squares scheme, the stress field near the vicinity of the crack-tip is captured by a ten-parameter solution (table 4.3). Using the ten-parameter solution (ten Mode-I and ten Mode-II parameters), theoretically reconstructed fringe pattern in colour is obtained over the field for comparing the performance of the three methodologies (figure 4.12(c)).

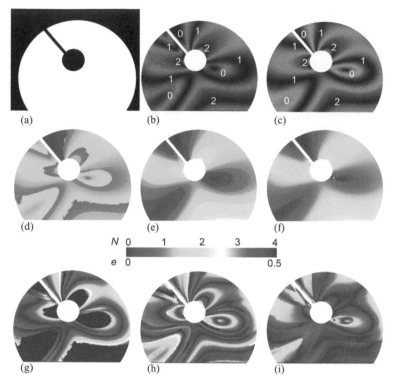

Figure 4.12. (a) Image mask. (b) Application image within the mask. (c) Theoretically reconstructed isochromatics in colour using the parameters in table 4.3. Colour map of refined fringe orders obtained by: (d) two-point colour adaptation, (e) three-point colour adaptation, (f) colour transfer method. Absolute error between theoretically reconstructed fringe pattern and (g) image 'd', (h) image 'e', (i) image 'f'. Courtesy: figures (a), (b), (d–i). Adapted from [31] with permission from Elsevier.

Table 4.3. Stress field parameters for theoretically generated fringe pattern in figure 4.12(c). Reprinted from [31] with permission from Elsevier.

Mode I parameters	Mode II parameters
$A_{I1} = 7.375\ 967\ \mathrm{MPa}\sqrt{\mathrm{mm}}$	$A_{II1} = -0.546\ 962\ \mathrm{MPa}\sqrt{\mathrm{mm}}$
$A_{I2} = 0.288\ 365\ \mathrm{MPa}$	$A_{II2} = 0.000\ 000\ \mathrm{MPa}$
$A_{I3} = -0.199\ 015\ \mathrm{MPa(mm)}^{-1/2}$	$A_{II3} = 0.068\ 140\ \mathrm{MPa(mm)}^{-1/2}$
$A_{I4} = 0.012\ 779\ \mathrm{MPa(mm)}^{-1}$	$A_{II4} = -0.035\ 680\ \mathrm{MPa(mm)}^{-1}$
$A_{I5} = -0.001\ 984\ \mathrm{MPa(mm)}^{-3/2}$	$A_{II5} = 0.005\ 148\ \mathrm{MPa(mm)}^{-3/2}$
$A_{I6} = -0.000\ 003\ \mathrm{MPa(mm)}^{-2}$	$A_{II6} = -0.001\ 107\ \mathrm{MPa(mm)}^{-2}$
$A_{I7} = 0.000\ 039\ \mathrm{MPa(mm)}^{-5/2}$	$A_{II7} = 0.000\ 047\ \mathrm{MPa(mm)}^{-5/2}$
$A_{I8} = -0.000\ 002\ \mathrm{MPa(mm)}^{-3}$	$A_{II8} = -0.000\ 017\ \mathrm{MPa(mm)}^{-3}$
$A_{I9} = 0.000\ 001\ \mathrm{MPa(mm)}^{-7/2}$	$A_{II9} = -0.000\ 004\ \mathrm{MPa(mm)}^{-7/2}$
$A_{I10} = 0.000\ 000\ \mathrm{MPa(mm)}^{-4}$	$A_{II10} = 0.000\ 001\ \mathrm{MPa(mm)}^{-4}$

The results show the efficacy of the over-deterministic least squares method to evaluate the fringe field and it can be noted that the comparison with figure 4.12(b) is extremely good. The stress intensity factors evaluated are $K_I = 0.585$ MPa$\sqrt{m}$ and $K_{II} = 0.043$ MPa$\sqrt{m}$. The convergence error obtained is 0.025 [31].

The whole field fringe order is obtained for the three cases when two-point adapted, three-point adapted, and the colour transferred calibration tables are made use of in the least squares analysis of the application image. The whole field fringe order thus obtained is refined by a sophisticated scanning, the details of which are discussed in the next section. The refined fringe orders are represented as figure 4.12(d–f).

The absolute fringe order error plots of the refined results with respect to the analytical solution are shown in figure 4.12(g–i). The mean absolute error (MAE) and standard deviation (SD) (in terms of fringe orders) after refining, for the colour transfer method are the least and they are 0.09 and 0.16, respectively. They are the highest for the two-point scheme (MAE = 0.36 and SD = 0.38), which is due to the presence of the red spot (hot pixel) in the application image (figure 4.11). These values are intermediate (MAE = 0.12 and SD = 0.17) for the three-point colour adaptation as it also involves the use of mean intensity in the calculations

4.7 Scanning schemes

Incorporation of the fringe order continuity criteria removes abrupt jumps in the fringe order estimated by the least squares method. Though this has helped to improve the quality of results in general, this has also made the method vulnerable to be influenced by the scanning scheme adopted for refining. Over the years, various scanning schemes have been proposed for refining the fringe order data such as, horizontal and vertical scanning schemes [16, 17], flood-fill scanning approach [12], four sub-image scheme [16], advancing front scanning scheme [7] and fringe resolution-guided scanning (FRSTFP) [8]. One of the important requirements of the scanning schemes is that it should be able to scan the complete model domain of any arbitrary shape. It is also desirable that the scanning scheme is sensitive to minimise noise propagation.

The scanning schemes can be generally classified into two—fixed seeding and flexible seeding schemes. Figure 4.13 shows the various scanning schemes used for refining the fringe order data in TFP/RGB photoelasticity. Simple schemes use a fixed seeding approach in which the first pixel of a row/column of the rectangular region of interest (ROI) is taken as the seed point for refining [16, 17]. Flexible scanning schemes give flexibility for the user to select seed points anywhere in the model domain. Flood-fill based scanning [12], four sub-image scanning [16], advancing front scanning (AFSTFP) [7] and fringe resolution-guided scanning (FRSTFP) [8] come under this category. The last two scanning schemes are abbreviated with TFP at the end to indicate that these scanning schemes are meant exclusively for TFP.

The horizontal, vertical or a combination of these two scanning approaches are restrictive methods, which work only on carefully selected ROI [32]. On the other

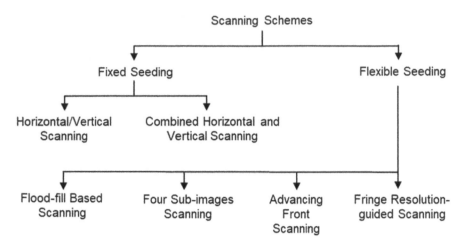

Figure 4.13. Classification of scanning schemes used for refining in TFP/RGB photoelasticity

hand, flood-fill, four sub-images, and AFSTFP are useful for a certain class of domains. The most advanced scanning scheme is FRSTFP as it incorporates the spatial resolution in scanning to avoid the propagation of noise, which is discussed in section 4.9.

4.7.1 Flood–fill scanning scheme

The flood-fill based scanning approach proposed by Madhu and Ramesh [12] in 2007 (hereafter called flood-fill scanning) in conjunction with RTFP, enables the user to choose any correctly resolved point within the model domain as the seed point for starting the refining process. The use of flexible seeding in conjunction with simple horizontal and vertical scanning has made it possible to completely refine simply connected model domains having arbitrary boundaries.

Figure 4.14(a) illustrates the flood-fill scanning for a model having an arbitrary geometry. To account for arbitrary shaped models, new data structures for boundary information handling, viz., XBN (x-boundary) for vertical scanning and YBN (y-boundary) for horizontal scanning have been developed [12]. Let A be the primary seed point selected for refinement. Horizontal scanning along the line AB yields secondary seed points. Vertical scanning using these will refine pixels in the regions I and II. To refine fringe order data in the regions III and IV, the primary seed point is re-assigned to point C, which is the midpoint of BD. This process of re-assigning the primary seed point is repeated until the new seed point coincides with the boundary pixel. This way, the entire model domain is scanned. In this approach, depending on the location of the pixels, the scanning scheme adopted is either horizontal or vertical. The application of this approach is demonstrated for a circular disc under diametral compression. The seed point is deliberately selected off-centre (figure 4.14(b)) to demonstrate the automatic generation of additional seed points to scan the complete model domain. The XBN and YBN files are primitive forms of using a boundary mask. It is now commonplace to have a boundary mask

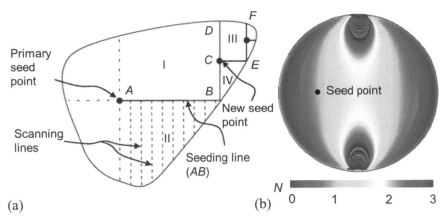

Figure 4.14. (a) Progressive flood-fill approach proposed by Madhu and Ramesh [12] for arbitrary boundaries. (b) Contour plot of the fringe order results obtained for the circular disc under diametral compression. Adapted from [32] with permission from Elsevier.

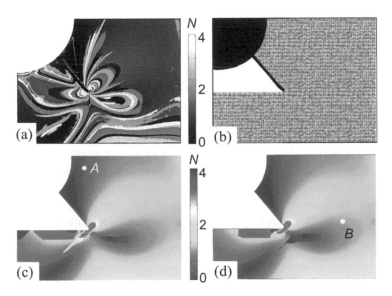

Figure 4.15. (a) Fringe order estimated using equation (4.12), (b) area of the model scanned using the scheme. Contour plot of the fringe order results obtained using (c) seed point A, (d) seed point B. Seed points used for refining are shown in yellow. Adapted from [32] with permission from Elsevier.

to define the region of interest conveniently. In fact, this scanning scheme is quite useful to generate the boundary mask, based on geometric primitives (section 3.7).

Next, the methodology is applied to resolve fringe fields near the tip of a crack (figure 4.11). Figure 4.15(a) shows the fringe order obtained using the colour difference formula (equation (4.7)). This needs to be refined. Figure 4.15(b) shows the area of the model scanned by the flood-fill scanning scheme. It is unable to scan the complete model domain owing to the presence of the crack, which is a discontinuity. Nevertheless, the influence of seed point location is studied for the problem using two seed points at *A* and *B*. The contour plots of the fringe order results obtained are shown in figure 4.15(c and d), respectively. The two contour plots are not completely identical, and some differences are seen in the neighbourhood of the crack-tip. Further, in both the cases, abrupt jumps are seen in the contour plot that indicate erroneous zones.

4.7.2 Four sub-images scanning scheme

The four-sub image scanning scheme was proposed by Ajovalasit *et al* [16] in 2010 as part of their work on the improvements in the window search method. The image is divided into four sub-images delimited by the row and column of the seed point selected (figure 4.16(a)). Hence, in this scheme, the area of the model scanned is dictated by the location of the seed point selected. Each sub-image is then scanned row-wise for fringe order refinement. The scanning scheme uses a boundary mask to identify the domain of interest. Figure 4.16(b) illustrates the progression of a four sub-images scanning scheme for a sample sub-image 4. Due to its dependence on the selected seed point location, the scheme would be unable to scan even a simply connected circular disc completely if the seed point were not selected appropriately [32].

The inclined crack problem (figure 4.11) is then considered. Figure 4.17(a–c) shows the area of the model scanned for various seed points selected in the model. The presence of the crack adds to the geometric complexity of the model. Figure 4.17(d and e) shows the contour plot of the fringe order values obtained using the seed points shown in figure 4.17(b and c), respectively. The difference in the results shows that the consistency and quality in fringe order results obtained using this method is highly dependent on the position of the seed point. In both the cases, the fringe orders below the crack-tip are found to be erroneous.

The use of seed point shown in figure 4.17(c) scans the maximum area for this problem, which is the same as the zone covered by the flood-fill scanning. The progression of the scanning scheme in sub-image 4 (figure 4.17(c)) is analysed at an interval of every 60 000 refined pixels. To accommodate more images, the overall

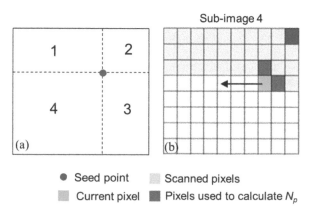

Figure 4.16. Four sub-images scanning scheme (a) four sub-images based on the seed point selected, (b) progression of the scanning scheme for sub-image 4. Reprinted from [32] with permission from Elsevier.

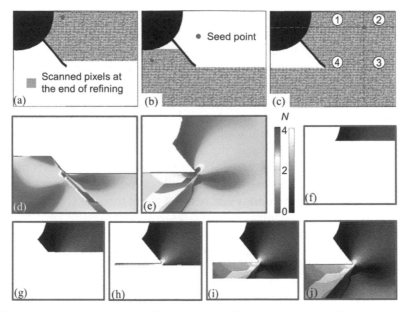

Figure 4.17. (a)–(c) Area of the model domain scanned using the selected seed point. Contour plot of the fringe order results obtained using: (d) seed point shown in (b), (e) seed point shown in (c). (f)–(j) Progression of the scanning scheme for sub-image 4 shown in (c) at an interval of 60 000 steps. Adapted from [32] with permission from Elsevier.

image is shrunk and presented in figures 4.17(f–j). It can be observed that the fringe orders are determined correctly till the scanning front reaches the region near the crack tip (figure 4.17(h)). The noise in the fringe order results originates from this zone and later propagates to the subsequent pixels. Though this method has not worked, the source of noise is identified as the crack-tip that has high density of fringes (low fringe resolution).

The four sub-images scanning does not involve progressive re-assigning of secondary seed points and progresses only by row-wise scanning. The flood-fill scanning scheme proposed by Madhu and Ramesh [12] makes judicious use of row and column scanning approaches and is comparatively superior in that respect.

4.7.3 Advancing front scanning scheme

Kale and Ramesh [7] in 2013 proposed an advancing front scanning (AFSTFP) scheme, which progresses by resolving a pixel that is surrounded by the maximum number of resolved pixels within a 3 × 3 window centred on it at every step. For the first time, the use of multiple simultaneous user defined seed points has been suggested for refining models having complex shapes and geometries. The process of refining could be achieved either by RTFP (equation (4.9)) or by window search method (equation (4.12) or equation (4.16)) in practice.

The major difference between AFSTFP and flood-fill method is that the flood-fill method will just take the first pixel in the queue for resolving, while the AFSTFP method searches for pixel having maximum resolved neighbours. Unlike the previous scanning schemes, the AFSTFP scheme can scan the complete model domain and the seed point location does not affect this aspect. This is its significant advantage. However, the fringe order results are dependent on the location of the seed point, which is not desirable.

Figure 4.18(a and b) shows the results for seed points A and B, respectively. In both the cases, erroneous results are obtained after the scan has crossed the crack-tip. Figure 4.18(c) shows the contour plot of the fringe order obtained using A and B as seed points simultaneously and the zone of noise has reduced considerably compared to the result shown in figure 4.18(a and b). The scanning fronts from the two seed points A and B have restricted the propagation of noise to some extent. To improve the results further, refining is carried out using three seed points A, B, and C (figure 4.18(d)). There is a significant improvement in the results near the crack-tip region compared to figure 4.18(c). The fringe order results near the crack-tip are found to improve further by slightly varying the location of seed points from A, B and C to A', B', and C' (figure 4.18(e)).

Hence, to obtain accurate results after refining using AFSTFP, the seed points must be selected carefully such that the noise that originates from the low-resolution zones is encapsulated within as small a region as possible. However, careful selection of multiple seed points at appropriate locations is tedious and time-consuming.

It is found that all the scanning schemes discussed so far lead to the propagation of noise from regions of low spatial resolution of the isochromatic fringe patterns. This leads to a large area of erroneous results after refining. Careful selection of multiple seed points using an advancing front scanning scheme can minimise noise propagation. However, the method involves extensive user interactions and requires

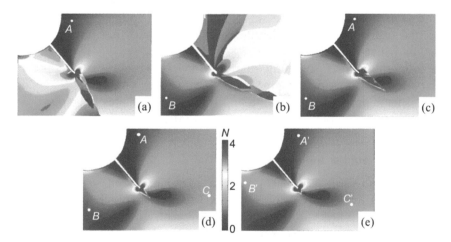

Figure 4.18. Contour plot of the fringe orders obtained after refining by advancing front scanning scheme using (a) seed point A, (b) seed point B, (c) simultaneous use of two seed points A and B—in this case leaving a small zone near the crack-tip other zones are resolved correctly, (d) simultaneous use of three seed points—in this the unresolved zone near the crack-tip is further reduced, (e) a slight change in the location of three seed points has only a marginal influence. Adapted from [32] with permission from Elsevier.

users with experience in photoelastic analysis and image processing for a quality output. The details of FRSTFP are discussed in the subsequent sections, which addresses these shortcomings and is the best scanning method available as of now.

4.8 Influence of spatial resolution

It is reported [18] that for accurate fringe order estimation by TFP, the spatial resolution of the fringe pattern must be at least 10 pixels per fringe order. Figure 4.19 shows the variation of the red, green, and blue intensity values along line PQ taken such that point P is far away from the crack-tip and point Q is closer to the crack-tip. The intensity modulation drastically reduces towards the crack-tip and the colour information is not adequately captured in this zone. Since TFP uses colour information to assign fringe orders, the results in this zone are erroneous. The noise originating from these low fringe resolution zones propagates and leads to a large area of erroneous results. This is because in equations (4.12) and (4.16) if the N_p value is erroneous and noisy, then the noise propagates to the current pixel due to false estimation.

4.9 Fringe resolution guided scanning in TFP (FRSTFP)

4.9.1 A new measure for identifying gradients in the model domain [8]

Figure 4.20(a) shows the dark field isochromatics theoretically simulated for a circular disc under diametral compression (load = 492 N, dia. = 60 mm) recorded in

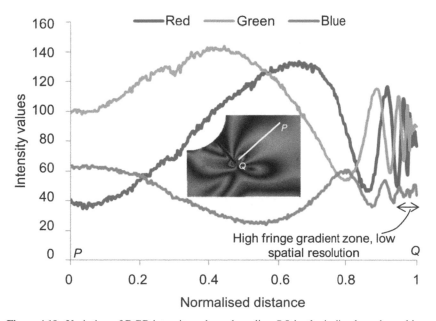

Figure 4.19. Variation of RGB intensity values along line PQ in the inclined crack problem

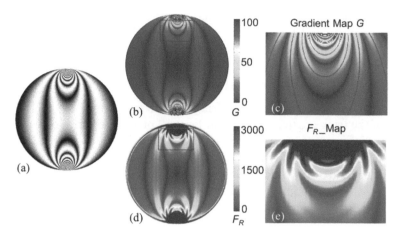

Figure 4.20. (a) Theoretically simulated fringe pattern in a circular disc under diametral compression, (b) the intensity gradient map G, (c) magnified view of a portion in (b), (d) whole field map resembling the fringe gradient (F_R_Map), (e) magnified view of a portion in (d). Adapted from [8] with permission from Elsevier.

monochromatic light. It can be observed that the fringe gradient is extremely high near the loading points. The objective here is to create a whole field map whose values are representative of the fringe gradient in this model.

For each pixel in the domain, initially the intensity gradients are computed by the central difference formula in the horizontal and vertical directions.

$$G_x = \frac{I(i, j + 1) - I(i, j - 1)}{2}; \; G_y = \frac{I(i + 1, j) - I(i - 1, j)}{2} \tag{4.23}$$

where, G_x and G_y are the intensity gradients in the horizontal and the vertical directions, respectively, and (i, j) represent the pixel coordinates of the point under consideration. Next, the absolute gradient is calculated for each pixel from the directional gradients as,

$$G = \sqrt{G_x^2 + G_y^2} \tag{4.24}$$

Figure 4.20(b) shows the absolute gradient map G for the circular disc. Although in the overall image, it gives a semblance that the intensity variation is indicative of the fringe gradient, in the magnified image (figure 4.20(c)), there are still pixels near the centre of the fringes in which the values are low (appearing as blue). Hence, the quantity G is not adequate to capture the fringe gradient. Next, by trial and error, another quantity designated as F_R (fringe resolution) is obtained by taking the sum of the absolute intensity gradient over a $k \times k$ pixel window centred around the pixel under consideration, which is given by,

$$F_R(i, j) = \sum_{m=i-p}^{m=i+p} \sum_{n=j-p}^{n=j+p} G(m, n) \quad \text{where } p = \frac{1}{2}(k - 1) \tag{4.25}$$

where (i, j) represents the global coordinates of the pixel under consideration and (m, n) represents their local coordinates in the $k \times k$ neighbourhood. A whole field plot of F_R is labelled as F_R_Map and is shown for the whole disc in figure 4.20(d). In the magnified view (figure 4.20(e)), consistently it shows only increasing values of F_R (indicated by the colour) suggesting that it can effectively be used as a measure to capture the fringe gradients in a problem domain.

The value of F_R is dependent on the choice of the kernel size 'k' and figure 4.21 shows its variation as a function of the kernel size along the line AB of the disc in which the fringe gradient increases from A to B. It can be seen from the figure that the values of F_R are indicative of the gradient of the fringes and the higher its value, the higher is the gradient. A kernel size of 11×11 is found to work well for a variety of problems.

This idea is extended for use in handling colour images. Figure 4.22(a–c) shows the absolute gradient plot of the image planes of red, green, and blue for the problem of bi-axially loaded cruciform specimen with an inclined crack (figure 4.11). The sum of the gradients of the three image planes is shown in figure 4.22(d). Using these, a F_R_Map is generated and shown in figure 4.22(e). It is clear from the figure that the crack-tip area being a zone of stress concentration (low fringe resolution zone), it is identified as a zone of increasing values of F_R. The F_R_Map identifies even minute marks in the model domain. Although it is not of any advantage to solve the current problem, it comes in handy for solving industrial problems with unwanted marks and pores on the specimens. Utilisation of this aspect will be dealt with and highlighted by solving a model made from rapid prototyping in section 4.9.3.

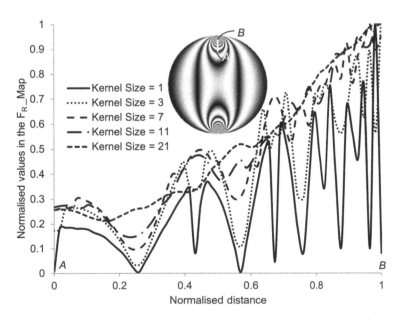

Figure 4.21. Variation of the values in the F_R_Map along a line AB at the high fringe gradient region of the simulated fringe pattern in a circular disc under diametral compression.

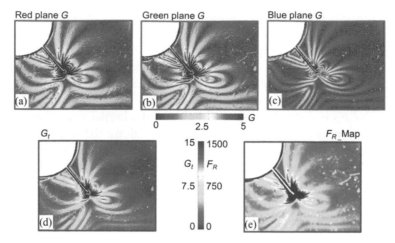

Figure 4.22. Absolute gradient for different colour planes of the bi-axially loaded cruciform specimen having an inclined crack (a) red, (b) green, (c) blue. (d) Sum of the absolute gradients (G_t). (e) F_R_Map contour plot generated by processing image (d). Adapted from [8] with permission from Elsevier.

4.9.2 Development and application of FRSTFP

It has been discussed earlier that while solving problems, noise from low spatial resolution zones (high stress gradient zones) gets propagated while refining the data using the window search method or RTFP. Since the zones of low spatial resolution can be identified by high values of F_R, a new scanning scheme is devised to take advantage of that, which is shown as a flow chart in figure 4.23. To start the refining process, a point that is correctly resolved by the least squares method is selected. At each step, all the unresolved adjacent pixels corresponding to the resolved pixels are considered for deciding the progression of the scanning scheme. Initially, the eight neighbouring pixels corresponding to the seed point are stored in a list of unresolved pixels. Subsequently, this list of adjacent unresolved pixels is updated. The algorithm searches for the least value of F_R amongst the list of adjacent unresolved pixels at each step so that all the high spatial resolution zones are resolved before resolving the pixels from low spatial resolution zones. If the pixel under consideration has more than one neighbouring resolved pixel, an average value of the fringe order is taken as N_p to maximise the use of neighbouring resolved pixel data. Hence, the FRSTFP scheme retains the advantages of AFSTFP scheme while addressing the issue of varied spatial resolution of the isochromatic fringes.

Refining is carried out by FRSTFP in conjunction with the modified window search method (equation (4.16)). The F_R_Map corresponding to the isochromatics fringe patterns is shown in figure 4.22(e). Figure 4.24(a–e) shows the progression of the scanning scheme for the inclined crack problem at an interval of every 100 000 refined pixels. The area near the crack tip which has the least spatial resolution is refined only at the end of the scan. Hence, the noise in the results due to inadequate resolution is encapsulated within this region and its propagation is restricted by

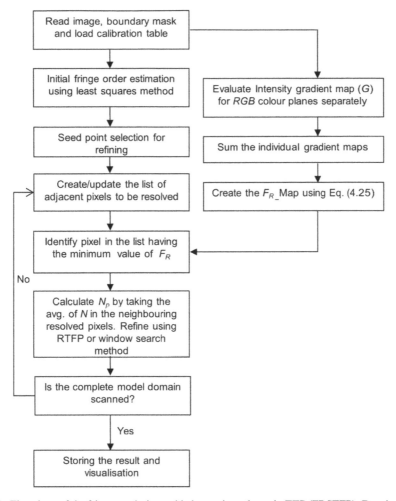

Figure 4.23. Flowchart of the fringe resolution guided scanning scheme in TFP (FRSTFP). Reprinted from [8] with permission from Elsevier.

moving it to the end of the scanning process. When applied to the problem of a thick ring subjected to internal pressure (figure 4.6(a)), it is shown in reference [8] that the refining by FRSTFP happens near the inner boundary only at the end, thus greatly minimizing the noise propagation from low spatial resolution zones.

Figure 4.25(a) shows the smoothed fringe order results obtained by multi-directional progressive smoothing scheme with seed point 1. Figure 4.25(b) shows the fringe order variation along the line AB. The fringe order results are compared with the integer fringe order values obtained by skeletonising the fringe pattern in the green plane image. The results obtained are consistent with the integer fringe orders. The mean errors in the fringe order values at the fringe skeleton points are only 0.05 for the inclined crack problem. Smoothed results obtained by two other seed point locations are shown in figure 4.25(c and d).

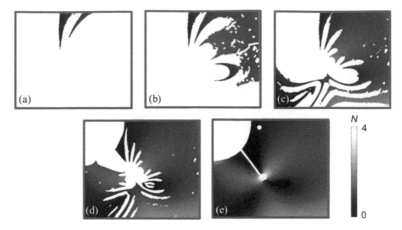

Figure 4.24. (a)–(e) Progression of the FRSTFP scheme in the interval of 100 000 pixels for the bi-axial fracture problem. The seed point is shown as yellow dot. The images of the models are shrunk for better representation and optimal utilisation of space. Adapted from [8] with permission from Elsevier.

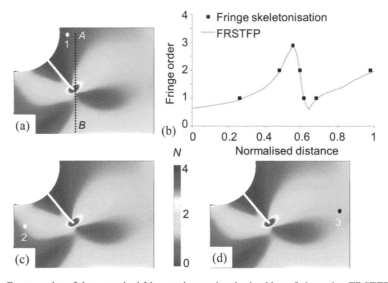

Figure 4.25. Contour plot of the smoothed fringe order results obtained by refining using FRSTFP scheme for the inclined crack in the bi-axially loaded cruciform specimen: (a) seed point 1, (b) fringe order variation along line AB that compares well with integer fringe orders obtained from fringe skeletonization, (c) using seed point 2, (d) using seed point 3. Fringe order variations obtained by employing different seed points are the same thus eliminating the dependence on seed point. Adapted from [8] with permission from Elsevier.

4.9.3 Application to a stereo-lithographic model

The recent developments in stereo-lithography have enabled one to make photoelastic models of desired shapes through layer-by-layer manufacturing. Stress analysis in such 3-D models is performed by stress-freezing followed by slicing. However, these models have dark spots due to porosity; randomly spread across the domain, which are potential sources of noise for photoelastic analysis. It is shown that more noise shows up in the isochromatic image as the thickness of the slice decreases [33]. However, to capture the variation of stresses using 2-D modelling, use of thin slices is recommended.

A circular disc (diameter = 60 mm and thickness = 6 mm) made using stereo-lithography is subjected to a diametral load of 64 N during stress freezing. Due to the layered manufacturing process, some levels of micro-porosity exist in each layer of the model. A slice of 2 mm thickness is then cut from the central portion of the disc carefully. Figure 4.26(a) shows the dark field isochromatics in a quarter of the disc recorded using an incandescent lamp. There are intermittent spots spread across the model domain, which makes the creation of boundary mask to encapsulate the noise exceedingly difficult. The F_R_Map generated for the problem is shown in figure 4.26(b). Apart from the high fringe gradient zones near the loading point, the map also captures the small spots owing to the sudden intensity jumps. This aspect is used to advantage in extracting the fringe orders over the domain. The fringe order is initially estimated by the colour difference equation involving the combined use of RGB–HSV followed by refining using FRSTFP. The progression of the scan after refining the first 210 000 pixels shows that the scheme refines the areas excluding the noise spots first and then gradually proceeds into the pores (figure 4.26(c)). The white spots in figure 4.26(c) correspond to the noisy spots in figure 4.26(a) and hence, removing the possibility of noise propagation from these spots.

The fringe order result obtained after refining using the FRSTFP scanning scheme is shown in figure 4.26(d). Isochromatic demodulation is carried out accurately only till four fringe orders as the intensity modulation is weak beyond this owing to the use of an incandescent lamp. The fringe order data is then smoothed by the multi-directional progressive smoothing scheme and the results obtained are shown in figure 4.26(e). The variation of fringe order values along line XY (figure 4.26(e)) is plotted in figure 4.26(f). The fringe order values obtained closely follow the corresponding theoretical values. Thus, this demonstrates that FRSTFP scanning scheme is also capable of handling models with random noise of varied sizes introduced during model preparation.

4.9.4 Application to interacting cracks in edge heating [34]

A 5.5 mm thick rectangular (125×100 mm^2) photoelastic specimen made of polycarbonate (PSM-1, Vishay Measurement Group, USA) with two interacting cracks initially at room temperature is suddenly subjected to heating at one edge. A crack of length 5 mm is cut near the bottom edge of the specimen using a jewel saw of 0.2 mm thickness. An additional crack of 5 mm length is cut asymmetrically near the existing crack [34] (figure 4.27). Thermal stress developed in the specimen is visualized as photoelastic fringes using a circular polariscope arrangement. It has been reported that PSM-1 is the most suitable material for photo-thermoelastic analysis as its properties remain constant in the temperature range of -10 °C to 55 °C [35]. The material stress fringe value is found to be 8.1 N/mm/fringe corresponding to the discrete fluorescent light source used in this study. In this problem, the fringes evolve as

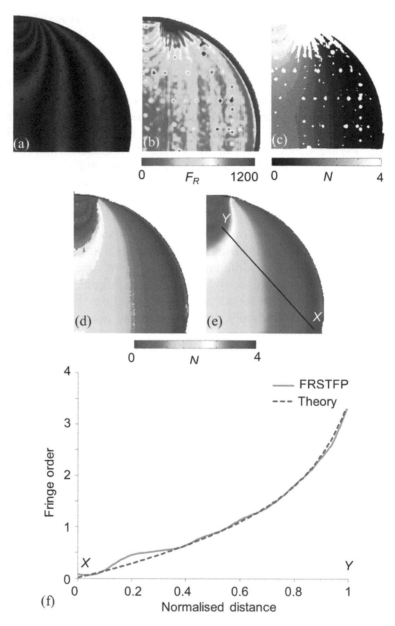

Figure 4.26. (a) Dark-field isochromatics in the circular disc under diametral compression made by stereo-lithography, (b) F_R_Map, (c) progression of the scanning scheme after 210 000 steps. Contour plot of the fringe order results obtained after: (d) refining, (e) refining and smoothing, (f) fringe order variation along the line XY compared with the corresponding values from theory. Reprinted from [8] with permission from Elsevier.

a function of time, and it is ideal to use TFP, as it needs only one image for evaluating the fringe orders at any given instant of time.

A hot plate made of steel with a temperature controller is used for heating the edge of the plate. It is maintained at 53 °C throughout the experiment and kept in between

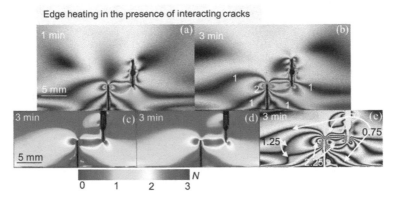

Figure 4.27. Experimental isochromatic fringe patterns near the interacting crack-tips at two different time intervals after the specimen is brought into contact with the hot plate (a) 1 min, (b) 3 min. The fringe patterns show slow evolution of fringes as a function of time making them ideally suitable for the use of TFP. Data extraction steps using TFP for edge heating at 3 min for interacting cracks: (c) refined fringe order data, (d) smoothed fringe order data, (e) mixed-field plot of fringe order. Courtesy: [34]. Adapted by permission from Springer Nature.

the elements of a circular polariscope arrangement. The specimen is initially at room temperature of 25 °C. Then, one edge of the specimen is brought in contact with the hot plate. As soon as the specimen is brought into contact with the hot plate, a stopwatch is started, and images of the isochromatic fringes are captured at an interval of 60 s for up to 15 min using a Canon DSLR camera with a zoom lens of 55–250 mm. The camera can provide image resolution varying from 10 to 27 pixels mm^{-1}. The individual images reported have the scale marked on them. The isochromatic fringe patterns near the interacting crack-tips at two different time intervals when the specimen is in contact with the hot plate are shown in figure 4.27(a and b).

Figure 4.27(c) shows the refined and figure 4.27(d) shows the smoothed fringe order data extracted from the dark-field isochromatics near the crack-tip at 3 min (figure 4.27(b)). The fringe order data thus obtained contains the fringe order at each pixel of the image. From this, the data required for the least squares algorithm to find stress field parameters are to be collected. It has been established that if data depicts the geometric shape of the fringe orders, the convergence is better [36]. As the dark-field image has only a few fringes, a *mixed-field* isochromatic image (fringe order varies as 0.25, 0.75, 1.25...) is used to extract fringe orders along fringe contours. The mixed-field plot of the fringe order data is shown in figure 4.27(e).

Using the fringe order data and the non-linear least squares algorithm as discussed in section 2.7, the SIFs are calculated for the crack-tips. In order to improve the accuracy of SIF evaluation, a semi-automated method of crack-tip refinement is also performed [37, 38]. The fringe features are quite complex and a simple solution with two parameters such as Irwin's solution [39] is not possible for this case. The converged values of the parameters obtained based on fringe order minimization criteria are used to reconstruct the crack-tip fringe patterns and the data points are echoed back on the reconstructed images as an additional check on the solution obtained. The choice of the number of parameters influences the nature

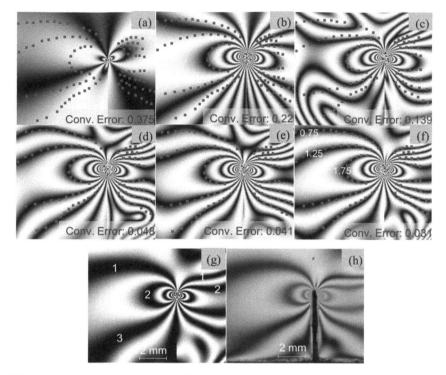

Figure 4.28. Theoretically reconstructed magnified mixed-field isochromatics with experimental data extracted from figure 4.27(e) echoed as red points on the image as a function of number of parameters: (a) two parameters, (b) four parameters, (c) six parameters, (d) eight parameters, (e) ten parameters, (f) twelve parameters. At twelve parameters, note that the experimental data match well with the theoretically reconstructed patterns. (g) Theoretically reconstructed dark-field image using the twelve-parameter solution, (h) green plane image of figure 4.27(b) at 3 min with the same magnification as the reconstructed image. Courtesy: [34]. Reprinted by permission from Springer Nature.

of the reconstructed isochromatic fringe field. To illustrate this, starting from two parameters, a theoretically reconstructed mixed-field isochromatic image (magnified) with the echoed data points at different number of parameters in steps of two are presented in figure 4.28. As the number of parameters is increased from two to twelve, one could see that the reconstructed fringe field matches well with the echoed data points. The convergence error has reduced from 0.375 for a two-parameter solution to 0.031 for a twelve-parameter solution. To illustrate that the results obtained correspond to the experimental image of figure 4.27(b) (3 min), the monochrome dark-field image is reconstructed and shown in figure 4.28(g). This compares well with figure 4.28(h), which is the enlarged view of the green image plane of the experimental image shown in figure 4.27(b) (3 min).

4.10 Image normalization methods

In TFP, uniform illumination is desirable. In experiments involving poor accessibility to the specimen, when there is non-uniform illumination over the model domain, the method of image normalization helps to handle such situations. Normalization has been shown to be quite useful in fringe thinning in section 2.3. In an industrial scenario for complicated structures coated with suitable photoelastic coatings, where uniform illumination is not possible due to practical considerations, one may think of the utility of normalization to interpret the data recorded in the best possible manner. Ramesh and Pandey [40] proposed a novel normalization scheme that increases the dynamic range in both low and high fringe gradient zones.

One of the issues in experimentation is the difficulty in ensuring a high dynamic range of intensities over the field due to several experimental factors. Normalization [41] transforms an image with intensity values in the range (*Min*, *Max*) into a new image with intensity values in the range (*Min*$_{new}$, *Max*$_{new}$).

Any arbitrary fringe field I_{xy} having matrix size $m \times n$ can be defined as

$$I_{xy} = a(x, y) + b(x, y) \cos \phi(x, y) \qquad (4.26)$$

where $a(x, y)$, $b(x, y)$ and ϕ represent background noise, amplitude modulation, and phase, respectively.

Before normalization, the image is first averaged to reduce noise in the image. This is done by convoluting the image matrix $I_{xy}(m \times n)$ by a filter matrix A (3×3), which is as follows:

$$A = \begin{bmatrix} 0.1111 & 0.1111 & 0.1111 \\ 0.1111 & 0.1111 & 0.1111 \\ 0.1111 & 0.1111 & 0.1111 \end{bmatrix} \qquad (4.27)$$

The equation for filtering image matrix I_{xy} with filter matrix A for each pixel (x, y) is defined as

$$I_A(x, y) = \sum_{j=-1}^{1} \sum_{i=-1}^{1} A(i, j) I_{xy}(x + i, y + j) \qquad (4.28)$$

A normalized image [41] is obtained by removing low frequency background noise and then making the amplitude modulation as unity, which can be defined as $\cos\{\phi(x, y)\}$. This is achieved by resorting to processing the image in the frequency domain. By calculating a 2-D discrete Fourier transform of the image, the image I_{xy} is converted to the frequency domain. Unwanted frequencies are filtered out by filtering (appendix A). The frequency filter H (u, v) has a pair of frequency filters [42], i.e., high-pass filter (HPF) and low-pass filter (LPF) on each channel. A normalization algorithm is applied on each channel separately. Later, the three channels are combined to get a normalized RGB image.

The high pass filter selected after a systematic study is defined as:

$$H_{HP} = 1 - e^{\left(-\frac{(u^2+v^2)^{0.8}}{2R_1^2} \right)} \qquad (4.29)$$

where R_1 is the radius of the HPF, u and v are the matrices defined as

$$u = \begin{bmatrix} 1 - u_0 & 2 - u_0 & \dots & n - u_0 \\ 1 - u_0 & 2 - u_0 & \dots & n - u_0 \\ \dots & \dots & \dots & \dots \\ 1 - u_0 & 2 - u_0 & \dots & n - u_0 \end{bmatrix} \quad v = \begin{bmatrix} 1 - v_0 & 1 - v_0 & \dots & 1 - v_0 \\ 2 - v_0 & 2 - v_0 & \dots & 2 - v_0 \\ \dots & \dots & \dots & \dots \\ m - v_0 & m - v_0 & \dots & m - v_0 \end{bmatrix} \quad (4.30)$$

where u_0 and v_0 are taken as: $u_0 = \frac{n}{2} + 1$, $v_0 = \frac{m}{2} + 1$.

Size of the variables u and v is the same as the matrix H_{HP}, i.e., $(m \times n)$. This filter when applied to image I_A in frequency domain, allows frequencies higher than a certain cut-off frequency. Next, a low pass filter (LPF) is applied on the high-pass filtered image matrix I_{HP}, which is defined as:

$$H_{LP} = e^{\left(-\frac{(u^2 + v^2)^{0.5}}{2R_2^2} \right)} \quad (4.31)$$

where R_2 is the radius of the LPF. The matrix H_{LP} has the same size as that of u and v, i.e., $(m \times n)$. This filter when applied to the image I_{HP} allows frequencies lower than a certain cut-off frequency. The low pass filtered matrix is represented by I_{LP} and is obtained as

$$I_{LP}(x, y) = F^{-1}[F(I_A) \times H_{HP}(u, v)] = b(x, y) \cos\{\phi(x, y)\} \quad (4.32)$$

where, F and F^{-1} represent 2-D Fourier and inverse Fourier transforms, respectively.

Hilbert transform [43] of the image I_{LP} is taken to get the quadrature I_Q of the image (appendix B). The phase of the normalized image $\phi(x, y)$ is obtained by dividing I_Q with I_{LP}.

$$\phi(x, y) = \tan^{-1}\left(\frac{I_Q}{I_{LP}} \right) \quad (4.33)$$

$$\text{Normalized Image} \quad I_N(x, y) = \cos \phi(x, y) \quad (4.34)$$

As one of the basic requirements of TFP involves constructing a calibration table, there have been efforts to use normalization schemes to simplify the process. In 2015, Swain et al [29] proposed a method for constructing a calibration table theoretically using the idealized material stress fringe value for each image plane. Later, they [44] processed the experimentally recorded image by normalization and extracted the fringe orders using the theoretical calibration table. Ramesh and Pandey [40] had brought out the lacuna in the theoretical calibration table, and established the importance of experimental calibration table generation. They recommended using normalization to both the calibration and the experimental application specimens as the way forward to handle poorly recorded images.

4.10.1 Performance of the normalization approach to a practical problem

One of the problems in industrial scenarios is that the application image may have acquired external marks on some portions of the problem domain. It may be difficult to repeat the experiment and one may still want to get some insight into the problem with the available images. This is demonstrated in the context of an *embedded* microcapsule subjected to a tensile load.

Polymeric composite materials are used extensively in engineering and industrial applications due to their high design flexibility. These polymeric composite materials are sensitive to damage in the form of cracks. Vivek *et al* [45] observed that debonding and crack initiation that occurs near a microcapsule in a tensile loaded epoxy matrix is due to high stress concentration. The embedded microcapsule in the epoxy matrix shown in figure 4.29(a) is taken at a high magnification. At such magnifications, in some of the regions, even light or tiny external marks become visible. In view of the higher level of fringe orders with the lowest fringe order being 5, the basic appearance of the coloured isochromatics is different. The merged calibration table (figure 4.3) shows a similar colour variation for this fringe order range. Only the RGB–HSV model is suitable to resolve the fringes. The domain excluding the microcapsule embedded at the centre is used for fringe order determination. The FRSTFP scanning can handle the multiply connected domain with noise points randomly scattered. The role of two-point colour adaptation and normalized calibration table (CT) approach for their performance in this case is evaluated. The isochromatic image is initially resolved using two-point colour adaptation (figure 4.29(b)). Next, figure 4.29(a) is normalized with $R_1 = 8$ and $R_2 = 10$ (figure 4.29(c)). The normalization distorts the colour totally, and the noisy regions are accentuated. Fringes resolved using the normalized CT are shown in figure 4.29(d). The first look of it shows that it compares well with figure 4.29(b). The colour map of absolute error (figure 4.29(e)) between figures 4.29(b and d) brings out the regions where the error is high. Figure 4.29(f) shows the plot of fringe order along lines *EF* and *GH* drawn in figure 4.29(b). The fringe order resolved by colour adaptation and normalized CT are comparable along line *GH* since the region of line is noise-free, but along line *EF*, the normalized CT shows repetitive jumps compared to smooth and continuous variation of fringe order by colour adaptation. This is because a part of line *EF* is passing through the region of external marks and the normalized CT is unable to demodulate fringe orders.

The two-point colour adaptation has worked well for this example problem. Unlike the problem of inclined crack where the $R_{\max}$ coincided with the hot pixel (figure 4.11), here there are no hot pixels, but noisy regions that appear with shades of grey. The minimum values of R, G, and B have coincided with one of these points but have not affected the results in any way. This may be because, if there had been a zeroth fringe order in the domain, the minimum values would be like that. The two-point colour adaptation has been successful in solving a variety of problems.

4.11 Five-step/ Four-step methods

It is established in the previous sections that using a single colour dark-field isochromatics, it is possible to extract the total fringe order at any point in the

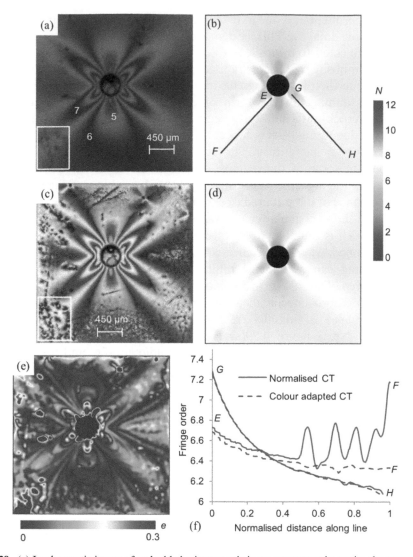

Figure 4.29. (a) Isochromatic image of embedded microcapsule in an epoxy matrix, notice the external marks on the specimen at the left corner, (b) colour map of fringe order obtained by colour adaptation, (c) image (a) normalized with $R_1 = 8$ and $R_2 = 10$, (d) colour map of fringe order obtained by normalized CT, (e) colour map of absolute error between (b) and (d), (f) fringe order plot along lines EF and GH in (b) and (d). Reprinted from [40] with permission from Elsevier.

model domain. Although, decisions on design can be arrived at for a variety of problems using isochromatic data alone, in certain problems, the knowledge of isoclinics is also desirable. In chapter 3, it has been discussed how high quality isoclinics can be extracted by processing four polarisation stepped images (first four arrangements of table 3.2 and equation (3.9)) recorded in colour. If this methodology is integrated with TFP, one can get both isochromatics and isoclinics by processing the data conveniently in colour domain. Since this involves just five steps, the method is christened the five-step method [46].

If four polarisation stepped images are available in colour domain (table 3.2), the dark-field isochromatic intensity can also be obtained by post-processing the four isoclinic fields as [9, 47]

$$I_j = \sqrt{(I_{3,j} - I_{1,j})^2 + (I_{4,j} - I_{2,j})^2} \ (j = R, \ G, \ B) \tag{4.35}$$

Where $I_{1,j}, I_{2,j}, I_{3,j}, I_{4,j}$ represents the intensity value of the different image planes (j) of the corresponding polarisation stepped image. $Digi$TFP® [24] is capable of generating such an isochromatic image if the four isoclinic images are available. Using the isochromatics thus generated, one can find the total fringe order by TFP. The four polarisation stepped images can be used to get the isoclinic field. Thus, digital photoelasticity has come a long way and it is enough to have a plane polariscope and a white light source to evaluate both isochromatics and isoclinics. There is no need of investing huge money on dedicated digital polariscopes. In addition, the four-step method is totally immune to quarter-wave plate errors, which has dominated the research in photoelasticity including digital photoelasticity for a long time. The applicability of the four-step method to a dental application is discussed next.

4.12 Digital photoelasticity applied to orthodontics

The use of dental implants in the case of missing natural dentition is now common in clinical dentistry. Implantology is one of the promising areas in dentistry, which poses challenging problems for both clinical practitioners as well as for design engineers. The development of improved designs of implant systems in dentistry have necessitated the study of stress fields in the implant regions of the mandible/maxilla (lower jaw/upper jaw) to understand better the biomechanics involved. The contacting interface of implant—abutment systems are crucial in the longevity and for the proper functioning of the implant system.

Figure 4.30(a) shows a typical human mandible with different regions indicated. The mesial and distal refer to the direction towards and away from the symmetric midline, respectively. The coronel region refers to the direction towards the crown of the teeth and apical refers to the region towards the root. The objective is to simulate the loads due to mastication and observe what happens at the implant region in a concept like All-On-Four® design [48]. The All-On-Four® concept is different from other implant retained prosthetic concepts, as it advocates using two longer implants in the posterior region. However, placing implants in this region of the edentulous arch is often complex owing to the anatomical structure and critical nervous system present in those areas [49, 50]. To circumvent this situation, the molar implants are usually tilted such that their placement will not disturb the critical nervous system. In clinical situations, generally, a range of angulations from 15° to 30° are used depending on the patient and in this work, an inclination of 30° is selected as it is the upper limit. The masticatory load depends upon several factors such as age, type of

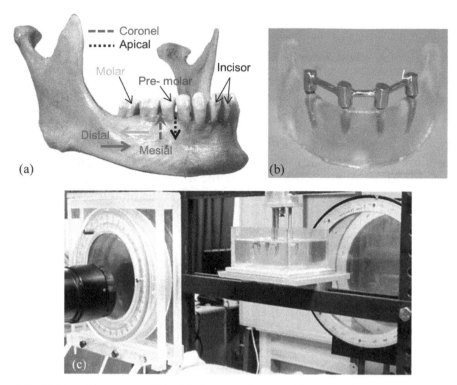

Figure 4.30. (a) Typical human mandible with the nomenclature and the zones of distal, mesial, coronel, and apical indicated, (b) stress free photoelastic model depicting All-On-Four® concept of implants, (c) photoelastic model loaded and kept in an immersion bath of epoxy resin having the same refractive index as the model to ensure normal incidence when viewed in a digital polariscope. Adapted from [53] with permission from Elsevier.

food, etc. Previous studies have shown that the masticatory forces in dentate patients vary from 175–350 N in normal situations [51, 52].

Preparation of a suitable model for analysis that retains the essence of the problem on hand is quite challenging and a mandible without soft tissue attachments and tongue is made carefully to minimize the formation of residual stresses [53]. The photoelastic model is then fitted with the bar attachments (figure 4.30(b)), which are cast using Ni–Cr alloy to connect the implants along the mandible using the standard procedure adopted by dentists. The photoelastic model in an immersion tank along with the loading fixture is shown in figure 4.30(c). A load of 237 N which is close to the average normal mastication load is applied at L_1 position indicated in figure 4.30(b), using a lever-arm loading system with a dead weight attachment. It is to be noted that the Young's modulus of the model material is found to be 3300 MPa. The Young's modulus of bone does not have a constant value and it may vary depending on several biological factors like age, quality/type of bone, etc. Studies on the cortical bone of a mandible have shown that the Young's modulus is in the range of 3200 ± 570 MPa [54]. Since the modulus and geometric aspects of mandible and the photoelastic model used in this study are comparable, load applied on the model is selected on a one-to-one scale.

Four polarisation stepped images are recorded, which are shown in figures 4.31(a–d). From these, using equation (4.35), the isochromatics are generated. Using the primitives available in *Digi*TFP® [24] software, a mask is carefully generated to limit the processing within the model domain (figure 4.31(f)). Initially, the isochromatic image is processed using TFP to get the fringe order data over the model domain. The use of equation (4.15) has resulted in figure 4.32(a) having discontinuities, which are removed by using equation (4.16) along with FRSTFP scanning (figure 4.32(b)). This has been possible using simultaneous multi-seeding with eight seed points, as shown in figure 4.32(a). The fringe order data obtained after refining may have minor jumps due to the localized colour variations in the isochromatic data, which is smoothed using the multi-directional smoothing

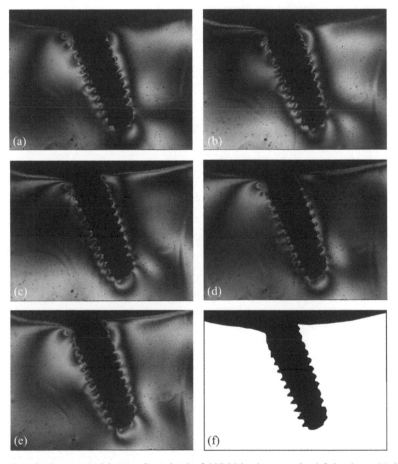

Figure 4.31. Polarisation stepped images for a load of 237 N in the posterior left implant: (a) 0° isoclinic, (b) 22.5° isoclinic, (c) 45° isoclinic, (d) 67.5° isoclinic. (e) Dark-field isochromatics generated from using equation (4.35). (f) Carefully generated boundary mask using the primitives of *Digi*TFP® [24] software. Adapted from [53] with permission from Elsevier.

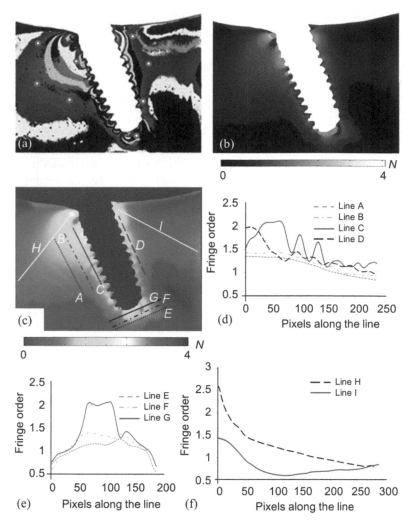

Figure 4.32. Isochromatic fringe order: (a) using equation (4.15)—seed points shown as yellow dots, (b) after refining using equation (4.16) along with FRSTFP scanning, (c) smoothed fringe order data after progressive multi-directional smoothing with the sequence 135°, 90°, 45°, and 0° using the NLR scheme with a span of 15 pixels. (d) Variation of fringe order along the lines *A, B, C, D*. (e) Variation of fringe order in the apical region along the lines *E, F, G*. (f) Variation of fringe order in the distal and mesial regions along the lines *H, I*. Adapted from [53] with permission from Elsevier.

algorithm [26] using a scanning sequence of 135°, 90°, 45°, 0°. The fringe order data obtained after smoothing is shown as a contour plot in figure 4.32(c).

Figure 4.32(d) shows the variation of the fringe order along the lines *A, B,* and *C* parallel to the implant length in the distal region. The corresponding fringe order variation along line *D* on the mesial side is also shown. It can be observed that the retardation levels increase as one moves closer to the implant. The isochromatic results capture the localized fringe order variation due to the influence of the implant threads (line *C*). The variation of fringe order due to the adjacent threads is close to 0.5. A point-by-point analysis using conventional photoelasticity would not have captured this variation resulting in wrong conclusions unless the correct line was identified *a priori*. Moreover, the implant interface in the distal zone (line *C*) experiences higher retardation when compared to the mesial side (line *D*).

Figure 4.32(e) shows the fringe order variation along lines *E*, *F* and *G* in the apical region of the implant. The maximum fringe order in this region is found to be 2.2. The fringe order along line *G* has the maximum variation from 0.5 to 2.2. The distal side (figure 4.32(f)) has a maximum fringe order of 2.6 at the coronal region. These fringe order variations are consistent with the clinical observations where the bone resorption is higher in the coronal region of the implant. However, the previous point-by-point study [55] on the All-On-Four® concept using photoelasticity had reported a contradicting result with the clinical observation. Accurate and robust methods employed in this analysis have helped in resolving this ambiguity.

Figure 4.32(f) raises an important issue as one can notice that, even though a reduction in fringe order is observed towards the distal region of the implant (line *H*), it is different along line *I*. The fringe order along line *I* is found to decrease initially, but after some distance, it gradually increases towards the mesial zone, which is due to the interaction effects of the incisor implants [53].

Isoclinics are obtained by processing the intensity data using equation (3.9), and are in wrapped format, i.e., they are in the range $-\pi/4$ to $+\pi/4$ (figure 4.33(a)) The isoclinics are unwrapped using a multi-seeded parallel unwrapping algorithm.

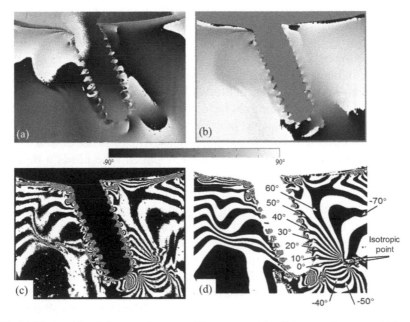

Figure 4.33. (a) Wrapped isoclinics as a phasemap, (b) unwrapped isoclinics as a phasemap, (c) isoclinics as a binary plot in steps of 10°, (d) smoothed isoclinics with the sequence 135°, 90°, 45°, and 0° by using RNLR for 135° and 0° scans and NLR for the other two scans. The span widths used were 10, 20, 10, 20 for each scan directions, respectively. Courtesy: figure (d). Adapted from [53] with permission from Elsevier.

A phasemap of the unwrapped isoclinic data is shown in figure 4.33(b). It is desirable to plot the isoclinics over the domain as binary contours of 10° steps for visualization. Binary representation of the isoclinics data for this case is shown in figure 4.33(c). The presence of noise in the phasemap can be identified from figure 4.33(c) and these are eliminated by smoothing the isoclinic data. Isoclinics are smoothed using the multi-directional progressive scanning method with a scanning sequence of 135°, 90°, 45°, 0°. Binary representation of the smoothed isoclinics is shown in figure 4.33(d). The *isotropic* points are marked in figure 4.33(d). In structural applications, *isotropic* points are used to remove material to reduce weight as it leads to less stress concentration if holes are drilled at these points [56]. With *isotropic* points identified, one can generate patient specific data. It can act as a guiding parameter in clinical surgical practices, to decide the area for material removal, especially for creating micro voids and the placement of surgical tools without affecting the structural stability of the bone.

4.13 Closure

In this chapter, it is demonstrated that the use of colour information in digital photoelasticity has become quite sophisticated and fringe orders up to 12 could be resolved comfortably. Extending the range of resolvability of fringes from three has been initially attempted by using a bulky tri-colour light source [57] and later research has revealed that the use of an HSV colour model in conjunction with RGB colour model could achieve this even with a simple fluorescent light source. In this chapter, one would find images recorded using incandescent, fluorescent as well as discrete fluorescent light sources successfully processed by the *Digi*TFP® software [24]. Such a versatility is needed in data processing to help customers from industry who may have different light sources. The focus should be to have many methods to process the available data as effectively as possible and at the same time set standards on what an ideal experiment should have.

In addition, in this chapter, images have been recorded by 3CCD cameras as well as recent cost-effective cameras that have Bayer colour filter array (CFA), as shown in figure 3.24. In Bayer CFA, a single CCD sensor records the colour information selectively in the pixels as dictated by the CFA panel. One needs a demosaicking algorithm to get the image planes of R, G, and B. Juan Briñez *et al* [58] conducted a systematic study on how demosaicking algorithms influence different parameters in PST and TFP. To minimize the error in TFP, the recommendation is to use an image resolution greater than 512×512 and use a fluorescent light source for illumination.

The development of FRSTFP has greatly enhanced the capabilities of TFP and complex fringe patterns could be quantitatively evaluated. The results obtained have been compared with the result of PST for the circular hole problem (figures 4.8 and 4.9) and for the inclined crack problem (figure 4.12) comparison has been made with the fringes reconstructed in colour based on the multi-parameter solution discussed in chapter 2. The reconstruction in colour has brought out visually the applicability of the over-deterministic least squares analysis in modelling the crack-tip stress fields. The FRSTFP scanning has been shown to process the stereo-lithographic models with pores effectively. Although with new developments in rapid prototyping, improved models are possible without such pores (section 5.7.3), the essence of the example chosen is to demonstrate that such randomly distributed noise can be handled effectively by FRSTFP. This will be especially useful for several industrial problems where one may make marks on the model for convenience—which do not hinder with normal visual analysis—this is now possible in digital domain with FRSTFP.

The use of just a plane polariscope to evaluate both isochromatics and isoclinics to solve a problem in orthodontics has demonstrated that a conventional polariscope with appropriate software is enough for a digital photoelastic analysis. The use of quality measure applied to TFP has also been attempted [59]. The future research should focus on combining the use of both the quality-guided approach and FRSTFP to solve problems of practical interest.

Appendix A: Applying a frequency filter to an image

The averaged image I_A in equation (4.28) is converted to frequency domain by calculating a 2-D discrete Fourier transform of the image before applying a filter.

The 2-D discrete Fourier transform $F(u_1, v_1)$ of the image I_A is:

$$F(u_1, v_1) = \sum_{x=0}^{m-1}\sum_{y=0}^{n-1} I_A(x, y)e^{-2\pi i\left(\frac{u_1 x}{m} + \frac{v_1 y}{n}\right)} \tag{A.1}$$

for $u_1 = 0, 1, 2,..., m - 1$ and $v_1 = 0, 1, 2,..., n - 1$

The zero-frequency components at the centre of the filter matrix (equation 4.29) are shifted away from the centre, let the shifted frequency filter matrix be $H_{HP_{\text{shifted}}}$. Applying $H_{HP_{\text{shifted}}}$ filter to the Fourier transformed image $F(u_1, v_1)$:

$$F(u_{1f}, v_{1f}) = F(u_1, v_1) \times H_{HP_{\text{shifted}}} \tag{A.2}$$

The 2-D discrete inverse Fourier transform of the high pass filtered matrix $F(u_{1f}, v_{1f})$ is:

$$I_{HP}(x, y) = F^{-1}\left(F(u_{1f}, v_{1f})\right) \tag{A.3}$$

$$I_{HP}(x, y) = \frac{1}{mn}\sum_{u_1=0}^{m-1}\sum_{v_1=0}^{n-1} F(u_{1f}, v_{1f})e^{2\pi i\left(\frac{u_1 x}{m} + \frac{v_1 y}{n}\right)} \tag{A.4}$$

for $x = 0, 1, 2,..., m - 1$ and $y = 0, 1, 2,..., n - 1$

Similarly, the low pass filter H_{LP} is applied on high pass filtered image $I_{HP}(x, y)$ to get the low pass filtered image $I_{LP}(x, y)$.

Appendix B: Applying Hilbert transform to an image

The notation $H_{2D}[I_{LP}(x_0)]$ indicates that each column of the function $I_{LP}(x_0) = I_{LP}(x, y)$ is independently replaced by its Hilbert transform [43], which is evaluated as

$$H_{2D}[I_{LP}(x_0)] = \int_{-\infty}^{\infty} \frac{f(x, y - x')}{\pi x'} dx' = \int_{-\infty}^{\infty} \frac{f(x, x')}{\pi(y - x')} dx' \tag{B.1}$$

Let the 2-D Fourier transform of image I_{LP} be $F(u_2, v_2) = F(\nu)$. The Hilbert transform in terms of 2-D Fourier transform $F(u_2, v_2)$ is:

$$H_{2D}[I_{LP}(x_0)] = \int_{-\infty}^{\infty} \int_{-\infty}^{\infty} -i \, \text{sgn}(v_2) F(\nu) e^{2\pi i (x_0 \nu)} du_2 dv_2 \tag{B.2}$$

The Signum function (sgn) used in calculation is defined as:

$S(u, v) = \dfrac{u + iv}{\sqrt{u^2 + v^2}}$; $S(u_0, v_0) = 0$ where u, u_0, v, and v_0 are the variables defined in equation (4.30)

The Signum function $S(u, v)$ is applied to $F(u_2, v_2)$ and inverse discrete Fourier transform (F^{-1}) is taken, which is called Quadrature I_Q.

$$I_Q = F^{-1}(F(u_2, v_2) \times S(u, v))$$
$$I_Q = b(x, y) \sin\{\phi(x, y)\} \tag{B.3}$$

Phase of the normalized image $\phi(x, y)$ is calculated as

$$\phi(x, y) = \tan^{-1}\left(\frac{I_Q}{I_{HP}}\right) \tag{B.4}$$

Cosine of the phase $\phi(x, y)$ is called normalized image.

$$\text{Normalized image:} \quad I_N = \cos(\phi(x, y)) \tag{B.5}$$

After normalization, an adaptive low pass Wiener filter [42] with the size of $[15 \times 15]$ is applied to each of the channel greyscale images to reduce the noise levels.

Exercises

1. For a load $P = 800$ N, calculate the maximum fringe order obtainable theoretically for the three calibration specimens shown in figure 4.2 by taking the material stress fringe value as 12 N/mm/fringe. The relevant geometric parameters marked in the figure are thickness = 6 mm, $h = 60$ mm, $a = 30$ mm, $c = 120$ mm, $d = 15$ mm. Which is the ideal calibration specimen among the three and why?
2. Why is green wavelength used as the reference wavelength in TFP?
3. What are the benefits of using a merged calibration table for isochromatic evaluation in TFP?
4. It is observed that the modulation of blue channel intensity in the calibration table is incredibly low after the third fringe order as compared to red and green channels, and this affects the quality of isochromatic evaluation. How is this issue addressed in TFP?
5. The R, G, B values from a row of a calibration table reads 145, 162, 87, respectively. Find the corresponding normalized HSV values.

6. The result after least squares analysis employing **RGB** colour difference formula for the biaxially loaded cruciform specimen having an inclined crack (figure 4.11) using a 0–4 fringe order calibration table (figure (a)) is shown in figure (b). Figure (c) represents the evaluated fringe orders along line segment *PS*. What is the reason for the sudden rise and dip in fringe order values at points *Q* and *R*, respectively?

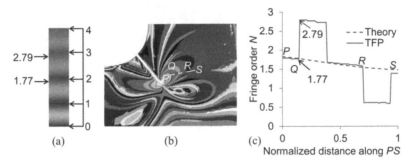

| (a) | (b) | (c) |

7. Figure (a) shows a circular disc of 60 mm diameter and 6 mm thickness subjected to a diametral compression of 855 N. The least squares evaluation of figure (a) using **RGBHSV** colour difference formula yields figure (b). The total fringe order at point *A* with coordinates (2.95 mm, 7.89 mm) with centre of the disc as the origin has to be determined by TFP. The RGB values at *A* are 89, 173, and 65, respectively. Also find the fringe order of point *B* (3.36 mm, 10.10 mm) next to *A* whose RGB values are 110, 152, and 69, respectively. Verify the results obtained with theory. Take the material stress fringe value to be 12.5 N/mm/ fringe. If necessary, refine the results obtained with RTFP using a *K* value of 100. Use the calibration table given for some selected data to base your computations.

R	G	B	N
90	98	35	0
174	173	74	0.50
117	101	64	1.00
156	178	67	1.50
152	108	73	2.00
119	176	69	2.50
81	175	65	3.00
175	114	73	3.50
184	115	77	4.00
80	167	69	4.50

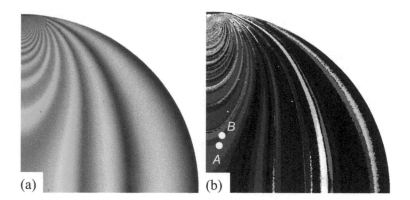

(a)   (b)

8. Solve Problem 7 using modified window search method. Generate the calibration table in steps of 0.05 fringe orders from 0 to 3.5 using a 15th order polynomial for RGB values.

$$I = P_{15}.N^{15} + P_{14}.N^{14} + P_{13}.N^{13} + P_{12}.N^{12} + P_{11}.N^{11} + P_{10}.N^{10} + P_9.N^9 + P_8.N^8$$
$$+ P_7.N^7 + P_6.N^6 + P_5.N^5 + P_4.N^4 + P_3.N^3 + P_2.N^2 + P_1.N^1 + P_0$$

The coefficients of the polynomial are summarized in the following table.

Coeff.	R	G	B
P_{15}	1.188 710	−0.984 520	−0.308 081
P_{14}	−35.568 042	33.350 095	7.717 427
P_{13}	474.926 809	−505.249 175	−78.764 087
P_{12}	−3726.497 798	4519.232 961	381.760 935
P_{11}	19 040.302 711	−26 530.463 462	−408.070 067
P_{10}	−66 327.542 787	107 475.800 725	−5711.725 959
P_9	160 302.040 926	−307 495.741 525	37 799.171 681
P_8	−268 263.128 651	625 330.126 404	−122 492.713 887
P_7	304 796.203 683	−897 988.920 148	245 934.065 606
P_6	−225 430.800 277	894 028.672 316	−319 938.648 640
P_5	99 547.939 111	−598 892.187 622	266 630.543 855
P_4	−19 900.861 314	259 240.267 303	−135 299.142 878
P_3	−2331.844 063	−69 046.042 206	37 947.174 956
P_2	1999.634 150	10 358.480 886	−5125.276 409
P_1	−121.112 986	−525.516 634	388.415 087
P_0	90.822 983	103.277 250	30.628 053

9. An epoxy disc (figure(a)) and a channel specimen for calibration are made from the same cast sheet for photoelastic evaluation. The tint of the disc, represented by the average R, G, B value was 132, 152, 185, respectively, which matched with the calibration channel specimen. To relieve the machining stress in the disc, it was subjected to an annealing thermal cycle that changed its tint as shown in figure(b). The average R, G, B value for the

annealed specimen is 156, 156, 161, respectively. Can the calibration table generated from the channel specimen be used directly to evaluate the fringe field in the annealed specimen? If no, what methods would help to do this?

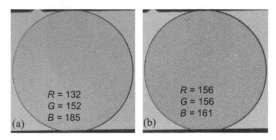

10. The calibration table for TFP is available for epoxy material. It was found that due to ageing, the tint of the model is slightly changed. The no load RGB values of the calibration and model specimens are, respectively: $R_z^c = 146$, $G_z^c = 135$, $B_z^c = 94$ and $R_z^a = 183$, $G_z^a = 165$, $B_z^a = 119$. The RGB values for a row of the calibration table are 129, 59, and 29, respectively. Determine the modified values after performing one-point colour adaptation.

11. The maximum and minimum R, G, B values from the calibration table of epoxy material are 169, 200, 105 and 22, 25, 18, respectively. A model made of epoxy material is subjected to stress freezing and, in the process, the tint of the material has changed. The maximum and minimum R, G, B values of the stress frozen specimen are 168, 166, 81 and 17, 20, 12, respectively. It is proposed to use the same calibration table available to generate the modified calibration table by employing colour adaptation methods. One of the R, G, B entries in the calibration table is known to be 25, 48, and 53. Determine the corresponding modified R, G, B values using two-point colour adaptation.

12. In Problem 11, the mean and standard deviations of R, G, B values for the calibration table were found as 102, 90, 40 and 81, 35, 16 respectively, whereas the same parameters for the model material were evaluated to be 125, 132, 78 and 48, 30, 22, respectively. Find the modified calibration table entry (given in Problem 11) using three-point colour adaptation and colour transfer methods.

13. The figure shows the dark-field isochromatics of a thick ring subjected to internal pressure. After least squares analysis, fringe order refining is set to be started from the seed point marked as yellow dot in the figure. Which of the available scanning schemes in TFP would be able to scan the entire model domain starting from the selected seed point? Discuss the algorithm behind those schemes in detail.

14. A circular disc of 60 mm diameter and 6 mm thickness is subjected to diametral compression of 492 N. Using a coordinate system whose origin is at the centre of the disc, three points A, B, and C are marked in the figure which are to be independently used as single seed points to refine the fringe order data after least squares analysis. If one uses four sub-images scanning for refining, which of these seed points would be able to scan the entire model domain and why?

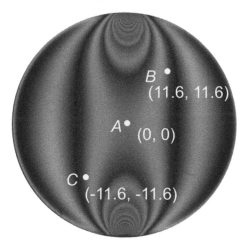

15. A seed point is put at the center of the 5×5 pixel window at the start of refining using advancing front scanning (AFSTFP) scheme. The first three instances in the scan propagation computed using a 3×3 window are marked in the figure. Find the correct location for the missing entries from 4 through 24.

?	?	?	?	?
?	1	3	?	?
?	2	0	?	?
?	?	?	?	?
?	?	?	?	?

16. A 5 × 5 pixel window from red, green, and blue intensity planes of an experimental isochromatics is shown in the figure. Compute the sum of absolute gradients (G_t) map and the F_R_Map. Use a kernel size $k = 3$ for F_R calculation.

	Red					Green					Blue			
136	136	135	133	132	38	37	36	34	34	47	47	49	49	49
139	135	136	133	131	39	37	37	36	34	46	46	48	48	48
140	138	136	134	133	41	38	37	36	35	46	46	47	48	49
139	137	137	136	134	41	39	38	36	37	45	46	46	47	48
141	140	138	136	135	43	41	40	38	37	45	44	46	47	47

17. The given figure is obtained by removing the low frequency background noise from figure 4.8(a) and making its amplitude of intensity modulation as unity. What is this process called in digital photoelasticity? What are its advantages and limitations?

18. '*Four polarization stepped images (I_1 to I_4 in table* 3.2*) recorded in colour are sufficient for the whole field isoclinics and* isochromatics *evaluation*'. Substantiate the statement.

19. For the problem of a plate with a hole, the *R, G, B* intensities corresponding to the four polarization stepped images (figure 3.14) at a point with coordinates (14.44, 1.69 mm) with hole centre as the origin are tabulated. Calculate the *R, G, B* intensity for the dark-field isochromatics at the same point.

	R	*G*	*B*
I_1	91	80	50
I_2	17	22	18
I_3	135	144	102
I_4	162	207	143

References

[1] Ramesh K and Deshmukh S 1996 Three fringe photoelasticity—use of colour image processing hardware to automate ordering of isochromatics *Strain* **32** 79–86

[2] Barone S and Petrucci G 1995 Towards RGB photoelasticity: full-field automated photo-elasticity in white light *Exp. Mech.* **35** 193–200

[3] Ajovalasit A, Barone S and Petrucci G 1995 Automated photoelasticity in white light: influence of quarter-wave plates *J. Strain Anal. Eng. Des.* **30** 29–34

[4] Ramesh K, Ramakrishnan V and Ramya C 2015 New initiatives in single-colour image-based fringe order estimation in digital photoelasticity *J. Strain Anal. Eng. Des.* **50** 488–504

[5] Voloshin A S and Burger C P 1983 Half-fringe photoelasticity: a new approach to whole-field stress analysis *Exp. Mech.* **23** 304–13

[6] Madhu K R, Prasath R G R and Ramesh K 2007 Colour adaptation in three fringe photoelasticity *Exp. Mech.* **47** 271–6

[7] Kale S and Ramesh K 2013 Advancing front scanning approach for three-fringe photo-elasticity *Opt. Lasers Eng.* **51** 592–9

[8] Ramakrishnan V and Ramesh K 2017 Scanning schemes in white light photoelasticity—part II: novel fringe resolution guided scanning scheme *Opt. Lasers Eng.* **92** 141–9

[9] Ramesh K 2000 *Digital Photoelasticity Advanced Techniques and Applications* (Berlin: Springer) https://www.springer.com/gp/book/9783642640995

[10] Ramesh K 2009 *e-book on Experimental Stress Analysis* 1st edn (Chennai, India: IIT Madras) https://home.iitm.ac.in/kramesh/ESA.html

[11] Ramesh K and Deshmukh S 1997 Automation of white light photoelasticity by phase-shifting technique using colour image processing hardware *Opt. Lasers Eng.* **28** 47–60

[12] Madhu K R and Ramesh K 2007 Noise removal in three-fringe photoelasticity by adaptive colour difference estimation *Opt. Lasers Eng.* **45** 175–82

[13] Ramesh K 2008 Photoelasticity *Springer Handbook of Experimental Solid Mechanics* ed J William and N Sharpe (New York: Springer) pp 701–42 https://link.springer.com/referenceworkentry/10.1007/978-0-387-30877-7_25

[14] Neethi Simon B and Ramesh K 2011 Colour adaptation in three fringe photoelasticity using a single image *Exp. Tech.* **35** 59–65

[15] Quiroga J A, García-Botella Á and Gómez-Pedrero J A 2002 Improved method for isochromatic demodulation by RGB calibration *Appl. Opt.* **41** 3461–8

[16] Ajovalasit A, Petrucci G and Scafidi M 2010 RGB photoelasticity: review and improvements *Strain* **46** 137–47

[17] Swain D, Philip J and Pillai S A 2014 A modified regularized scheme for isochromatic demodulation in RGB photoelasticity *Opt. Lasers Eng.* **61** 39–51

[18] Ajovalasit A, Petrucci G and Scafidi M 2015 Review of RGB photoelasticity *Opt. Lasers Eng.* **68** 58–73

[19] Sonka M, Hlavac V and Boyle R 2015 *Image Processing, Analysis, and Machine Vision* ed H Gowans (Boston, MA: Springer) https://link.springer.com/book/10.1007/978-1-4899-3216-7

[20] Ramji M and Ramesh K 2008 Stress separation in digital photoelasticity part A—photoelastic data unwrapping and smoothing *J. Aerosp. Sci. Technol.* **60** 5–15

[21] Ramji M and Ramesh K 2008 Whole field evaluation of stress components in digital photoelasticity—issues, implementation and application *Opt. Lasers Eng.* **46** 257–71

[22] Ramji M and Ramesh K 2010 Adaptive quality guided phase unwrapping algorithm for whole-field digital photoelastic parameter estimation of complex models *Strain* **46** 184–94

[23] Grewal G S, Dubey V N and Claremont D J 2006 Isochromatic demodulation by fringe scanning *Strain* **42** 273–81

[24] Ramesh K 2017 *Digi*TFP®-software for digital twelve fringe photoelasticity *Photomechanics Lab. IIT Madras* https://home.iitm.ac.in/kramesh/dtfp.html accessed Apr. 06, 2021

[25] Cleveland W S 1979 Robust locally weighted regression and smoothing scatterplots *J. Am. Stat. Assoc.* **74** 829–36

[26] Ramesh K, Hariprasad M P and Ramakrishnan V 2015 Robust multidirectional smoothing of isoclinic parameter in digital photoelasticity *Opt. Eng.* **54** 081205.1–9

[27] Martínez-Verdú F, Chorro E, Perales E, Vilaseca M and Pujol J 2010 Camera-based colour measurement *Colour Measurement* (Amsterdam: Elsevier) pp 147–66 https://doi.org/10.1533/9780857090195.1.147

[28] Neethi Simon B, Kasimayan T and Ramesh K 2011 The influence of ambient illumination on colour adaptation in three fringe photoelasticity *Opt. Lasers Eng.* **49** 258–64

[29] Swain D, Thomas B P, Philip J and Pillai S A 2015 Novel calibration and color adaptation schemes in three-fringe RGB photoelasticity *Opt. Lasers Eng.* **66** 320–9

[30] Reinhard E, Adhikhmin M, Gooch B and Shirley P 2001 Color transfer between images *IEEE Comput. Graph. Appl.* **21** 34–41 https://ieeexplore.ieee.org/document/946629

[31] Sasikumar S and Ramesh K 2020 Applicability of colour transfer techniques in twelve fringe photoelasticity (TFP) *Opt. Lasers Eng.* **127** 105963

[32] Ramakrishnan V and Ramesh K 2017 Scanning schemes in white light photoelasticity—part I: critical assessment of existing schemes *Opt. Lasers Eng.* **92** 129–40

[33] Ashokan K, Prasath R G R and Ramesh K 2012 Noise-free determination of isochromatic parameter of stereolithography-built models *Exp. Tech.* **36** 70–5

[34] Vivekanandan A and Ramesh K 2020 Study of crack interaction effects under thermal loading by digital photoelasticity and finite elements *Exp. Mech.* **60** 295–316

[35] Miskioglu I, Gryzagorides J and Burger C P 1981 Material properties in thermal-stress analysis *Exp. Mech.* **21** 295–301

[36] Vivekanandan A and Ramesh K 2019 Study of interaction effects of asymmetric cracks under biaxial loading using digital photoelasticity *Theor. Appl. Fract. Mech.* **99** 104–17

[37] Simon B N, Prasath R G R and Ramesh K 2009 Transient thermal stress intensity factors of bimaterial interface cracks using refined three-fringe photoelasticity *J. Strain Anal. Eng. Des.* **44** 427–38

[38] Hariprasad M P and Ramesh K 2018 Analysis of contact zones from whole field isochromatics using reflection photoelasticity *Opt. Lasers Eng.* **105** 86–92

[39] Irwin G R 1958 Discussion of the dynamic stress distribution surrounding a running crack-a photoelastic analysis *SESA Proc.* **16** 93–5

[40] Ramesh K and Pandey A 2018 An improved normalization technique for white light photoelasticity *Opt. Lasers Eng.* **109** 7–16

[41] Larkin K G, Bone D J and Oldfield M 2001 Natural demodulation of two-dimensional fringe patterns. I. General background of the spiral phase quadrature transform *J. Opt. Soc. Am.* A **18** 1862–70

[42] Gonzalez R C and Woods R E 2008 *Digital Image Processing* (Upper Saddle River, NJ: Pearson) https://www.pearson.com/us/higher-education/program/Gonzalez-Digital-Image-Processing-3rd-Edition/PGM197080.html

[43] Noo F, Clackdoyle R and Pack J D 2004 A two-step Hilbert transform method for 2D image reconstruction *Phys. Med. Biol.* **49** 3903–23

[44] Swain D, Thomas B P, Philip J and Pillai S A 2015 Non-uniqueness of the color adaptation techniques in RGB photoelasticity *Exp. Mech.* **55** 1031–45

[45] Ramakrishnan V, Mallikarjunachari G, Amrita R, Pijush G and Ramesh K 2014 Photoelastic and optical study of polymeric microcapsules dispersed in epoxy matrix *Int. Conf. on Experimental Mechanics (University of Cambridge, UK)*

[46] Kasimayan T and Ramesh K 2010 Digital reflection photoelastcity using conventional reflection polariscope *Exp. Tech.* **34** 45–51

[47] Petrucci G 1997 Full-field automatic evaluation of an isoclinic parameter in white light *Exp. Mech.* **37** 420–6

[48] Maló P, Rangert B and Nobre M 2003 'All-on-Four' immediate-function concept with Brånemark System® implants for completely edentulous mandibles: a retrospective clinical study *Clin. Implant Dent. Relat. Res.* **5** 2–9

[49] Federick D R and Caputo A A 1996 Effects of overdenture retention designs and implant orientations on load transfer characteristics *J. Prosthet. Dent.* **76** 624–32

[50] Elias C N 2011 Factors affecting the success of dental implants *Implant Dentistry—A Rapidly Evolving Practice* ed P I Turkyilmaz (Rijeka: InTech) pp 319–64 https://www.intechopen.com/chapters/18426

[51] Anderson D J 1956 Measurement of stress in mastication. I *J. Dent. Res.* **35** 664–70

[52] Regalo S C H, Santos C M, Vitti M, Regalo C A, de Vasconcelos P B, Mestriner W, Semprini M, Dias F J, Hallak J E C and Siéssere S 2008 Evaluation of molar and incisor bite force in indigenous compared with white population in Brazil *Arch. Oral Biol.* **53** 282–6

[53] Ramesh K, Hariprasad M P and Bhuvanewari S 2016 Digital photoelastic analysis applied to implant dentistry *Opt. Lasers Eng.* **87** 204–13

[54] Odin G, Savoldelli C, Bouchard P and Tillier Y 2010 Determination of Young's modulus of mandibular bone using inverse analysis *Med. Eng. Phys.* **32** 630–7

[55] Begg T, Geerts G A V M and Gryzagoridis J 2009 Stress patterns around distal angled implants in the all-on-four concept configuration *Int. J. Oral Maxillofac. Implants* **24** 663–71 https://pubmed.ncbi.nlm.nih.gov/19885406/

[56] Umezaki E and Shimamoto A 2000 Application of isotropic points extracted using computerized photoelastic experiment to design structural members *J. Strain Anal.* **35** 415–21

[57] Yoneyama S, Shimizu M and Takashi M 1998 Higher retardation analysis in automated white light photoelasticity *Proc. of XI Int. Conf. on Experimental Mechanics (University of Oxford, UK)* pp 527–32

[58] Briñez-de León J C, Restrepo-Martínez A and Branch-Bedoya J W 2019 Computational analysis of Bayer colour filter arrays and demosaicking algorithms in digital photoelasticity *Opt. Lasers Eng.* **122** 195–208

[59] Ashokan K and Ramesh K 2008 Quality assisted three fringe photoelasticity for autonomous evaluation of total fringe order *J. Aerosp. Sci. Technol.* **60** 139–45

Chapter 5

Diverse applications of photoelasticity

Diverse applications of photoelasticity are brought out in this chapter. Photoelasticity has been traditionally used for stress analysis in mechanical sciences and in the recent past, the utility of photoelasticity has increased phenomenally in diverse areas such as neurobiology, plant biology to determine root stresses, biomaterials etc. Even in traditional areas, use of photoelasticity to solve complex problems such as stresses induced due to manufacturing, assembly and residual stresses are discussed as these cannot be easily modelled either numerically or analytically. Further, as problems are becoming more and more complex, either numerical or experimental work alone cannot solve a given problem. How photoelasticity is essential to model the boundary conditions realistically has been brought out through a simple example. Other examples where photoelasticity has ably contributed, such as, the process simulation of precision glass lens moulding and stress analysis of 3-D electronic devices involving through silicon vias (TSVs) are also discussed. Rapid prototyping technologies have revolutionised photoelastic model making of complex 3-D structures for more realistic experimental modelling. An application related to stress field in heterogeneous concrete is presented. Applications involving granular materials and ready-to-use stress field equations for determining the stress field in arbitrarily loaded self-equilibrated forces acting on a circular disc is presented. The utility of photoelasticity in teaching subtle concepts of solid mechanics and the future course of photoelastic research is finally indicated[1].

5.1 Introduction

The technique of photoelasticity is now being applied to a wide variety of problems on a scale and complexity hitherto not explored in several emerging areas that are breaking new ground. This is propelled by advancements in manufacturing technologies and data acquisition through modern approaches. For the sake of

---

[1] Parts of this chapter have been adapted with permission of Elsevier from the author's own work in [158].

brevity, the focus of this chapter is to bring out the effectiveness of photoelasticity in problem solving through a few selected examples in various domains.

In the earlier chapters wherein various digital data acquisition methods have been discussed, several case studies have been presented as an application of using those methods. Each of the methods such as fringe thinning, fringe multiplication, carrier fringe analysis, phase shifting techniques and colour image processing methods are suitable for a class of problems and these were effectively brought out. For example, carrier fringe analysis is a useful tool in evaluating residual stresses in float glass (sections 2.14 and 2.15), which can be used in the production line to monitor the manufacturing process of glass. Phase shifting technique, which provides the most accurate results for photoelastic parameters, has been applied to solve the complex three-dimensional problem of a ball and socket joint using the multi-seeding approach (section 3.12.2). The use of four-step method has been demonstrated for data extraction in orthodontic dentistry employing All-on-four® concept (section 4.12) for implants. Apart from these, evaluation of residual stress in polycarbonate due to the manufacturing process (section 2.12), influence of residual stress on the values of stress intensity factors (SIFs) and T-stress, evaluation of stress field parameters for interacting cracks in biaxial loading (section 2.9) as well as those resulting from mild thermal shock (section 4.9.4) have also been discussed. These examples demonstrate the practical utility of photoelasticity and give the right choice of the methodology of photoelasticity to be adopted in such situations.

In this chapter, the focus is mainly on problems from diverse fields and in each of these cases, sufficient information is provided to appreciate the actual problem in hand and how this has been ably addressed by photoelasticity. The domain specific terms are also briefly explained to comprehend the problem better. Additional details can be obtained from the references listed. The problems chosen are quite complex and, in some cases, they are not easily adaptable for either analytical or numerical modelling and the results from experiments can go a long way in improving those models for effective parametric study. The case studies presented can be grouped in many ways and a single case study may well fit into more than a single group. The case studies are presented in an order that would better reflect their role and that choice is subjective.

Usually, stresses generated during the manufacturing and assembly processes, problems related to residual stresses, are difficult to model either analytically or numerically. The examples where the focus is mainly on manufacturing stresses include stresses due to the manufacturing process of plastic beverage bottles as well as 3-D printed electronic chips with sophisticated interconnection methods. The evaluation of residual stresses is the main concern in precision glass moulding and swelling of polymer composites (a multi-physics domain problem). Contributions from the photoelastic studies can help to model such problems better. Photoelasticity assisted finite element (FE) modelling is brought out in precision glass lens moulding and in the problem of mild-thermal shock given to one of the edges of a rectangular polycarbonate plate.

It is photoelasticity that has emphasised the need for higher order terms in fracture mechanics, while evaluating the crack-tip stress fields. The higher order terms play a vital role, even in crack growth prediction, and this has been brought out with the study of turbine blade root dovetail joint subjected to cyclical loading.

Photoelasticity is to be used as an effective tool to solve a given problem judiciously without complicating it. Engineers would like to solve most problems reliably with simple 1-D or 2-D modelling and avoid 3-D modelling as far as possible. This aspect is well brought out in the design assessment of an aerospace component and the failure analysis of chains. Insights on how photoelasticity can be used for solving complex 3-D problems is illustrated through the analysis of seals whose loading itself is a challenge for stress freezing studies. In both chains and seals, the stresses are developed during assembly, which are difficult to model either analytically or numerically. The 3-D analysis of chain offset plate is discussed to bring out the role of stress relieving holes in reducing stress concentration.

Rapid prototyping has now matured and with advancements in materials, photoelastically sensitive 3-D models could be easily made, that can be stress-frozen and sliced for revealing the 3-D stresses. In conjunction with computerised tomography (CT) scanning, it has provided an excellent opportunity to mimic complex structural systems and its use in modelling heterogeneous concrete to study its behaviour is taken up to illustrate the potential of the approach.

Understanding the complex behaviour of granular materials has caught the attention of physicists recently. They have literally resurrected the large-scale use of photoelastic studies in physics departments, which prompted easy availability of large-sized polaroid sheets, a multitude of methods to make consistent photoelastic models in large volume (the number of models range from 1000 to 15 000 per experiment!) and several software tools for force extraction from such studies. In civil structures, masonry has been used from time immemorial, yet the analysis is quite complex due to its heterogeneous nature. Only recently have there been studies on masonry structures using both transmission and reflection photoelasticity and these are presented in this chapter.

With ever-increasing population and shrinking agricultural lands, efficient use of available land for cultivation becomes paramount for food security. It is interesting to see that by exploiting the granular nature of the soil, photoelasticity finds its use in measuring the stresses exerted by the plant roots on the soil. This information in turn can be used to prepare the lands available for efficient agricultural use.

Over the last decade, there have been several applications of photoelasticity in neurobiological studies to determine the forces developed by different organisms for biomimetics. In these, the major challenge has been to measure exceedingly small forces exerted by those organisms and in some cases, identifying the mechanism of locomotion and how photoelasticity has been used ingeniously to solve such problems is discussed in this chapter. In both neurobiology and plant biology studies, the researchers have developed very elaborate experimental rigs with sophistication that reflects their confidence and significance of photoelasticity for such studies.

There has been an ever-increasing use of photoelasticity in bio-medical applications and development of suitable bio-compatible materials to be used as tendons and development of suitable surrogate materials to study needle insertion problem finds a place in this chapter.

Infrared (IR) photoelasticity has been used for defect detection in silicon chips. Phase shifting methods developed in visible domain have been extended to the infrared domain too. These are used to determine stresses developed during the manufacturing of 3-D connected chips, which is briefly described in this chapter. The use of photoelasticity for improving solid mechanics education is discussed towards the end of this chapter.

5.2 Photoelasticity impacting everyday life

The use of poly(ethylene terephthalate) (PET) bottles for beverages is quite common mainly due to its strength, light weight and recyclability. The material used is birefringent and easily lends itself to be analysed directly on prototypes by photo-elasticity. When they were initially introduced, they consisted of two pieces: a PET blown bottle and a bottom cap for standing. In the mid-eighties, single piece blown PET bottles with a petaloid bottom were made to allow them to be self-standing, thereby reducing the manufacturing processes as well as the material [1]. Cracking of the petaloid bottom has been an issue in such bottles (figure 5.1).

Usually, one selects a preform, and the preforms are tube shaped, typically 100 mm long and about 30 mm in diameter with threads and a compression ring at the thread end [2]. The preform is blown until stretched to the walls of the cold mould, where it freezes to the shape of the bottle and is then removed. During the blowing operation, molecular re-orientation takes place that causes the formation of residual stresses in the bottle. The earliest design approach [1] envisaged FE analysis,

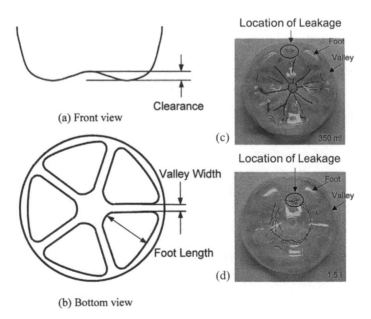

Figure 5.1. Petaloid bottom (a) front view, (b) bottom view, (c) cracks in radial direction, (d) cracks in circumferential direction. Reprinted from [1] with permission from John Wiley and Sons.

and the study has focussed only on the influence of the shape with respect to the internal pressure of the beverage without any consideration to the residual stresses introduced during its manufacture. Although it has been studied for actual pressures in carbonated bottles, due to the blowing-up process, the thickness of the bottle varies considerably over the shape and in FE simulation, it has been assumed to be only uniform [1] due to practical difficulties.

Residual stresses due to blow moulding can interact with service loads significantly and its understanding in terms of distribution is easily possible through photoelasticity by working on prototypes as the material used is birefringent. The thickness variation, which is difficult to predict is easily embedded in a prototype analysis. The ingenuity in applying the technique also includes how to view the bottle in a polariscope. Isochromatics due to the blowing-up process are easily recordable for the base of the bottle by keeping the bottle horizontal and sending the light through the neck region in a circular polariscope [3]. This has been done for different bottle samples of 0.5 and 1 litre bottles, randomly selected from supermarkets. The isochromatics differ even for the bottles of the same volume studied from Publix Supermarket in Atlanta [3] (figure 5.2) as they are influenced by various process parameters. The yellow points marked on the isochromatics indicate biaxial points and the fringe patterns can be idealised to depict cyclic symmetry as the petaloid. The maximum shear stress developed lies in the range 11–30 MPa [3]. For the cylindrical portion of the bottle, the isochromatics are recorded by cutting it in half so that the relieved stresses due to the cutting operation are recorded [4] (figure 5.2(c)). As preforms play an important role, a preliminary study has also been done to classify these using photoelasticity. To see the isochromatics in the neck region, the preform is positioned at an acute angle relative to the optic axis of the polariscope to enable viewing fringes from only one side of the wall of the preform [5] (figure 5.2(d)).

For a realistic FE simulation, inputs from photoelasticity are essential. With such inputs, one may be able to arrive at optimal values of process parameters such as selection of preform, pre-blow and blow pressures, temperature of the mould and its cooling curve for a given shape of the blowing mould. Even a small saving in material per bottle can go a long way in saving our environment, particularly due to the large volume of consumption.

5.3 Photoelasticity in solving a problem in multi-physics

Thermoset epoxy resins are used in matrices of advanced composites, anticorrosive pipe coatings, electronic packaging etc. In such applications, the epoxies used should have the ability to maintain their properties within a fixed range during their operative life and show significant resistance to external ageing factors such as thermal cycles, solvent absorption, and desorption such as water etc. The principal concern in terms of environmental ageing for such materials is mainly water uptake as epoxies are generally prone to solvent penetration. Hygrothermal ageing that happens naturally is usually studied by hydrothermal ageing (conditioning in hot water) to simulate and accelerate the long-term consequences of water uptake in

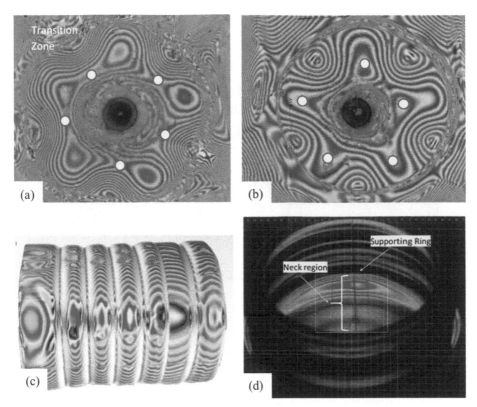

Figure 5.2. Residual isochromatics recorded for base, cylindrical wall, and neck of PET bottles. Base of one litre bottles showing high levels of stress that are functions of the several manufacturing parameters such as selection of preform, blow pressure, shape of the mould, cooling temperature etc. (a) Sample 1, (b) Sample 2. Adapted from [3] with permission from Elsevier. (c) Typical fringes observed for semi-cylindrical cut of bottle at rib geometries. Courtesy: [4]. Adapted by permission from Springer Nature. (d) Isochromatics in the preform neck region recorded by oblique incidence. Courtesy: [5]. Adapted by permission of Taylor & Francis Ltd, http://www.tandfonline.com.

composites. Hygrothermal effects in epoxy are quite complex as water diffusion mode and related mechanisms are not fully understood yet [6, 7].

Swelling due to water uptake is non-uniform resulting in induced internal stresses [8–12]. To visualise these, one requires sensitive and real time whole-field stress or strain measurement techniques. There have been efforts to measure swelling strain fields by local or whole-field deformation measurements [13–15] that have not helped in deciphering the phenomenon. However, photoelasticity is ideal for visualising these stress fields for the entire cycle of absorption, saturation, and desorption.

Two epoxy systems DGEBA (2,2-bis[4-(glycidyloxy) phenyl]propane) and DGEBF (bis(4-glycidyloxyphenyl)methane) with curing agent 4,4′ diamino-diphenyl sulfone (DDS) have been extensively studied under various conditions [8–12]. The epoxy DGEBA has a higher cross-linking density than DGEBF and DGEBA is more brittle with a fracture toughness of 0.57 MPa$\sqrt{m}$ than DGEBF, which is tougher with a fracture toughness of 0.64 MPa$\sqrt{m}$ [10]. Glass transition

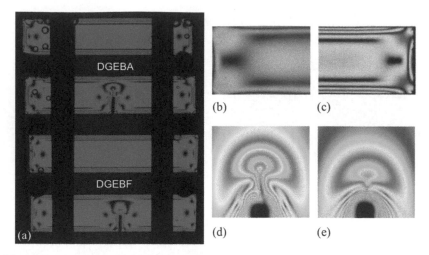

Figure 5.3. (a) Epoxy beams with and without crack in ionised water bath at 80 °C inside the furnace. (b) Residual stresses developed in a glass plate due to thermal gradients, (c) segment of water absorbed epoxy beam with residual stresses developed due to non-uniform swelling. Close-up of crack-tip fringe fields with fringes parallel to the flank indicating crack-closure: (d) DGEBA at 50 °C after 40 h, (e) DGEBF at 50 °C after 65 h. Courtesy: figures (a) and (c)–(e), adapted from [12] and [10] with permission from Elsevier.

temperature (T_g) is one of the important parameters of an epoxy and it is governed by the choice of epoxy, curing agent, the curing conditions and the filler used. Unlike the melting temperature where a phase change from solid to liquid happens, glass transition temperature specifies a temperature within which the amorphous polymer changes from a hard or glassy state to gradual softening and this happens over a temperature range. Glass transition is a complex process and is a reversible transition that occurs when an amorphous material is heated or cooled in a particular temperature range. The T_g is higher for DGEBA (226 °C) than DGEBF (202 °C).

Hydrothermal ageing is carried out at 50 °C and 80 °C using ionised water and desorption is done at room temperature ensuring a dry environment. It is found that DGEBA absorbs more water. The absorbed water is in part filling the free volume and in part chemically reacting with the epoxy network forming polar bonds that are usually bulkier than normal interchain hydrogen bonds causing swelling during absorption. Comparatively more isochromatic fringes are seen in DGEBA than DGEBF (figure 5.3(a)). Figure 5.3(b) shows self-equilibrated internal stresses seen in a glass plate under thermal gradients [16]. Like thermal gradients, the swelling is non-uniform; the outer material in direct contact with water swells first and the inner material is un-swelled, which results in the outer layers being subjected to compression and the inner layers in tension (figure 5.3(c)). The material stress fringe value for epoxies is much lower than glass and hence more nicely coloured fringes are seen in the swelled epoxy (figure 5.3(c)). Kinetics of the stress evolution is influenced by the network structure of the polymer and DGEBF exhibits slower kinetics of formation due to lower cross-linking density than DGEBA [10]. Swelling stresses during absorption increase with increasing material T_g. It is interesting to note that an initially stress-free specimen again becomes stress-free at saturation [8].

During desorption, the outer edges are subjected to tension, which can lead to the growth of edge cracks due to traction at the edges [12]. In desorption, higher stresses are observed for materials with lower T_g [11]. Modifications in the fracture toughness of the polymer during hydrothermal ageing have also been studied. It is interesting to note that during absorption, the fracture toughness reaches a peak [12] due to the formation of a compressive wake behind the crack causing crack-closure (figure 5.3(d and e)). Even at saturation, the fracture toughness is higher due to a different mechanism. The water bonds formed due to absorption of water determine a weakening of interchain van der Waals forces resulting in increased chain mobility called plasticisation [12] that is achieved at saturation. This is translated into a significant increase of fracture toughness observed only in DGEBA.

Improved processing of composites is now contemplated. Curing of epoxy resin by ionising radiation results in lower working temperatures and hence, the mechanical properties can be improved. Irradiated and thermally post-cured DGEBF exhibits lower stresses in desorption, slow stress kinetics and low stresses in absorption while keeping a high T_g and a lower water absorption [11]. Hence, it is an ideally suitable polymer for critical applications.

It is seen that photoelasticity is very well suited to solving problems in domains where a multi-physics approach is needed. Although in general, one would prefer an epoxy with a higher T_g, a systematic study on swelling stresses has recommended the use of DGEBF that has a comparatively lower T_g.

5.4 Photoelasticity assisted FE modelling

5.4.1 Role of photoelasticity in assisting process simulation of precision glass moulding

Precision glass moulding (PGM) is an environment friendly manufacturing process for fabricating high volume aspherical lenses, which are used for high-end precision optical applications such as imaging, laser pointing and aiming, medical devices, mobile phone cameras etc. Over the past two decades, it has gradually become a competitive hot-replicative manufacturing technology [17]. In a PGM process [18, 19], a gob of glass is heated above its glass transition temperature (T_g) and moulded to lens shape against moulds with mirror finished surfaces (figure 5.4). The lens is slowly cooled while it is still under small load. Whenever the temperature of glass falls below the T_g, the moulding tools move apart, and the glass lens is then subjected to a fast cooling process. The lens is taken out of the moulding machine when it is cooled to ambient conditions and is expected to be used as such without further finishing operations.

At the end of the glass moulding process, residual stresses and shape deviations get induced in the lens, which degrades its optical properties. Presence of load during cooling, kinetic phenomena of structural relaxation in glass and spatially non-uniform cooling rate of the specimen contribute to the formation of residual stresses in the lens [20]. Conventionally, the selection of a suitable thermal cycle, which optimizes the process time, residual stress, and shape deviation, is achieved experimentally by trial

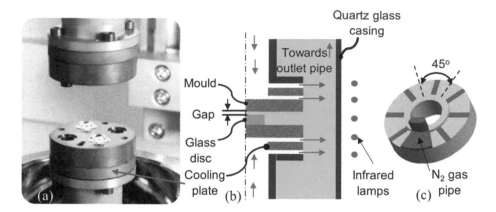

Figure 5.4. (a) PGM machine with gobs. Courtesy: Department of Fine Machining & Optics, Fraunhofer Institute for Production Technology IPT, Aachen, Germany. (b) The blue arrows indicate the nitrogen flow in the PGM machine for cooling and the green colour indicates the gas flow domain. (c) Schematic showing the provisions for cooling of the cooling plates. A sector of the enclosure for cooling gas entry is shown in deep brown colour. The red patches show the slotted portions to cool the plates. On scrutiny, sectorial slot openings are visible in figure (a). Courtesy: figures (b and c), adapted from [19] with permission from John Wiley and Sons.

and error. These are time-consuming and cost intensive. Shape deviation occurs due to thermal shrinkage, which requires a thorough understanding of the thermal behaviour of glass for numerical modelling. Evolution of FE simulation of lens moulding is desirable as it greatly helps in minimising the iterations involved in developing a new lens with a suitable glass for a specific application.

FE simulation of the replicative process of fabricating an aspherical lens is a coupled temperature-displacement problem [21]. In this problem, thermal as well as mechanical boundary conditions (BCs) are necessary (figure 5.5). In the literature, attempts have been made to experimentally obtain the parameters necessary for modelling the mechanical BCs [22, 23]. The PGM process occurs in a closed chamber with movable components, which makes it difficult to directly measure all the thermal boundary conditions of glass. In view of this, these were usually assumed in the FE simulation [21, 24–31].

Thermal BCs include heat transfer between (1) the mould and the N_2 gas, (2) the glass and the mould, and (3) the glass and the N_2 gas. Thermal BC for mould and N_2 gas interaction is approximated by measuring the temperature of the mould experimentally [20]. Heat transfer between glass and mould is modelled using contact conductance [20]. The heat transfer between glass and N_2 gas is usually modelled as convection BC in the FE model. Yi and Jain [31] had assumed natural convection during heating with heat transfer coefficient (HTC) as 20 W $(m^2 K)^{-1}$ and forced convection during cooling with HTC of 200 W $(m^2 K)^{-1}$. Such an assumption needs to be assessed and it is desirable to know the mode of heat transfer from glass for this configuration to fine-tune the FE simulation [19].

Thermal cycling experiments of circular glass disc specimens without pressing stage are normally recommended to validate various aspects of the FE simulation [19, 20]. Circular P-SK57™ (supplied by Schott AG) glass discs are subjected to

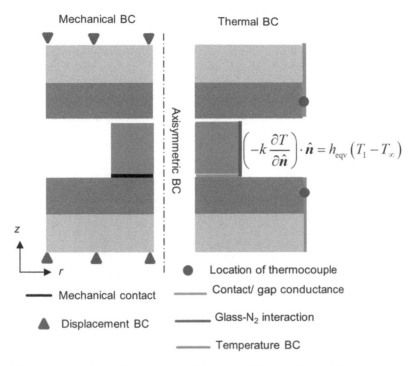

Figure 5.5. Mechanical and thermal boundary conditions of the PGM machine for FE analysis. Adapted from [19] with permission from John Wiley and Sons.

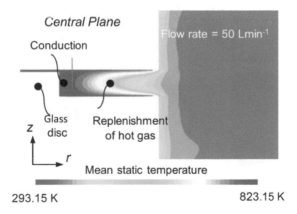

Figure 5.6. Result of the ANSYS FLUENT study that established that glass-N_2 interaction is a combination of conduction and replenishment of hot gas that can be modelled by an equivalent heat transfer coefficient. Adapted from [19] with permission from John Wiley and Sons.

thermal cycling for three different cooling rates in a TOSHIBA GMP 207HV glass moulding machine [19]. Computational fluid dynamics (CFD) analysis of N_2 gas flow in PGM machine is done to investigate the heat transfer mechanism from the glass surface (figure 5.6). Residual birefringence of glass disc after thermal cycling is measured using the latest developments in digital photoelasticity [32] and used as a useful input to improve FE modelling. Figure 5.7(a) shows the modified version of the automated polariscope AP-07 (GlasStress, Tallinn) developed for analysing lenses conveniently. Figures 5.7(b–g) show the six-phase shifted images obtained using the AP-07, which has a resolution of 2 nm in view of the use of high-quality optical elements that are pre-aligned and brought before the light source to record the phase shifted images by a stepper motor. The resulting images are processed by

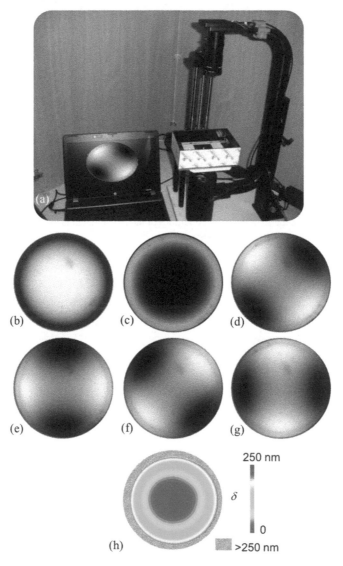

Figure 5.7. (a) Modified AP-07 with immersion tank for lens analysis. (b–g) Six phase shifted images of the thermally cycled disc. (h) Phase retardation obtained by AQGPU using *Digi*Photo software [33].

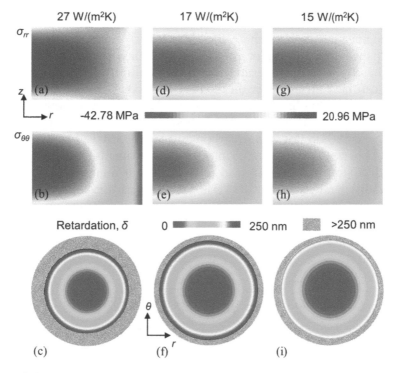

Figure 5.8. Variation of radial and tangential stresses over the depth of the circular disc and variation of integrated retardation over the complete lens for various values of h_{eqv}: (a)–(c) 27 W $(m^2K)^{-1}$, (d)–(f) 17 W $(m^2K)^{-1}$, (g)–(i) 15 W $(m^2K)^{-1}$.

using AQGPU (section 3.10) in-built in the *DigiPhoto* software [33]. The retardation pattern obtained is axisymmetric that justifies the use of axisymmetric modelling in the FE simulation.

Figures 5.8(a)–(i) show the contour map of simulated radial and hoop components of stress tensor as well as the distribution of retardation in the thermally cycled disc. Figure 5.8(c) shows the simulated birefringence distribution by modelling the glass–N_2 interaction by an equivalent heat transfer coefficient (h_{eqv}) of 27 W $(m^2K)^{-1}$. It can be observed that the residual birefringence obtained from FE simulation is high as compared to the experimental values (figure 5.7(h)). The values of h_{eqv} are lowered and FE simulation is repeated for each of these and compared with the experimental results. For h_{eqv} of 15 W $(m^2K)^{-1}$, it is in complete agreement with digital photoelastic experimental results (figure 5.7(h)). Thus, the appropriate h_{eqv} value is finally taken to be 15 W $(m^2K)^{-1}$ for a flow rate of 50 1 min^{-1}. Similarly, for the flow rates of 30 1 min^{-1} and 80 1 min^{-1}, the parametric study is repeated. The final equivalent HTC values for these cases are found to be 14 W $(m^2K)^{-1}$ and 18 W $(m^2K)^{-1}$, respectively.

The study has been successfully extended to the moulding of plano-convex lenses by precision glass moulding [34].

5.4.2 Thermal stress evolution in a plate subjected to a mild thermal shock

In many engineering components, thermal stresses generated due to differential thermal expansion play a major role in precipitating cracks leading to eventual failure of the component. In certain situations, multiple cracks develop, and their interaction can have an influence on the life of the component. One of the simplest examples of thermal loading is the stresses that develop in a rectangular plate when one of its edges comes in contact with a hot surface suddenly (figure 5.9(a)). This is modelled in ABAQUS for a polycarbonate plate (PSM-1, Vishay Measurement Group, USA) of size $125 \times 100 \times 5.5$ mm^3 at room temperature of 25 °C coming into contact with a steel plate maintained at 53 °C. Four-node plane stress thermally coupled quadrilateral element, CPS4T is used to mesh the model. A fine mesh of element size of 1 mm is maintained near the contact region and an element size of 3 mm is adopted in other zones [35].

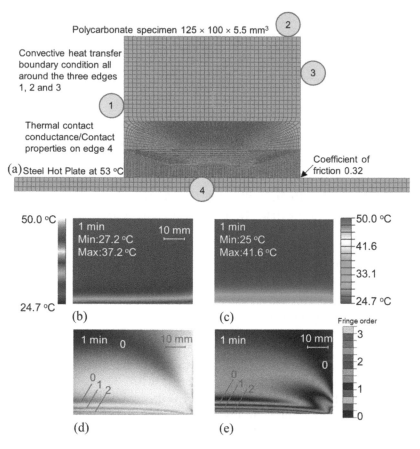

Figure 5.9. (a) Discretisation of polycarbonate plate resting on a hot steel plate. Plot of different data on a portion of the plate that includes the right corner of the plate: (b) IR camera record of temperature variation, (c) temperature variation simulated using FE, (d) isochromatics recorded experimentally, (e) FE simulated isochromatics. Courtesy: figures (a), (d) and (e), [35]. Adapted by permission from Springer Nature.

The ultimate focus of the study is to determine which of the interacting crack configurations in the bottom edge of the plate are significant when it is exposed to a thermal shock. Obviously, it is easier to do a parametric study numerically for various crack configurations, thereby, reducing the experimental verifications to the minimum. From a numerical point of view, a first look at the problem without any cracks in the plate would appear to be very trivial as the geometry of the problem domain is simple, and the boundary conditions appear not quite difficult to specify. The first choice will be to apply convection boundary conditions on the three edges, as well as a simple conduction boundary condition at the bottom edge in contact with the hot plate (edge-4 in figure 5.9(a)). The convective heat transfer coefficient is taken as 50 W $(m^2K)^{-1}$ and the conductive heat transfer (contact conductance) coefficient is taken as 120 W $(m^2K)^{-1}$ [35].

Being a thermal problem, one of the possible ways to verify FE modelling is to compare the prediction with the results of an IR camera (FLIR T530 of 320 × 240, 76 800 pixels resolution) to see the evolution of the temperature profile. A closeup view of the temperature profile at one minute, after the plate is in contact with the hot surface, recorded by IR camera (figure 5.9(b)) and those obtained by FE modelling are shown (figure 5.9(c)). The comparison gives the impression that all is well with the FE modelling. However, it is not so in reality.

Polycarbonate is birefringent and is the most suitable material for photothermoelastic analysis as its properties remain constant in a temperature range of −10 °C to 55 °C [36]. Figure 5.9(d) shows the experimentally recorded fringes near the right corner of the plate at one minute. Most FE packages do not have a facility to plot contours that mimic photoelastic fringes. A simple procedure has been developed to post-process FE results of ABAQUS to mimic photoelastic fringes (section 1.15) [37]. Figure 5.9(e) shows the isochromatics obtained by post-processing the FE results. There is a striking geometric change of the fringe features near the right corner between the two. In fact, in the early development of fracture mechanics, unlike other techniques like holography, Moiré or caustics, the ability of photoelasticity to show significant geometric changes in the isochromatic fringe field has prompted the researchers to recognise the importance of higher order terms in the stress field equations near the crack-tip [38].

This difference in the geometry of the fringe patterns prompted a closer scrutiny of the FE analysis and results. The displacement of the edge-4 is plotted as a function of time and it showed micro-lifting of the plate (figure 5.10(a)) due to differential thermal expansion leading to a violation of the assumed boundary condition along edge-4 as being simple conduction throughout the analysis. As there is no uniform contact along the length, gap-conductance must be brought in to specify the boundary condition. A study by Fuller and Marotta [39] on the pressure-dependent contact conductance values between the metal-polymer interface reveals that at low pressures, the contact conductance value at the interface ranges between 30 and 120 W $(m^2K)^{-1}$. After a few iterations of trial and error, the contact conductance (h_c) value is specified based on the level of micro-lifting. If it is fully in contact, h_c value of 120 W $(m^2K)^{-1}$ is taken and if there is a micro-lifting of 10 microns, it is reduced to 110 W $(m^2K)^{-1}$ and for values of 20 microns and above, a uniform value of 90 W $(m^2K)^{-1}$ has been found to be

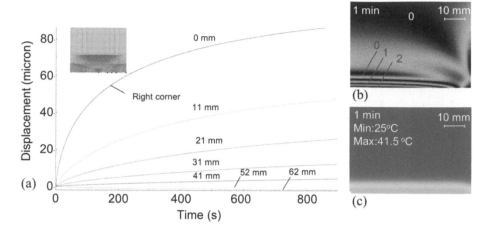

Figure 5.10. (a) Graph showing the micro-lifting of the various points lying on edge-4 as a function of time. The distance of the node from the right end of the specimen is marked on the graph. FE simulated results by modelling the influence of micro-lifting: (b) isochromatics, (c) temperature variation. Courtesy: [35]. Adapted by permission from Springer Nature.

suitable to capture the phenomenon better. The post processed FE results in the form of isochromatics shown in figure 5.10(b) are in good agreement with figure 5.9(d). For completeness, the temperature profile predicted by FE is also plotted and shown in figure 5.10(c), which is quite like figure 5.9(b). Thus, there is no clear prompting signal from the IR camera. Hence, a need even to rethink about the validity of the FE model does not arise, which is deceptive.

In the presence of a single edge crack, the isochromatics observed experimentally are shown in figure 5.11(a), and with the improved modelling of micro-lifting, the post-processed FE results closely match with those from the experiment (figure 5.11(b)). On the other hand, if a simplistic FE analysis ignoring micro-lifting is used, the isochromatics from FE results shown in figure 5.11(c), have larger fringes indicating a higher value of SIF. Hence, the study has brought out the elusive nature of boundary conditions even in seemingly simple problems.

With such improved FE modelling, edge-crack configurations that are parallel, collinear, and asymmetric were considered for a parametric analysis. The study revealed that the asymmetric configuration is the one most influenced by crack interactions. Figure 5.11(d) shows the variation of SIF considering micro-lifting and without micro-lifting for the asymmetric crack configuration. Missing out the physics can lead to overestimated results for SIF, which is not desirable in the new context where the design offices have a competitive need for more optimal solutions.

5.5 Importance of higher order terms in crack growth prediction

Aeroengine compressor blade and disc assemblies are subjected to thermo-mechanical loads due to high-speed rotation and fluid flow over the blades. With new advancements in aeroengines, several researchers have devoted their attention to

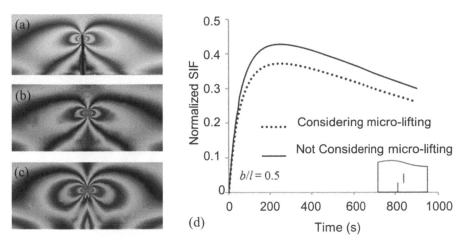

Figure 5.11. (a) Experimental isochromatics for a single crack. FE simulated isochromatics for a single crack: (b) influence of micro-lifting modelled, (c) influence of micro-lifting ignored, (d) graph showing the comparison of normalized SIF as a function of time by including and ignoring micro-lifting. Courtesy: figures (b), (c) and (d), [35]. Adapted by permission from Springer Nature.

understanding the stress distribution in the dovetail joint region. In most cases, cracks get initiated at the dovetail joint due to fretting action at the blade–disc interface [40]. Identification of critical crack orientations and evaluation of SIFs would be of interest to study the crack growth. A study by Abhishek Gandhi in 2017 [41] on the possible accuracy of SIF evaluation by various experimental methods such as strain gauges, transmission photoelasticity, coherent gradient sensing (transmission and reflection modes) and digital image correlation established that among all these methods, photoelasticity guarantees the most accurate evaluation of SIFs.

Shi *et al* [42] showed that low cycle fatigue (LCF) in blade–disc interface is precipitated by the action of centrifugal force causing a relative motion between the contact surfaces and high cycle fatigue (HCF) is caused by the vibrational loads. The focus of the study is to experimentally simulate LCF. Prior to that, static experiments are carried out to identify critical crack orientations. The significant forces acting on the blade and disc assemblies are radially outward centrifugal force due to blade rotation and the bending of the blade due to gas pressure. Following the work of Forshaw *et al* [43], the ratio of axial to bending load of 25:1 is used to conduct static experiments on the dovetail assembly with crack orientations varying from 45° to 120° (measured with respect to the dovetail edge) in steps of 15° (figure 5.12(a)). Figure 5.12(b) shows the close-up view of the isochromatics for a crack at 90°. The colour image is used for identification of the fringe orders and the monochromatic image is used for data processing. The fringes are thinned using the software *Digi*Photo developed in-house [33] and the data points are collected (figure 5.12(c)). Using these, as discussed in section 2.7, the stress field parameters are evaluated in an over deterministic sense using the software PSIF developed in-house [44]. An interesting observation from the close-up view of the fringes for a crack at 90° is that

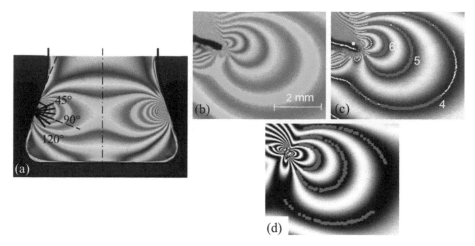

Figure 5.12. (a) Isochromatics on a loaded dovetail without a crack. Crack orientations selected for further experiments are marked in the stress concentration region. (b) Close-up view of the isochromatics for a 90° crack, (c) fringe skeletonised image with data points echoed, (d) theoretically reconstructed image with a 7-parameter solution. Courtesy: figures (b)–(d) adapted from [45] with permission from ASME.

the stress field is Mode-II dominated ($K_I = 0.222$ MPa $\sqrt{m}$, $K_{II} = 0.293$ MPa $\sqrt{m}$, T-stress $= 14.133$ MPa) and the T-stress values are also significant [45]. A similar feature has been observed for all the crack angles chosen. The values reach a peak for a crack at 120° followed by for a crack at 90°. The gradient of the SIFs has been the highest for a crack at 90°. The SIFs for a crack at 45° are comparable to those for cracks at 90° and 120°.

Hence for fatigue studies, edge cracks of 2 mm at angles 45°, 90° and 120° are considered. Crack growth due to fatigue for each of these cases has been examined until complete failure of the specimens. For fatigue testing, the maximum and minimum forces necessary for epoxy specimen are evaluated based on Paris law considering the initial crack length as 2 mm with an expected life of 10 000 cycles using the Paris law constants given in reference [46]. The calculation gave the maximum and minimum forces as 707 N and 27 N, which are taken as 750 N and 20 N for convenience with the corresponding value of $\Delta\sigma$ being 4.86 MPa. The dovetail assembly is subjected to sinusoidal load variation with a frequency of 2 Hz as it falls in the region of LCF [47]. The isochromatic fringe fields are recorded at selected intervals using a high-speed colour camera Delsa Genie (CR-GC00-H6400, HM series) with a framing rate of 50 fps. It is noted that the specimen with an initial crack at 45° undergoes 4058 cycles before catastrophic failure and the specimens with an initial crack at 90° and 120° fail at 3210 and 6916 cycles, respectively [45].

To get the crack growth angle, the isochromatic images where a significant change in crack length is observed, are processed to get the exact outline using the edge detection tools of digital image processing. The relevant angles are determined using the CAD software CATIA V5R20. These images are then superimposed on each other to get the crack path and figure 5.13(a) and (b) shows these for the 45° and 90° cracks, respectively [45].

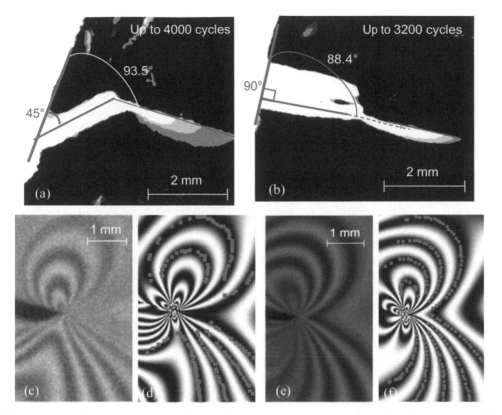

Figure 5.13. Crack growth superimposed images for a few instances up to an instant just before catastrophic failure showing initial crack growth angle with respect to the dovetail edge and the crack progression at discrete steps as different shades: (a) 45° crack, (b) 90° crack. Sample isochromatics just before catastrophic failure and the theoretically reconstructed images after using the nonlinear over-deterministic method of stress field evaluation with nine-parameter solution: (c) and (d) 45° crack, (e) and (f) 90° crack. Adapted from [45] with permission from ASME.

The crack growth angle of cracks at 45° and 90° are evaluated for two cases each, one at the initial stage of crack growth and another just before catastrophic failure, using three different criteria viz., MTS (maximum tangential stress), SED (strain energy density) and GMTS (generalized MTS). Tangential stress in a linear elastic cracked body can be written from equation (2.7) as

$$\sigma_{\theta\theta} = \frac{1}{\sqrt{2\pi r}} \cos\frac{\theta}{2}\left[K_{\mathrm{I}}\cos^2\frac{\theta}{2} - \frac{3}{2}K_{\mathrm{II}}\sin\theta\right] + T\sin^2\theta + O(r^{1/2}) \qquad (5.1)$$

where r and θ are the coordinates with respect to the crack-tip, K_{I} and K_{II} are the mode-I and mode-II SIFs, and T-stress is a constant and non-singular term independent of distance r from the crack-tip. The higher order terms $O(r^{1/2})$ are usually negligible near the crack-tip.

Unlike conventional MTS and SED criteria, the GMTS criterion also considers the effect of the non-singular T-stress term. As per GMTS criterion, crack growth initiates radially from the crack-tip along the direction of maximum tangential stress. Based on the GMTS criterion, the angle of maximum tangential stress θ_0 is determined as

$$\frac{\partial \sigma_{\theta\theta}}{\partial \theta}\bigg|_{\theta=\theta_0} = 0$$

giving,

$$[K_{\mathrm{I}} \sin \theta_0 + K_{\mathrm{II}}(3 \cos \theta_0 - 1)] - \frac{16T}{3}\sqrt{2\pi r_{\mathrm{c}}} \cos \theta_0 \sin \frac{\theta_0}{2} = 0 \qquad (5.2)$$

The crack growth angle θ_0 is obtained by solving equation (5.2) such that $\sigma_{\theta\theta}$ is maximum [48]. For numerical calculations in the GMTS criterion, an appropriate value for the critical distance r_{c} should be used [49]. As suggested in [49], several values of r_{c} were checked, and the results were compared with the experimental values. It is observed that the critical distance r_{c} as 2 mm can predict the crack growth angles that match with the values obtained from experiments.

The SIFs and the T-stress needed for the calculations are obtained from processing the respective isochromatics using the nonlinear least squares method discussed in section 2.7 by judiciously using the software *Digi*Photo and PSIF [33, 44]. Figure 5.13(c)–(f) shows the isochromatics just before catastrophic failure and the theoretically reconstructed image using a nine-parameter solution for the cracks initially oriented at 45° and 90°, respectively. The respective values of SIFs and T-stress are reported in table 5.1. The table summarises the experimental values of crack growth direction and those predicted by MTS, SED and GMTS criteria. Local tangent to the crack is taken as the reference to find the crack growth angle at each stage. A striking message from the table is that when the T-stress value is significant, the inclusion of T-stress is essential to correctly predict the angle of crack growth. In the turbine blade dovetail, the crack that is initially Mode-II dominant grows in a fashion that it becomes Mode-I dominant (table 5.1) before catastrophic failure.

The fringe patterns encountered in this study are overly complex geometrically and from mode-mixity perspective. A simple two-parameter solution is inadequate and only a multi-parameter solution is appropriate (section 2.7), and it has been demonstrated that one can successfully evaluate the stress field parameters reliably

Table 5.1. Comparison of crack growth angles evaluated experimentally and by theoretical models MTS, SED and GMTS.

				Crack growth angle			
Initial crack angle (deg)	K_{I} (MPa$\sqrt{m}$)	K_{II} (MPa$\sqrt{m}$)	T-stress (MPa)	Exp. (deg)	MTS (deg)	SED (deg)	GMTS (deg)
45	0.1380	0.1134	0.1345	−48.49	−49.93	−59.58	−53.28
45[a]	0.3931	0.0347	8.9655	3.35	−9.93	−5.10	1.74
90	0.1263	0.0658	13.544	1.63	−41.08	−41.25	1.92
90[a]	0.4129	0.1085	12.035	1.06	−26.34	−17.76	3.89

[a] Before catastrophic failure

using photoelasticity. The software *Digi*Photo and PSIF [33, 44] have been quite useful to handle such complex fringe fields with ease.

5.6 Ingenuity of solving problems by simplifying the problem

5.6.1 Design assessment of an aerospace component

It is well known that cut-outs and notches act like stress raisers. The focus of the current case study is to assess a component used for separation of stages in a rocket during its flight that had two stress raisers — a cut-out and a notch (figure 5.14(a–c)). A cylindrical shell known as frangible ring has a square cut-out for functional reasons on the axial-hoop plane interacting with a notch in the axial-radial plane to assist stage separation. An extended tubular assembly (XTA) sits on a bracket on the inner side of the frangible ring carrying explosives. The design assessment is to evaluate whether the component can withstand flight aerodynamic loads before separation [50].

Since both the stress raisers are an integral part of the structure, it is difficult to find out the contribution to the stress concentration by each of these from a single component test. The use of strain gauges to measure the strain concentration is difficult as the notch radius is only 1 mm, which is quite small. A 3-D photoelastic analysis may be better but replicating all the assembly features would be cumbersome and time-consuming (section 3.12.2). A 2-D photoelastic study has been envisaged to measure the effect of the notch alone (figure 5.14(d)). The size and load required are calculated from similitude relations.

$$\sigma_p = \sigma_m \left(\frac{F_p}{F_m} \right) \left(\frac{L_m}{L_p} \right)^2 \tag{5.3}$$

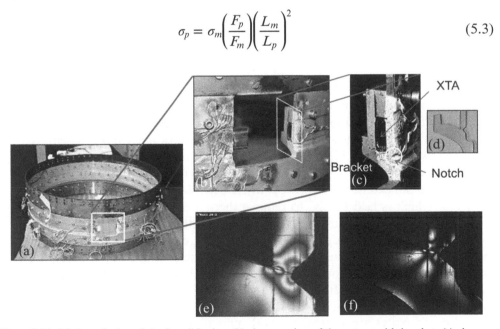

Figure 5.14. (a) Overall view of the frangible ring, (b) close-up view of the cut-out with bracket, (c) close-up view of the bracket with notch, (d) 2-D photoelastic model, (e) isochromatics observed, (f) zero-degree isoclinic passing through the root of the notch. Adapted from [50].

where, σ_p is the prototype stress, σ_m is the model stress, F_p is the load acting on the actual component, F_m is the applied load on the model, L_p and L_m are the characteristic lengths of the prototype and the model, respectively. The length ratio is taken as unity, and for a prototype load of 218.8 kN, (125% of the expected load) that acts on the entire circumference of the shell, the load that needs to be scaled and applied on the model that is only of 2 mm thickness, the corresponding prototype load is 251 N. Based on practical considerations of visualising enough fringes, the model load is taken as 16.8 N. The load ratio is thus close to 15. The isochromatics observed are shown in figure 5.14(e). Figure 5.14(f) shows that zero-degree isoclinic passes through the notch tip, thereby simplifying the data interpretation of the fringes. The fringe order at the notch tip is 3.95, which corresponds to a model stress of 18.6 MPa and a component stress of 280 MPa as per equation (5.3). The frangible ring is made of AA 7075-T7352 with a guaranteed minimum yield stress of 350 MPa. Considering the Young's modulus as 70 GPa, the measured strain would correspond to 4000 με.

Generally, the aerospace structures are ground tested for their flight worthiness and the actual component level test done at 135% of the actual load is utilised to measure the total strain in the component by using 3-D DIC (digital image correlation). The test revealed a total strain of 5200 με. When the result of photoelastic test is upgraded to the actual test load of 135%, the strain would be 4320 με. The difference of 880 με between the two can be attributed to the cut-out and thus the role of cut-out is not critical. This is attributed to the extra stiffness in the structure provided by the bracket and related assemblies [50].

Thus, a seemingly complex problem has been efficiently handled by judiciously combining two appropriate experimental techniques with intelligent modelling that has come about through experience in solving practical design problems.

5.6.2 Failure analysis of chain links

Most automotive and industrial chains are an assembly of five parts shown in figure 5.15(a) namely, outer plate, inner plate, bush, pin, and roller [51, 52]. Two inner plates are press-fitted with two bushes to form an inner block assembly. The bushes shown in figure 5.15(b and c) can be made using cold extrusion, cold forming, or cold drawing processes. The chain plates are made using blanking and piercing operations. Cold extrusion is a process where extrusion punches are pushed against a solid pin, which is held firmly, to remove the material to make it hollow. In cold metal forming process, a steel strip of specified width is cut into smaller lengths and rolled into the shape of a cylinder using a forming die set. There is a fine slit along the length of the formed bush (figure 5.15(c)). In the cold drawing process, a higher diameter tube is pulled through a die set to make it into a smaller diameter tube with increased wall thickness. In both cold extrusion and cold drawing processes, the geometric shapes are well-controlled.

The outer plates are press-fitted with pins after passing the pins through the assembled bushes of the inner block. The roller is a rotating member, which is placed over the bush during the inner block assembly. Inner block assembly is the load

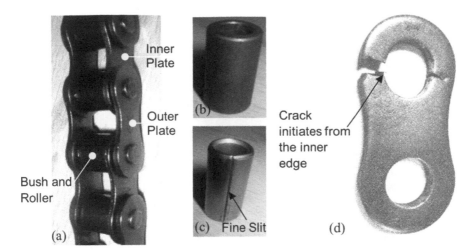

Figure 5.15. (a) General chain assembly, (b) extruded bush without slit, and (c) formed bush with fine slit along its length, (d) random fatigue failure of the inner plate. Courtesy: [52]. Adapted by permission from Springer Nature.

transfer member from the sprocket tooth and the outer block assembly helps in holding and pulling the inner block over the sprocket teeth. Thus, interference fits are the necessary features in a chain assembly to keep the parts together with sufficient holding force.

There were random fatigue failures of chain assemblies encountered in the field. Most of the failures were observed with the inner plate in the problem under study and the nature of the failure is as shown in figure 5.15(d). The reason for the random failure could not be found out through regular quality checks. The use of strain gauges could not give any satisfactory results. Even a 3-D FE analysis of the problem was not able to detect the cause of the random failures. A simple transmission photoelastic analysis has brought out the core of the problem that is associated with interference fits. Figure 5.16 shows the typical fringe patterns observed in an epoxy model of the chain plate inserted with a solid pin and a bush made by the forming operation containing a slit. A perfectly round pin produced fringes that are concentric to the hole and on the other hand, the slit bush produced a flowery pattern dictated by the geometric undulations of the formed bush leading to random fatigue failures in service. As there is no control on the outer surface in a slit bush, it has led to minute undulations. In interference fits, even a small increase in the interference level can lead to a spike in the stress levels locally and this has caused the formation of cracks leading eventually to fatigue failures. A detailed study by reflection photoelasticity [52] on the actual metallic inner plates reconfirmed this observation. This led the manufacturing company to change their production line to use only extruded bushes, whose outer geometry is well controlled and most suited for the chain assembly to improve its fatigue life.

Figure 5.16. Isochromatics observed for a chain link inserted with a machined pin (isochromatics in colour) and a slit bush (isochromatics in black and white). Courtesy: [52]. Adapted by permission from Springer Nature.

5.7 Three-dimensional photoelastic analysis

5.7.1 Chain offset plate failure

One of the aspects of a chain assembly is that if a chain length is of odd number of pitches, it requires an offset plate [51] as shown in figure 5.17(a) to connect the two ends of the chain together and to make the chain endless. The offset plate can be made using a cold cranking or hot cranking process based on the thickness and its offset depth. In cold cranking process, a steel strip of specified width is cut into a smaller length and cranked into an offset plate shape using a cranking die set. Hot cranking is a process where the plate is heated to a temperature, ranging between 800 and 900 °C, before cranking. An inbuilt thinning zone, free formed radius, and smaller cranking radius at the offset zone are unavoidable in both the processes of cranking. When the chain is assembled with an offset plate, the chain fatigue life is observed to be only 20% to 25% of the total life of a chain assembled without an offset plate. The reason for the premature failure could not be found through regular quality checks. Preliminary studies by numerical methods have only shifted the focus and suggested a true 3-D photoelastic analysis for problem identification.

An epoxy solid block was machined using a CNC machining centre to make a chain plate model like the actual chain plate. To get the correct size and shape of the offset plate (i.e., the cranking radius as well as the free form radius) in the photoelastic model, the prototype offset plate was scanned using a portable FARO laser scanning machine. The final CAD model was generated based on this and using this, an epoxy chain plate was machined out from a block using a CNC machine [53]. The model is then stress-frozen with a suitable tensile load as discussed next.

For exact similarity, the model and prototype should be geometrically similar when deformed by the respective loads. This necessitates that the corresponding strains in the model and prototype be equal. To achieve this, an appropriate force scaling is done. Assuming that both the model and prototype have the same Poisson's ratio, the stress in the model and prototype are related as in equation (5.3).

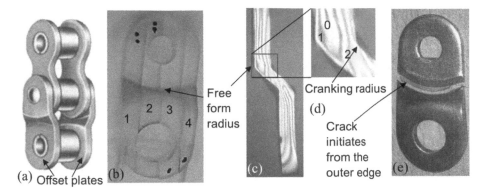

Figure 5.17. (a) Use of offset plates to make the chain with odd number of pitches endless, (b) stress-frozen and sliced model of the offset plate, (c) slice 1 viewed in a circular polariscope, (d) enlarged view of the slice showing higher fringe order at the cranking radius, (e) failure of the plate at the cranking radius with the crack emanating from the outer edge. Figures (a), (b) and (e) adapted from [53] with kind permission of Trans Tech Publications.

Replacing the stresses in terms of strains and Young's moduli in equation (5.3), the force scale obtained is given by,

$$F_m = F_p \frac{E_m}{E_p}\left(\frac{L_m}{L_p}\right)^2 \quad \text{or} \quad F_m = F_p \frac{E_m}{E_p}\left(\frac{L_m \times t_m}{L_p \times t_p}\right) \tag{5.4}$$

The chain plate is 4 mm thick and is made of steel (Young's modulus 210 GPa). The epoxy model is also 4 mm thick with Young's modulus of 3.3 GPa. In the present work, the model is an exact replica of the prototype ($L_m = L_p$ and $t_m = t_p$); equation (5.4) reduces to

$$F_m = F_p \frac{E_m}{E_p} \tag{5.5}$$

Considering 7848 N as the applied load on the actual chain plate, the equivalent load to be applied on the epoxy model is calculated to be 123.3 N. This would be the load required for room temperature test. To see the internal stresses, it was planned to stress-freeze the model suitably. The room temperature F_σ for the model material is 10.57 N/mm/fringe. The F_σ at stress freezing temperature is close to 0.3 N/mm/fringe. The load that is required to be applied for stress freezing is estimated as

$$F_{mf} = \frac{123.3 \times 0.3}{10.57} = 3.5 \text{ N}$$

The stress frozen sample was sliced into four pieces as shown in figure 5.17(b), to estimate the nature of stress distribution and its magnitude at the offset zone. The slices are viewed in a circular polariscope with white light and figure 5.17(c) shows the isochromatics for slice 1. It is observed that the highest fringe order is at the cranking radius side and zeroth fringe order is observed remarkably close to the formed radius side at the offset zone (figure 5.17(d)). This has supported the nature of failure observed in the field near the cranking radius (figure 5.17(e)). The nature of fringe pattern depicts that there is a bending moment acting at the offset zone due to its shape even though the external loading is only an axial pull, which is causing higher surface stresses at the vicinity of the cranking radius. The study has revealed the nature of the stress distribution and the stress concentration is relieved by putting a stress relief hole of appropriate shape at the offset region.

5.7.2 Seals

A seal is an integral packing element in many mechanical systems used to prevent fluid leakage and contain pressure. It is utilized in a wide range of devices ranging from household pressure cookers to rocket propulsion units. They are broadly classified based on their cross-section as O-ring, square/rectangular ring, X-ring, and D-ring (figure 5.18). The failure analysis of seals gained much importance once the investigation on the Challenger space shuttle found out that the failure of O-ring joints in the solid rocket booster led to the disaster. Square and rectangular cut rings are considered as an inexpensive, loose-tolerance alternative to moulded O-rings. They are used in static applications as a direct replacement for O-rings. However, they are inferior to O-rings in dynamic applications.

One of the most common failures of an O-ring in the field is a spiral failure usually found in long stroke hydraulic piston seals. This can be precipitated due to inadequate or improper lubrication, uneven surface finishes, too slow stroke speed, wide clearance combined with side loads or even the use of exceptionally soft rubber for the seal. In spiral failure, the seal becomes 'hung-up' at one point on its diameter and slides and rolls at the same time due to which, deep spiral cuts get developed on the surface of the seal. Any effort in design to prevent twisting of the seals is desirable to avoid this failure mode.

D-rings can typically be retrofitted into existing O-ring grooves with the size often based around the original O-ring size. The flat geometry on the base of the D-ring stops the seal from twisting and rolling, preventing spiral failure. The X-rings are four-lipped seals with a specially developed sealing profile that provides twice the number of sealing surfaces as a square cut ring. Further, the gap between the sealing surfaces does a better job of retaining the lubricant. The four-lobed design not only provides lower friction than an O-ring and a square cut ring, but also, due to its cross section, resists spiral twist. The X-rings are used in many dynamic applications where O-rings have failed to deliver satisfactory performance.

Seals made of rubber have a low elastic modulus, their Poisson's ratio is greater than 0.49 and they are nearly incompressible, which allows the seals to achieve

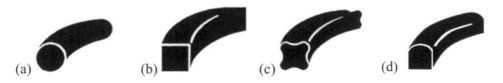

Figure 5.18. Various cross-sectional seals. (a) O-ring, (b) square-ring, (c) X-ring, (d) D-ring.

automatic sealing. Since their Poisson's ratio is close to 0.5, photoelastic analysis using stress freezing closely mimics this behaviour and the use of a simple one-to-one scale of the photoelastic model is feasible. In service, the O-rings are laterally constrained, squeezed, and pressurised (LCSP). A novel stress-freezing fixture for loading the seals under uniform squeeze rate has been designed and fabricated [54, 55] (figure 5.19). The squeeze rate is controlled by internal diameter variations of the groove. The O-ring of size 6.98 ± 0.15 mm with an internal diameter of 121.5 mm made of high-temperature Araldite B41 and hardener HT903 (Hutsman Advanced Materials, Switzerland) at weight ratio 10:3, cylinder and guide ring are heated for 40 min at 120 °C (close to the stress freezing temperature of the epoxy used) in the furnace. The O-ring, cylinder and guide ring are assembled in the furnace at 120 °C. Although the designed squeeze rate was 20%, the actual squeeze rate achieved is 18% possibly because the actual stress freezing temperature was about 130 °C. Nevertheless, the approach adopted by them is quite novel to mimic the initial squeezing of the seals in their assemblies.

The loading rig also had a provision to apply a variable hydraulic pressure [56]. The tests have been conducted for pressures of 0.98, 1.96, 2.94, 3.92 MPa and other suitable pressures as the seal design demands. For each of these pressures, a new model needs to be stress frozen. Slices of 2 mm thickness are cut at 90° intervals. These are then polished down to a thickness of 1 mm for recording the isochromatics. Using the same loading rig, the cross-sections of the seals are modified to study the nature of stress distribution due to the shape change of the seals [54–62]. The isochromatics observed for various cross-sections of the seals are summarised in figure 5.20. The material stress fringe value at stress freezing temperature for the resin is typically about 0.242 N/mm/fringe [61]. The isochromatics provide a wealth of information, for comparative studies on the role of cross-section of the seals in influencing the stress distribution.

The pressure of the sealed fluid modifies the contact stress of the seal inside the assembly and proper sealing is guaranteed only when the maximum contact stress

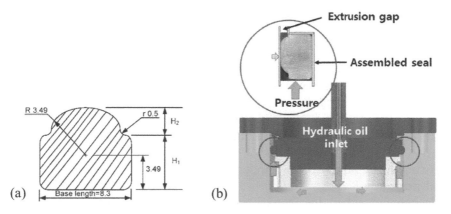

Figure 5.19. (a) Geometrical details of the D-ring in mm. Courtesy: [56]. Adapted by permission from Springer Nature. (b) Sectional view of the loading device with D-seal assembled with 20% initial squeeze. Adapted from [55] with permission of JSME.

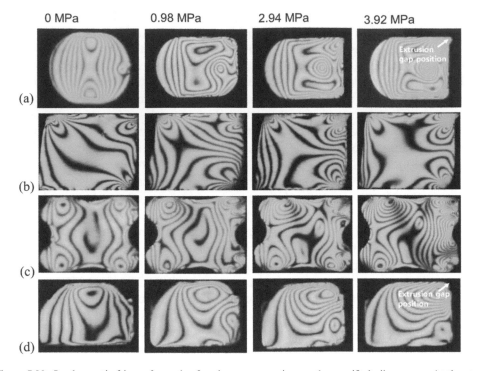

Figure 5.20. Isochromatic fringes for seals of various cross-sections under specified oil pressure. At the start, the seals have stress due to initial squeeze of 20% while assembling seals tightly (a) O-ring, (b) square-ring, (c) X-ring, (d) D-ring with $H_1/H_2 = 3.5$. Courtesy: figures (a) and (d) adapted from [55] with permission of JSME. Figures (b) and (c) from [61, 62] adapted by permission from Springer Nature.

remains greater than the hydraulic pressure [56]. The photoelastic experimental hybrid method (PEHM) [54] has been used in the contact and fracture analysis of various geometries of seals [54–62]. The PEHM uses the data points collected from experimental isochromatic fringes to get a consistent match with the theoretical solution obtained by Muskhelishvili's complex function and Airy's stress function by minimizing the error between them. Studies on O-rings [54, 55, 57, 59] have revealed that a significant portion of the seal gets sheared during assembly and they are also prone to spiral failure in dynamic applications [57]. Although the sparse iso-chromatic fringes in the upper and lower portion of the square and rectangular ring seals showed stability in stress distribution (figure 5.20(b)), the stress singularity at the corners was not desirable [56, 62]. The X-rings and D-rings were developed by combining circular and rectangular profiles for high-pressure applications. Bernard *et al* [58] and Mose *et al* [56] have studied the failure criteria for X-ring and D-ring using stress-frozen models, under uniform squeeze rate for various internal pressures P_i.

Seals can fail either due to extrusion of seal material into the assembly gap between sealing surface until a point where the contact stress surpasses the yield strength of the sealing material [56] or by fracture caused by high lateral pressure of the sealed fluid [57]. The extrusion pressures were found to be 1.96 MPa for O-ring seals, 3.92 MPa for rectangular seals, 5.88 MPa for X-ring seals and 6.87 MPa for D-ring seals [56]. In X-ring seals, even after extrusion, the isochromatics at the upper left portion showed symmetry which revealed its ability to still contain fluid [58]. The highest contact stresses also differed based on the cross-section; the values being 10 MPa for an O-ring, 15.64 MPa for X-ring and 26.86 MPa for square geometry. The fracture pressures are at much higher levels and for a D-ring, it is estimated to be 70 MPa [56] and hence the seal performance is mostly dictated by the extrusion pressures.

5.7.3 Rapid prototyping and photoelasticity

Ramesh *et al* [63] in 1999 foresaw that a fusion of rapid prototyping (RP) for model making, digital photoelasticity for experimental analysis, and finite element method (FEM) for numerical analysis as the way forward for production and analysis of complex structures. Fused deposition modelling (FDM) is one of the earliest methods to make 3-D models based on CAD data, which was used to make complex models like a turbine blade [63]. Using the FDM generated turbine blade as the model, moulds were made by rapid tooling (RT) to make the final photoelastic model with special quality RT resins like SG 95. The RT resins showed good stress freezing properties, but their optical response was one-third of the conventional photoelastic material. Models made this way can be analysed by conventional stress freezing and slicing. The only challenge is to minimize the residual stress while moulding, which can be accomplished by using ice bath to take away the heat. Dentists have successfully exploited this approach for making patient-specific models of mandible (lower jaw) and maxilla (upper jaw) obtained by CT scanning of the patient [64, 65] and subsequently extracting stress information by photoelasticity.

Stereolithography (STL) (3D systems, USA) is another RP technique to build prototypes using a monomer formulation usually of acrylic- or epoxy-based resins. The advantage of STL is that the model can be directly used for stress freezing and slicing. Due to layered manufacturing, the models can exhibit micro-porosity and show directional dependence. Nevertheless, complex models like gears have been made (SL5510 material) to extract the fringe data using phase shifting technique [66]. The issue of noise due to porosity in STL models, which results in discontinuity of photoelastic fringe contours was solved by Ashokan *et al* [66]. Vivek and Ramesh [67] demonstrated that such issues can also be handled by TFP with its advanced implementation of FRSTFP scanning, followed by multi-directional progressive smoothing (section 4.9.3). Properties of some of the STL resins are given in [68].

Photopolymers typically shrink up to 8% by volume during polymerization that leads to build-up of residual stress. Umezaki *et al* [69] have done the stress and flow analyses of photopolymer pentaerythritol triacrylate by considering a thick layer of 15 mm of the resin cured by exposure to UV light. They have considered an acrylic-based photopolymer as it is expected to show more shrinkage strain than epoxy-based photopolymers [70]. The UV-curable resin illuminated from above with UV rays was cured from the upper part, and the cured area spread in the downward direction with time. The residual stress is observed as isochromatics and the stresses in the upper part were larger than the lower part and there is hardly any dependence on the flow of the resin.

With advancements in 3-D printing technologies and newer materials, better photoelastic models that are nearly homogeneous and show low directional dependence are now possible [71–80]. The 3-D printer Objet Connex 500 3D (Stratasys ltd, USA) has been successfully used to create overly complex models of rock and concrete. The print resolution is $600 \times 600 \times 1600$ dpi in the x, y, and z directions with a dot accuracy of 10–50 μm and it can use 17 different types of basic photopolymers from which, hundreds of composite materials can be generated using specific mixing ratios. A photopolymer named VeroClear RGD 810 (Stratasys ltd, USA) is used as the basic matrix material, which is a unique epoxy resin material that is rigid and colourless and has excellent dimensional stability. Since the raw material used is in the form of a liquid, the typical layer thickness possible ranges from 16 to 30 μm. Use of a high-intensity UV radiation instantaneously hardens the sprayed layer thus significantly reducing the effect of layered structure on the macroscopic response of the printed sample [72]. The use of polyjet printing involving a liquid resin, makes the printed specimens more homogeneous than those fabricated by powder-based or filament extrusion 3-D printers.

Uniaxial compression tests revealed that VeroClear undergoes elastic to plastic deformation and displays favourable ductility in the post-peak stage. Further, it displays perfect elastic performance up to 80% of the peak load. An opaque photopolymer RGD 525 exhibits elastic to brittle behaviour. However, both the materials exhibit brittle fracture under tensile stress. Lattice support material recommended is Fullcure 705 (Stratasys ltd, USA), which has a loose structure and low strength. This material is ideally suitable for creating voids in the 3-D model wherever required [73]. The models made need to be carefully handled before testing. In addition to the normal precaution of keeping the model in a desiccator, it must be opaquely wrapped to prevent ambient light and moisture from affecting its mechanical and optical properties.

With the power of making complex 3-D printed photoelastic models, the group led by Yang Ju [71–80] have attempted to solve difficult problems in rock mechanics, porous solids as well as modelling heterogeneity of concrete. The details of their study on concrete will be discussed next.

This is the first pilot study on mimicking the heterogeneity of concrete [73]. To make photoelastic analysis easier, the mass fraction of aggregates is kept low at 15.7% and irregularly shaped limestones are used with sizes ranging from 5 mm to 10 mm with about 72% in the size range of 6–8 mm. Local Portland cement P.O. 42.5 along with coarse aggregates and river sand with a granular diameter ranging from 0.25 to 0.35 mm for fine aggregates are used to manufacture concrete mortar. Using these, a cylindrical concrete specimen of diameter 50 mm with a length of 100 mm is made for CT scanning.

The geometric details of the concrete sample are carefully identified by high-resolution micro-CT imaging that has a resolution of 4 μm. With a scanning interval of 0.1 mm, approximately 1000 images are taken for the specimen with each image

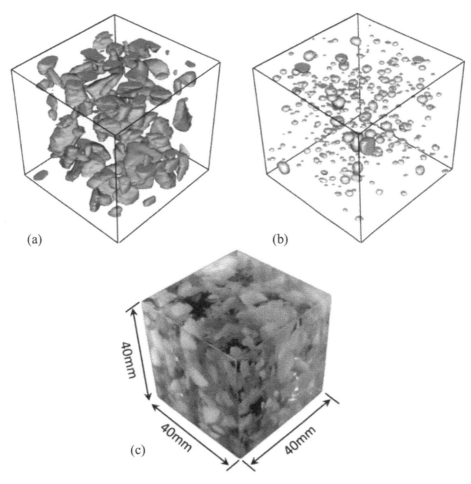

Figure 5.21. 3-D perspective view of (a) aggregates, (b) voids, (c) 3-D printed photoelastically sensitive concrete model with VeroClear as the matrix, RGD 525 as aggregates, and Fullcure 705 as voids. Reprinted from [73] with permission from Elsevier.

of size 650×650 pixels out of which, the central core of 400×400 pixels is culled out for mimicking a photoelastic model of concrete. Using VeroClear as matrix, Fullcure 705 to simulate voids and opaque printable material RGD 525 (Stratasys ltd, USA) for aggregates, a model of concrete cube of size $40 \times 40 \times 40$ mm^3 has been made [73] (figure 5.21). Although the individual elastic property values are different for the lightweight aggregate concrete and the photoelastic model, the requirement for the aggregate strength/stiffness to be lesser than that of the matrix for lightweight aggregate concrete is satisfied by the photoelastic model. The ratio of Young's modulus of aggregate to matrix in photoelastic model is 0.45 at 25 °C (0.53 at 125 °C) and for concrete, this ratio varies between 0.18 to 0.45 [73]. The ratio of Poisson's ratio of aggregate to matrix for photoelastic model material is 0.39 at 25 °C (0.45 at 125 °C) and for concrete, it varies from 0.36 to 0.40 [73]. Thus, the simulated model is expected to be elastically similar when subjected to equivalent loads.

Two sets of experiments have been conducted. One in which the cube is subjected to compression and stress frozen at 125 °C and slices at periodic intervals are cut and viewed in a polariscope with white light illumination (figure 5.22(a)). Second in which, from the already developed 3-D computerised model, 2-D slices of 6 mm are freshly 3-D printed and subjected to equivalent live loads and the isochromatics are recorded. The material stress fringe value of VeroClear at room temperature is 35.06 N/mm/fringe and 2.13 N/mm/fringe at 125 °C [81]. Figure 5.22(b) and (c) corresponds to slices cut from the stress frozen model and figure 5.22(d) and (e) are the isochromatics for the corresponding 2-D slices. One striking feature between the figures is that the fringe orders (by comparing the colours with the colour code) are less in the case of 3-D slices than the 2-D slices. This is attributed to a combination of structural heterogeneity and lateral inertial confinement of the 3-D model.

The approach adopted has the potential to study the influence of heterogeneity in the design of concrete very comprehensively. As the making of the model itself is quite challenging, the authors have confined to mere visualisation of the isochromatics in their first study. This group has also now adapted to using the ten-step phase shifting method (sections 3.3 and 3.4) for data interpretation [80]. In the years to come, more insights to the analysis of complex models are possible.

Usually, the Poisson's ratio of epoxy resin will approach a value of 0.5 at stress-freezing temperatures and this feature has been used to advantage in extracting stress fields in seals made of rubber (section 5.7.2) whose Poisson's ratio is about 0.5. Unlike the epoxy resins, it is reported that the Poisson's ratio of VeroClear reduces from 0.38 at room temperature to 0.295 at the stress-freezing temperature [73]. This is an interesting property since, for most metals the Poisson's ratio is around 0.25 at room temperature. As the stress distribution is influenced by the Poisson's ratio in 3-D specimens (section 1.13), stress-frozen experimental results from conventional epoxy models are suitably re-interpreted currently, which may not be necessary if models are made of VeroClear. Further, there is scope for improving the 3-D printable material to have modified elastic properties by various means to satisfy varied requirements. As the models are grown using digital means, a numerical model featuring all the geometrical and physical features is quite possible that has also been effectively used by the researchers to compare their results [73, 74, 76]. Thus, a true synergy of rapid prototyping for model making, 3-D photoelasticity for experimental analysis and parametric studies by numerical modelling such as FE is now a reality.

5.8 Phenomenological studies on granular materials and structures

5.8.1 Granular materials

Granular media are encountered everywhere in our daily lives, from sugar and flour to construction materials, mining products and pharmaceuticals. The granular materials are quite different from the ordinary solids, liquids, and gases that one

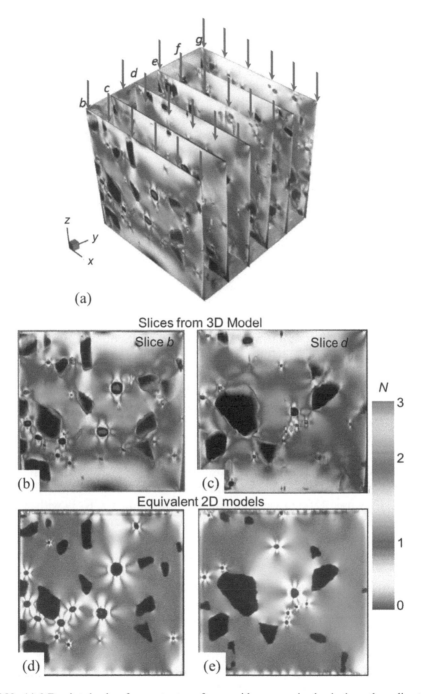

Figure 5.22. (a) 3-D printed cube of concrete stress frozen with compressive loads along the z-direction with slicing locations. Isochromatics observed in slices cut from the 3-D cube: (b) slice b, (c) slice d. Isochromatics observed in equivalent 2-D slices subjected to compression corresponding to (d) slice b, (e) slice d. Adapted from [73] with permission from Elsevier.

learns from physics textbooks. They are a collection of discrete particles, the behaviour of individual particles is governed by classical mechanics, whereas a collection of particles exhibits complex and fascinating behaviour. Understanding how granular materials behave and fail is important to predict phenomena such as avalanches, mudslides, and earthquakes.

Grains in a granular system can span a large range of sizes from a few hundred microns in powders to several meters in boulders. In a dry granular system such as dry sand, there are no attractive interactions between the grains. Unlike an elastic solid, they cannot support tension. However, they can support compression as one can walk over them. At the same ambient conditions of the temperature, it can also be poured as a liquid. This is quite different for ice—only when it is heated, can it be poured as a liquid. To sustain the flow of granular materials, external force is required as energy is lost through frictional interactions.

In the analysis of liquids or gases, one need not deal with individual molecules as thermodynamical principles allow one to use averaging techniques to predict the waves in a pond and the pressure of a gas. In granular materials, the particles are large enough that gravity and friction prevent the random motion induced by temperature. Hence, in granular systems, standard thermodynamic approaches break down and the systems are far from equilibrium and highly nonlinear. In view of it, the solid-like and the liquid-like states of granular materials often exhibit counterintuitive behaviour. Stirring of a granular system will not mix but will separate the constituent particles!

One of the earliest experimental studies on granular materials has been done by Wakabayashi [82] using crushed glass as granular medium immersed in a matching refractive index liquid. To evaluate the stresses developed, he has embedded phenolite rings (photoelastically sensitive material) at selected locations and viewed the system in a circular polariscope. One of the key observations is that the system showed hysteresis during loading and unloading cycles. Work of Dantu [83] kick-started the idea of using photoelastically sensitive discs themselves as the granular medium and was first experimented by Drescher and Josselin de Jong in 1972 [84]. In their experiment, they used 1200 discs of thickness 6 mm made of CR-39 with six different diameters varying from 8 to 20 mm. The loading arrangement shows a fixed arm and a pivoted arm that can be rotated, and the discs are loaded by dead weight (figure 5.23(a)). Isochromatics recorded for a portion of the assembly at maximum anticlockwise position of the pivoted arm is shown in figure 5.23(b). The experiment revealed for the first time that the force is not distributed uniformly but gets transferred along chains of discs, which are later christened as force chains [85] and brought out the heterogeneous distribution of forces. They also determined the contact forces and plotted the force chains (figure 5.23(c)). The study also underlined the inadequacy of determining the granular system behaviour based on stress or strain as they are averages that blur the real physical entities responsible for the mechanical actions.

The mere observation of *force chains* [86] has given insight into phenomenological understanding of varied phenomena such as *meteor impact* [87, 88], faults in earthquakes [89, 90]. In missile or meteor impacts, the penetration is better at slow speeds than at higher impact speeds as the force chains grow more extensively causing higher resistance. It is interesting to note the extensive experimental

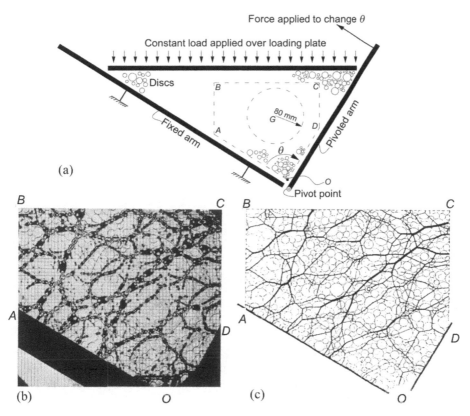

Figure 5.23. (a) Loading arrangement for about 1200 particles of six different diameters of circular discs ranging from 8 to 20 mm of thickness 6 mm made of CR-39 polymer. (b) Isochromatics observed for a portion of the tested model when the pivoted arm is at its maximum anticlockwise position. (c) Illustration of contact forces as chains, the line thickness indicates the magnitude of the forces. Figures (b) and (c) adapted from [84] with permission from Elsevier.

arrangement to study the phenomenon of seismogenic faults in earthquakes. Most earthquake producing faults contain a layer of granular fault rocks. Daniels and Hayman [89] in 2008 have used about 10 000 particles made of PS-3 polymer (Micro-measurements group, USA) of 60% circular (5.6 mm diameter) and 40% elliptical (major and minor axes of 6.8 mm and 4.7 mm) shapes of 3 mm thickness. The heterogenous mixture of two shapes prevents the granular medium into a tightly packed configuration of a crystal. The seismogenic fault is simulated using a single layer of particles with a stationary side and a pulled side, which creates a granular-on-granular shear zone (figure 5.24). The stationary side can have either a constant volume or a constant dilatational boundary condition. Starting from a randomised configuration with few force chains, when the pulling starts, the particles rearrange in response to shear. Based on the boundary conditions, force chains develop in opposition to the motion. They found that by a differencing technique, the force chains can be interpreted better. The image differencing technique uses a pixelwise

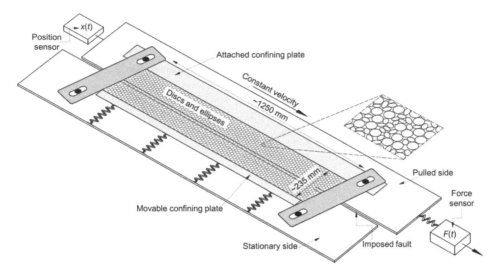

Figure 5.24. Isometric view of the particle layer consisting of about 10 000 particles made of PS-3 polymer (Micro-measurements group, USA) of 60% circular (5.6 mm diameter) and 40% elliptical (major and minor axes of 6.8 and 4.7 mm) shapes of 3 mm thickness with suitable shearing arrangement.

subtraction of the initial image from the final image for each event that helped to visually inspect the granular mechanisms associated with each event [89, 90].

Although quantitative evaluation of forces has been done by Drescher and Josselin de Jong in 1972 [84], automated and dramatic comparison of results was possible only with the work of Majmudar and Behringer in 2005 [91]. They devised a procedure for evaluating the contact forces, which are iteratively changed until the reconstructed fringe patterns closely match the experimental ones by comparison of 500 points in each particle (figure 5.25). To make the computations simple, they have used only circular cylinders of diameters 8 mm and 9 mm with a length of 6 mm in the ratio of 80% smaller and 20% larger cylinders to prevent crystallisation. The coefficient of friction for these cylinders is 0.8. In total, about 2500 to 3000 cylinders are used and the camera images about 1200 cylinders in the central zone. Except for the cylinders on the boundary, the contact forces inside have been calculated by an automatic procedure.

To load the system, the walls are moved by a stepper motor with a step of 0.0004 mm, which is approximately 0.005 times the diameter of the cylinders. The process of loading is quasi-static and after each step, tapping is applied to relax stress in the system and two images are captured: one with the polarizers to get the fringes and the other without the polarizers to be used for identifying the boundaries of the discs and their respective centres [91].

The walls are given uniform bi-axial compression (figure 5.25(a)) and their effect on the cylinders has been labelled as isotropic compression (figure 5.25(b)). Using their iterative procedure to calculate the forces (discussed subsequently) acting on each disc, they have reconstructed the fringe patterns (figure 5.25(c)) and the comparison is quite dramatic for a complex system such as this. When the walls

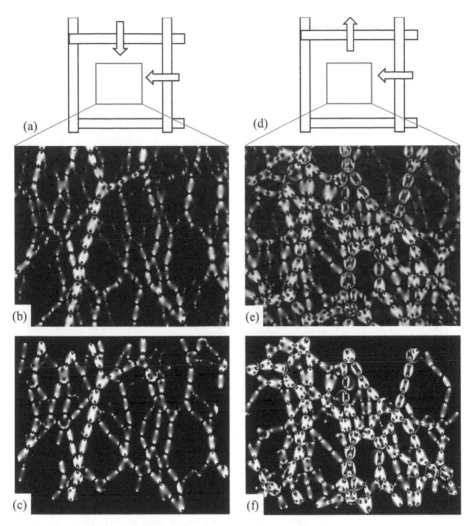

Figure 5.25. (a) Biaxial compression loading arrangement, (b) experimentally obtained fringe patterns of the discs in the central region, (c) theoretically reconstructed fringe patterns, (d) loading arrangement to generate shear, (e) experimentally obtained fringe patterns of the discs in the central region, (f) theoretically reconstructed fringe patterns. Courtesy: figures (b), (c), (e) and (f), [91]. Reprinted by permission from Springer Nature.

are moved as in figure 5.25(d), the cylinders are labelled to be subjected to shear and the experimental and theoretically reconstructed fringe patterns are shown in figure 5.25(e and f), respectively.

Later improvements by the students of Behringer have enabled the use of a single colour camera to record simultaneously, the position as well as the stress information by using two monochromatic light sources of red and green; the red image plane giving the positional information and the green channel providing the fringe patterns [92] (figure 5.26). It is interesting to note that the gradient information, which has been used to identify the zones of low spatial resolution in TFP (equation (4.23)) is

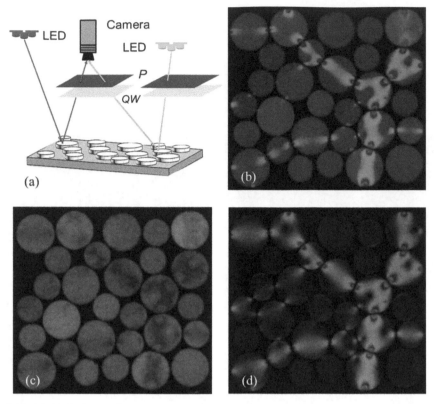

Figure 5.26. (a) Typical experimental arrangement of reflection polariscope with two monochromatic illuminations recorded by a colour camera, (b) image recorded by the colour camera, (c) red plane image is shown to identify the position of individual discs, (d) green plane image showing only the isochromatics. Figures (b)–(d) reprinted from [92] with the permission of AIP Publishing.

used in granular materials to evaluate the magnitude of the sum of forces acting on a particular disc [93]. In TFP, only gradients in the x and y directions are used for computation and here even the cross-diagonal terms are also used in computation. It is established that for a circular disc, the value of G^2 (equation (4.23) with additional cross diagonal terms) is found to be proportional to the sum of all the contact forces acting on the periphery of a disc [93].

The automated procedure for force determination has three steps: pre-processor, fringe-inverter, and a post-processor [92]. The pre-processor identifies the disc centres, identifies the neighbours based on proximity and validates which ones of these neighbours are in contact by using a threshold value of G^2 within small white and green circles of figure 5.27. The green circles are above the threshold and labelled as contacts and the white circles are below the threshold, which are labelled as neighbours. The value of G^2 obtained within the green circles is used as an initial guess in the fringe-inverter step that utilises the theory of elasticity solution to calculate the forces acting on each disc iteratively until the fringe information is satisfied in about 500 points within each disc. The choice of initial guess plays a

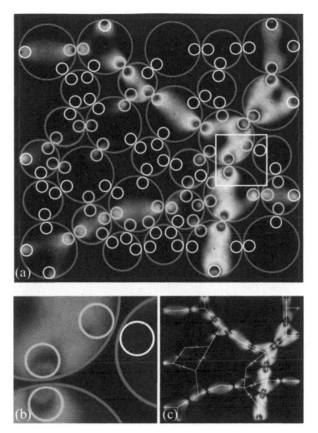

Figure 5.27. (a) Identification of big and small discs and their centres marked red. Small white and green circles indicate the possible contact points, (b) close-up view of the rectangle marked in yellow showing the classification of contact regions based on information analysed in the small circles. The green circles are above a threshold declared as contacts while white ones are below the threshold labelled as neighbours. (c) Shrunk image of the whole pattern showing active contacts represented as lines wherein both the sides are declared as contacts. Reprinted from [92] with the permission of AIP Publishing.

particularly important role in convergence. The post-processor plots the theoretically generated fringe patterns based on the converged iterative solution of the vectorial forces acting on the disc. The calculation of forces is checked for force and moment balance in static problems. To plot theoretically generated fringe patterns, one should have the necessary equations defining the stress field for an arbitrary set of self-equilibrated forces acting on the disc. Though there have been general discussions on this in the granular literature, and their software use this for fringe plotting, there has been no attempt to give a ready-to-use set of equations for this. This is presented in an elegant manner in the appendix.

Much software has been made available by researchers and a wiki has also been maintained by a group of scientists to help the research community in this area [94]. Ideally, a single run of the iterative process is sufficient to determine all the forces at contact points for all the particles [84]. If there is difficulty in convergence due to poor-quality initial guesses, there is a provision in the software to allow a sequential approach of scanning the packing in such a way that the information of the previously solved particle forces is used suitably as initial guess for the next particle. This process can be repeated multiple times until all contact forces with their magnitudes and directions are determined for each of the particles in the system [84]. The photoelastic grain solver (PEGS) [95] is quite flexible to analyse forces even in dynamic situations and has helped in the study of steady two-dimensional avalanches simulated by 2-D free-surface, gravity-driven, dry granular flows [96]. To record the rotational information of the discs, the use of drawing a line on each disc using a special pen, which makes the line visible in ultraviolet light has been proposed and used [85].

The relationship between inter-particle forces and the macroscopic behaviour of granular materials is an important area of research. Evaluation of collective statistical properties of inter-particle forces has been achieved through simple photoelastic analysis of 2-D systems. The forces above the average decay in an exponential manner, which indicates that the force is heterogeneously distributed, and the forces less than the average follow a power law distribution [91].

It is instructive to note that photoelastic technique has been very dominant in this area by noting the title of one of the recent research papers 'Extracting inter-particle forces in opaque granular material: beyond photoelasticity' [97]. Such studies have used just 34 discs [97] to make measurements using DIC! In comparison, even the very first study by Drescher and Josselin de Jong in 1972 used 1200 discs! In numerical analysis, after finite elements, discrete element method (DEM) came into picture and the current trend is towards the development of GEM (Granular element method). For all such numerical developments, photoelastic results can be quite crucial for comparisons and validations.

5.8.2 Photoelasticity applied to masonry

A large number of monuments such as temples in India (figure 5.28(a)), temples in Cambodia, City of Great Zimbabwe in Africa, Roman, Greek, and Medieval

Figure 5.28. (a) Dry stack stone constructions in India, Sri Kedarnath Temple, Uttarakhand, India (possibly 12th c. AD), (b) localised cracks in dry stack construction marked by red rectangles. Courtesy: Professor Arun Menon, IIT Madras.

structures in Europe, and many more structures in Asia, America and Australia built across different eras used different types of stones for construction [98]. Natural building stones were used to construct most of these historical monuments for reasons such as local availability of material, strength, and aesthetics, but primarily durability. Masonry in many of these structures is composed of stones interlocked with each other without any binding mortar such as mud or lime mortars. Even constructions originally built with mortar joints have suffered a significant or complete loss of bedding mortar due to chemical, environmental, and mechanical degradation over time. The load transfer behaviour of dry-stone masonry is different from that of the masonry with mortar, as the lateral resistance is solely due to the frictional resistance of these joints.

In dry stack construction, localized failure of stone blocks (figure 5.28(b)) is often observed in some of the stone units. The localized failure of the stone block cannot be explained from compression behaviour, which is true only in an average global sense, but not in a local sense. An in-depth understanding of the mechanical behaviour of these structures is essential to develop appropriate structural assessment criteria. Many of these structures are found in moderate-to-high intensity earthquake-prone areas; therefore, assessment of the stability and deformation conditions of the structure is required for a better insight on the conservation of historical structures. Since the possibility of carrying out destructive tests on historical constructions, either *in situ* or by extracting samples large enough to be representative, is generally not feasible [99], experimental research on masonry structures is largely based on laboratory tests. It is surprising to note that in view of the overly complex nature of the experiments, the first known credible experiment using photoelasticity was done only in 2010.

Bigoni and Noselli [100, 101] in 2010 used transmission photoelasticity for understanding the load path behaviour of brick masonry and noted that nominally identical masonry structures can be subjected to different stress states under the same loads, due to randomness of contact, causing arbitrary distribution of highly localised stress percolation. They also developed two ideas on modelling the force percolation: one using micromechanics and the other using orthotropy of the masonry. Baig *et al* [102] have employed digital photoelastic analysis of three fringe photoelasticity on a model of dry masonry wall of size ($180 \times 220 \times 6$ mm^3) with bricks ($20 \times 10 \times 6$ mm^3 and $30 \times 10 \times 6$ mm^3) made of epoxy, which are cut from epoxy sheets of thickness 6 mm. The longer bricks are placed at the ends of every alternate layer to ensure vertical stagger of masonry joints in the wall. The model is subjected to a load of 197 N at the top [102]. Figure 5.29(a) shows the nature of fringe patterns simulated using the generalised solution for flat punch contact reported in reference [103], if the wall is made of a single homogeneous sheet and figure 5.29(b) shows the experimental fringe patterns on the discretised wall made of photoelastic bricks. The domain is divided into three zones and images are recorded separately using a colour 3CCD camera (Sony XC-003P) having a resolution of 752×576 pixels and these are stitched appropriately to form a single image [102].

The isochromatics in figure 5.29(b) clearly bring out that the stress percolation shows a random pattern and one can identify the force chains that are formed (shown only for some segments to have appreciation of the isochromatics as the lines

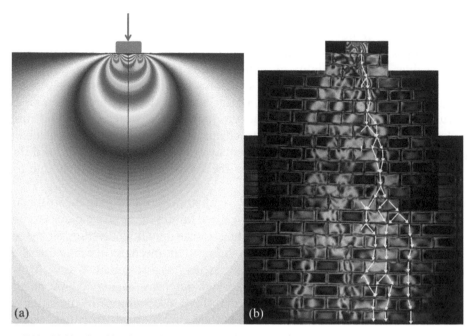

Figure 5.29. (a) Simulated isochromatics at a compressive load of 197 N on a wall of homogeneous plate (180 × 220 × 6 mm³) indicating continuous and smooth flow of force, (b) isochromatics experimentally recorded at a compressive load of 197 N on a model of dry stack masonry using epoxy blocks (20 × 10 × 6 mm³) indicating the influence of heterogeneity of the structure. White lines indicating force chain are marked on a segment of the wall.

hide the patterns). Though the discrete nature of the wall is captured, the blocks are identical as they are machine cut without any undulations as otherwise seen in natural stones and in that sense the model is incomplete. However, certain ideas related to influence of heterogeneity, influence of openings in such structures can be studied by using transmission photoelasticity.

The next step is to go for model studies involving stones using reflection photoelasticity. As stone is a heterogeneous material by nature, the presence of stones in irregular geometry with high variability in mechanical properties and natural planes of weakness results in a complex behaviour in structures built with stone. Due to this heterogeneous and composite behaviour of stone masonry, there are complex links and interaction between the stone components at their joints, which tend to disobey the generally expected behaviour as observed in a relatively homogenous material.

In 2017, Colla and Gebrielle [104] have done a preliminary study on the application of reflection photoelastic analysis on clay brick masonry column under eccentric loading but could not record meaningful fringe patterns. Pankaj *et al* [98] in 2020 used natural granite stones procured from Mahabalipuram, Tamil Nadu, India to match the stones (coarse-grained igneous rock—one of the three main rock types formed through the cooling and solidification of magma or lava) used in historical structures in India. Stone samples are prepared and tested for strength as per the procedure mentioned in IS code 1121. One of the key challenges is, how to apply the photoelastic coating to sample walls made of these stones. After a few trials, two sets of walls are prepared. The first set is made using granite

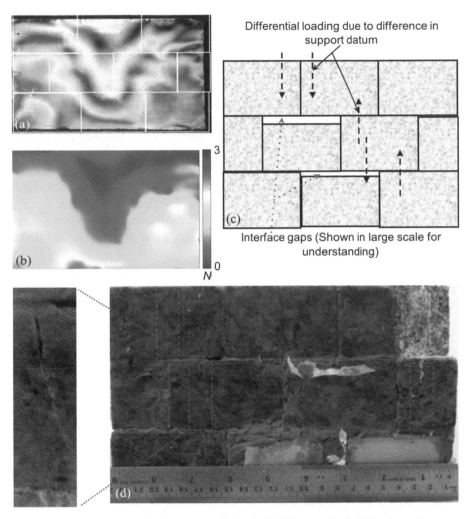

Figure 5.30. (a) Isochromatics recorded at a compressive load of 10 kN on a wall mimicking dry stack masonry, (b) fringe orders away from the edges extracted by TFP coupled with FRSTFP scanning, (c) schematic of the load transfer indicating the shear transfer mechanism, (d) rear view of the wall with the red rectangles indicating the cracks after the wall failed at a load of 120 kN. Adapted from [98] with permission from John Wiley and Sons.

stones of size $50 \times 40 \times 30$ mm^3 (a scaled model of 1:20 of the original structure), where the epoxy used for bonding the coating has seeped into the gaps between the stones. The thickness of the infiltration is not uniform across the interface gaps between the stones used for the wall and hence its behaviour lies between a dry-stack masonry in some regions and masonry with confinement offered by the adhesives in other regions. To have better control, another reduced sized wall with just three layers of stones of size $70 \times 50 \times 40$ mm^3 is made wherein the seepage is largely prevented by very thin threads of moulding clay filled in the interfaces. Thus, wall-2 prepared is closely modelling a dry masonry stack.

The isochromatics recorded for wall-2 under uniaxial compression of 10 kN are shown in figure 5.30(a) and the fringes are processed using *Digi*TFP® software [105] that employs TFP coupled with FRSTFP scanning (figure 5.30(b)) (section 4.9). As the stones are not bonded, they freely underwent differential movement and uneven and irregular gaps between stone units tend to close-in (figure 5.30(c)) with the increase of compressive load and the wall failed at 120 kN. Rear view of wall-2 after the failure is shown in figure 5.30(d). Inspection of the stone units in wall-2 showed that stone units marked in red rectangles split by the formation of vertical or sub-vertical cracks, which is exactly like the ones seen in the archaeological sites (figure 5.28(b)). There has been a tendency to undergo concentrated vertical shearing of stone units due to the differential loading caused by the differences in interface gaps of the supporting stone blocks (figure 5.30(c)). The significant variations in the stress levels within a dry stack stone masonry wall, with clearly identifiable high and low levels of stress zones have been brought out by the isochromatic fringe field recorded. Propagation of failure is dictated by the contact mechanics, which is governed by the non-uniformity of block geometry even in very regular dry stack masonry.

Numerical modelling of dry-stone masonry is a cumbersome task. Interface conditions across joints in dry-stack stone masonry are variable, and hence finite element analysis will not be appropriate or realistic unless the correct boundary conditions are given. Modelling proper boundary conditions that are closer to reality is not straightforward. Results of photoelastic analysis can contribute to improved numerical modelling of dry-stone masonry.

5.9 Photoelasticity for food security

Food security will be the focus of emerging economies. When roots grow into physically inhospitable soil, they send inhibitory signals to leaves, which inhibit growth of above ground biomass. This can happen despite the availability of water or mineral elements, thus, affecting the agricultural production. The changes in agricultural practices have brought in a new problem of soil compaction with constantly increasing size and weight of tractors that contribute towards impeding the growth of roots.

There is constant coupling between root development and soil reorganization. The plant roots exhibit gravitropism (roots grow in the direction of gravitational pull) and reorient if the root tip is placed in a direction that is not initially aligned with gravity. The classical view has been that the photosynthetic chemical energy accumulated within the plant tissues is converted into turgor pressure (it is caused by osmotic flow of water through a selectively permeable membrane to maintain the rigidity and sturdiness of plant cells), which then overcome the resistance to the displacement of the soil around the root [106]. This view has been based on the fact of considering the soil as a continuous and homogeneous medium, whereas in reality, it is granular. The level of mechanical stress at which growth is arrested is estimated to be above 5 MPa, whereas the common value of turgor pressure measured is below 1 MPa [106]. Thus, a relook at the biophysics involved is needed.

In compacted soils, roots grow preferentially within macropores (large soil pores, usually between aggregates, that are generally greater than 0.08 mm in diameter), cracks, biopores (continuous pores formed by plant roots and burrowing soil animals such as earthworms) etc, thereby sustaining the plant development [107]. Further, soil being granular, the force distribution is dictated by force chains that are inhomogeneous. It may be possible to create better soil processing procedures for agriculture to improve the health of the plants and ensure superior crop yields if one understands how the roots respond to the inhospitable conditions in which they grow. There is also interest to develop root inspired robotic devices (called plantoids with sensorized robotic roots) to penetrate granular media [108] that use less energy to penetrate, which may have applications in disaster areas covered in rubble, areas prone to avalanches etc.

Inspired by the work of Behringer on using photoelastic discs simulating grains, Kolb *et al* [107] in 2009 developed a simple device to measure the root stresses allowing the root to pass through a narrow gap of soil model made of cylinders of polyurethane (PSM4, Micro measurements, USA) of diameter 9 mm and thickness 7 mm. They selected the chickpea as a test plant, as it is the third most important pulse crop, which in unstressed conditions has a root diameter of approximately 1 mm that facilitates easy design of the test condition to maintain a gap of 0.5 to 1.5 mm between grains to study the root stresses. Further, its root network is not too dense, and the seed is large enough to feed the plant during the first cycle of growth. Two snapshots of the isochromatics developed over a period of 10 h are shown in figure 5.31(a) and (b) and figure 5.31(c) and (d) show the force developed and the growth of root diameter.

A more exhaustive study on the influence of granular nature of soil on the plant growth to determine the response of plant roots to different levels of force between grains and how the growing roots alter the force distribution in a granular system has been done by Wendell *et al* in 2012 [109]. They selected pinto beans as they have large roots, which are easy to grow and germinate in a variety of environments and grow in well-drained soil that has negligible soil cohesion stresses. This facilitates the simulation in the experiment to be closer to their natural environment with large grains. The apparatus is designed for long term behaviour of plants, wherein the grains are in random configuration and allowed system-level arrangements during root growth. They used polyurethane discs of 9.5 mm diameter cut by waterjet machining that are packed in an acrylic box of 100 mm^2. They have used seven boxes, of which six are with pinto seeds and the seventh one is a box that contained only photoelastic grains so that the forces in the grains remain static not influenced by root growth. The experiment is carried out for a period of 10 to 18 days. The roots exerting force on the grains is captured. Figure 5.32(a) and (b) shows the overall view and close-up view of the central grain stressed by the roots. Their study showed that the pinto bean roots tend to grow between grains when the force level is less than 0.5 N. The average force exerted by the roots on the grains is about 110 mN and they vary as a function of time.

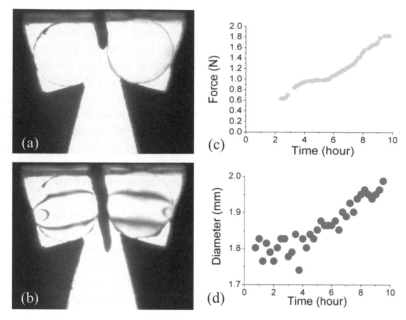

Figure 5.31. (a) Chickpea root just entering the gap, (b) after 10.25 h of growth, (c) development of force as a function of time, (d) development of root diameter as a function of time. Courtesy: [107].

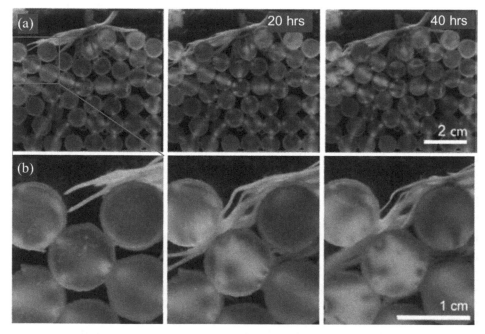

Figure 5.32. (a) Overall view of the roots interacting with the granular system taken at intervals of 20 h. (b) Close-up view of the rectangular portion showing the central grain experiencing force from the roots. Courtesy: [109]. Reprinted by permission from Springer Nature.

It is interesting to note that the experimental setup of plant root studies is fully automated with periodic photography of the roots every 70 min. The overall experimental system is well presented in figure 5.33 reported by Barés *et al* [110]. They have made further improvements on the shape of the grains used to facilitate root growth in between them by providing a cylindrical groove on the thickness (4 mm) of the grain with a depth of 0.5 mm and a width of 1 mm. In addition, they also attempted to study the influence of applying a shear on the granules to see the root response. In all, they have used 11 test cells, of which some are filled with bidisperse glass beads of 2 and 4 mm diameters and some with 7 and 9 mm photoelastic discs made of polyurethane and chickpea is germinated. They observed that the root system architecture (RSA) took more than three weeks to develop in photoelastic granular cell compared to glass bead cell that took just a

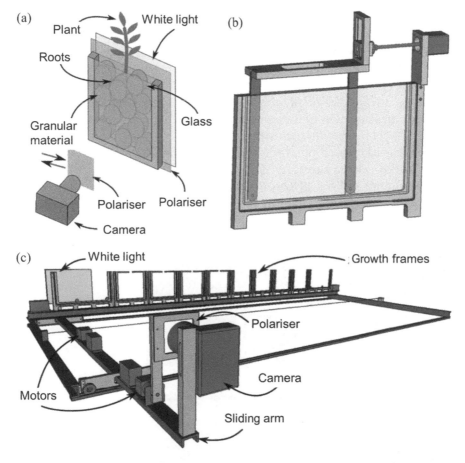

Figure 5.33. (a) Sample growth frame with photoelastic discs of thickness 4 mm with diameters of 7 mm and 9 mm each with a central groove of 0.5 mm depth by 1 mm width over its thickness to facilitate root growth in between. (b) Two of the cells are provided with a shearing arrangement to shear the granular medium by a stepper motor when the growth of roots has ended. (c) Complete setup with 11 growth cells to be photographed at periodic intervals to monitor growth of plants. Courtesy: [110].

week. Further, the tortuosity (degree of twists and turns) in grooved photoelastic cell is much less than in glass beads cell. They concluded that root–grain interaction is a dynamical process that needs to be followed over the long term. Their shear experiments revealed that RSA stabilizes the granular system. It has been a successful model experiment that has the potential to delve deep into the complex interaction between root growth and surrounding soil mechanical evolution.

5.10 Photoelasticity applied to neurobiology

Many organisms have adapted to live in their environments. For these organisms to survive in nature, an efficient strategy in locomotory behaviour has presumably developed. Locomotion results from complex, nonlinear, dynamically coupled interaction between an organism and its environment. Integrating studies of locomotor mechanics, neural control and isolated muscle function remains a challenge in the field of neurobiology. Development of models of how legs, joints, multiple muscles, and neural networks work together to produce locomotion requires calculation of forces developed by the legs during an event.

One of the advantages of photoelasticity is its whole field approach to enable simultaneous measurement of forces in multiple legs of an organism such as a cricket (figure 5.34(a)) [111]. The photoelastic setup consists of a thin slab (2 mm) of gelatin placed between the crossed polariser and analyser (figure 5.34(b)) and the organism is placed on the gelatin substrate. Gelatin substrate is prepared with 8% gelatin (weight to volume) and 5% glycerol (volume to volume) to maintain hydration of gelatin as it is prone to fluctuations in hydration, which leads to residual stresses affecting the photoelastic data interpretation. Further details on the preparation of

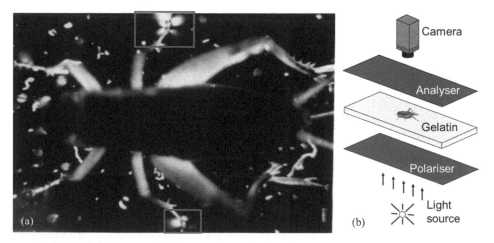

Figure 5.34. (a) Isoclinics revealed when a cricket adopts a posture prior to jumping on a gelatin substrate. The dirt on the substrate is deposited by the insect. The size and the skew of the isoclinics are calibrated to provide the force. The force and the stress developed by the leg are estimated to be 3.57 mN and 8.93 kPa. Reprinted from [111] with permission from John Wiley and Sons. (b) Schematic of the experimental setup.

gelatin substrate can be found in reference [111]. The size and skew of the isoclinics developed is empirically interpreted to the magnitude and direction of the force [111, 112].

As photoelasticity measures stress, the techniques for measuring force range are theoretically infinite as exceedingly small forces acting on small areas can generate sufficient stress for data reduction. The sensitivity of the system is dictated by the optics and although Harris [111] has claimed that his system is capable of measuring force as low as 98 nN, other researchers have generally reported the sensitivity of the system to be of the order of milli Newtons only [112, 113]. This technique has been used to find the force developed by a cockroach during the righting behaviour, which was eight times greater than those observed during high-speed running [112]. The use of a similar approach for the study of slithering motion is discussed next.

5.10.1 Slithering locomotion

Limbless locomotion on land necessarily relies on sliding and understanding it as a form of locomotion presents new challenges. One observes in Nature that snakes are long and flexible, which enables them to enter crevices of dimensions much smaller than their body lengths and they slither through mechanically complex environments such as sand, soil, or grass. Understanding and modelling such locomotion can inspire the design of limbless robots that can move through complex terrain, especially in confined areas.

According to the pattern of placements of their limbless bodies on the ground, one can identify four principal 'gaits' for snakes [114]. Most common of these is slithering known as lateral undulation in which the body propagates a 2-D travelling wave from head to tail. A simplistic understanding of 2-D slithering of snakes on flat surfaces (essentially above ground) has been modelled by Hu and Shelly [114] using resistive force technique that has been previously developed to model the locomotion of small organisms in fluids. The greatest challenge in the study lies in modelling the environments that are quite complex.

Unlike the earlier understanding that the locomotion of snakes is based on *push points* [115], Hu and Shelly have shown that snake motion is highly dependent on frictional anisotropy that is facilitated due to the nature of construction of their belly scales that are arranged like the overlapping shingles on the roof of a house (figure 5.35(a)). Hu and Shelly [114] have experimentally measured the directional dependence of frictional coefficient by putting the snakes to sleep for a few minutes using anaesthetic gas and straightening them out on various configurations on an inclined plane. Measurement of a milk snake's [114] (a class of non-venomous snakes with vibrant colour, docile and friendly creatures) frictional coefficient on cloth indicated that the frictional coefficient is the least for moving forward (0.1) compared to moving backward (0.2) and intermediate for sideward (0.14).

One of the verifications of their mathematical model was based on the prediction of the speed of the snake's centre of mass. In their initial model, they assumed that the weight of the snake is uniformly applied to the ground along its length. Although

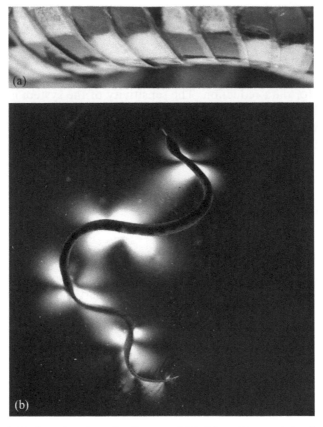

Figure 5.35. (a) Typical belly scales of a snake. Courtesy: [114]. Adapted by permission from Springer Nature. (b) Snake photographed in a polariscope on a sheet of gelatin—the luminescent areas show the contact of snake to the gelatin sheet. Credit: © Grace Pryor, Mike Shelley and David Hu, Applied Mathematics Laboratory, New York University, and Department of Mechanical Engineering, Georgia Institute of Technology.

the frictional anisotropy and the parameters of undulation kinematics such as the wavelength, amplitude and frequency are reasonably modelled, their speed prediction was just half of the actual speed observed in snakes! This only indicated that some key aspect of the snake's behaviour has not been modelled correctly.

This mystery has been solved by a phenomenological study employing photoelasticity. The snake is photographed in a polariscope while slithering on a sheet of gelatin (figure 5.35(b)). Although, the adhesiveness of gelatin can interfere with the snake's locomotion, the study brought out an important fact that the entire length of the snake's body is not in contact with the ground! There is perceptible dynamic distribution of body weight while slithering. Once this aspect was modelled, the study predicted an increase in the speed by 35% and increase in efficiency by 50%. Further understanding of this dynamic distribution of weight can explain how speedier snakes like black mamba can move quickly at speeds of about 5 m s^{-1}.

Although in other gaits, earlier researchers have noticed body lifting, in slithering it is quite subtle and photoelasticity has helped to model this effect emphatically.

The body of the snake that touches the ground (gelatin here) and its neighbourhood appears brighter due to stress induced birefringence on gelatin (figure 5.35(b)).

5.10.2 Burrowing

The locomotion studies now focus on understanding the mechanisms of burrowing. The burrowing animals contribute to bioturbation, which is the churning or reworking of sediments by living organisms. It may enhance or decrease the porosity and permeability of the sediments. Burrowing animals can have great environmental and ecological consequences. Sediments exhibit a dramatic variety of material properties, ranging from granular materials such as clean, monodisperse sands to linear elastic materials such as muddy sediments. Differences in material properties result in differences in burrowing behaviour in different sediments.

Bioturbation models have generally ignored the elastic behaviour of soils. The first breakthrough in understanding burrowing came from extension of the idea of models on bubble growth in double-strength gelatin (2× gelatin) blocks [116, 117]. Double-strength gelatin in seawater has similar ratio of fracture toughness to Young's modulus as elastic cohesive sediments [116]. This approximately has just 2.85% of gelatin so that the mud is soft enough for burrowing in comparison to a stiffer sheet (8%) used for measuring forces by organisms. As photoelasticity has always been in the forefront of phenomenological understanding of complex phenomenon, it has come to the rescue of the researchers to observe and establish that burrow extension can happen due to crack propagation! [118]. This is observed for *Alitta virens*, formerly *Nereis virens* [119] (WoRMS, 2020), which is an annelid worm (ringed worms or segmented worms) that burrows in wet sand and mud. Sandworms make up for a large part of the live sea-bait industry and are grown commercially. One difference between muddy sediments and gelatin is that gelatin displays a higher restoring force, hence it can prevent free movement of the worm. Compared to other worms, *Alitta virens* could burrow normally in gelatin [120].

One finds very sophisticated and elaborate photoelastic experimental preparation for capturing the burrowing phenomenon with live samples by introducing a crack using forceps and a worm placed within (figure 5.36) [120]. Experiments are conducted in a cold room at 11 °C that is within the living habitat of the worm. The experiments are tricky as the worm should continue to burrow in gelatin and worms sometimes stopped moving but would often continue again after the tail was gently touched. In some instances, the worms just do not burrow and these need to be carefully replaced to do the experiment! Also, a deviation of about 30° from the original axis is set as a limit for interpreting the behaviour.

The worm burrows in 3-D, but data analysis is generally restricted to 2-D. Quantification of the force generated is established empirically by measuring the size of the isochromatics selectively [120] instead of the isoclinics as proposed by Harris [111]. The force required to extend the crack is estimated to be of the order of 0.023 N that is less than 10% of the force required by a worm against the aquarium wall [118]. Based on the experimental results, the important finding

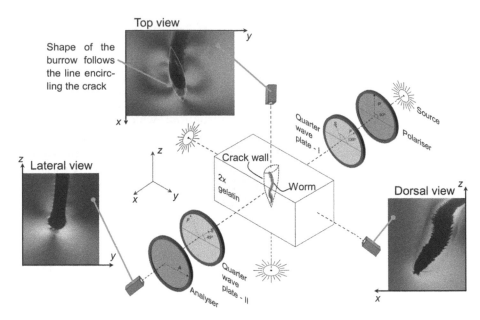

Figure 5.36. Schematic of the experimental setup to record burrowing kinematics and forces of *Alitta virens*, formerly *Nereis virens* worm. The fringe patterns correspond to using white light for illumination. The optical elements for recording the top view and dorsal view are not shown in the picture for clarity. As the burrow is made by the worm it is not straight at higher magnifications. Courtesy: Coloured isochromatics from [118]. Adapted by permission from Springer Nature.

is that the crack extension can be modelled as a wedge-driven stable crack growth.

Further studies with five different gels with variation of the ratio between fracture toughness to Young's modulus in the range 0.005 to 0.0312 (for 2× gelatin, it is 0.0092) have shown [121] that the burrowing mechanics depends on the relative stiffness and fracture toughness of the mud. In a material with high fracture toughness, *Alitta virens* everted its pharynx (protruded a muscular food pump in the head) to propagate the crack. In stiff materials with low fracture toughness, worms moved their head side-to-side to extend the crack [121]. A typical worm moves forward extending the crack then begins pharynx eversion and between eversions, moves its head from side-to-side within the plane of the crack thus extending the crack laterally [122]. The use of transparent birefringent gelatin has helped in simultaneous observation of burrowing kinematics and quantification of forces and stresses through photoelastic fringes during burrowing [122].

The hypothesis of burrowing by crack extension is seen in *Cirriformia moorei* (a species of segmented worms found in marine habitat) with distinct morphology (lacks an eversible pharynx but uses its hydrostatic skeleton to expand its anterior to apply stress on the burrow wall thereby precipitating fracture) and *Alitta virens* suggesting that this mechanism may be widespread among burrowers [123]. Thus, photoelasticity has helped in understanding certain aspects of burrowing mechanics emphatically.

5.10.3 Burrowing in granular media

To improve the sensitivity of the methodology and take it to simulate granular materials, Ceniceros [113] has proposed the use of gelatin (8% mass fraction of gelatin with no glycerol) micro-spheres of the size of 100 μm to measure the forces generated by *Caenorhabditis elegans* (*C. elegans* is a free-living transparent nematode), which is a worm about 1 mm in length and 100 μm in diameter [113] that lives in temperate soil environments. Many of the genes in the *C. elegans* genome have functional counterparts in humans which makes it an extremely useful model for human diseases. *C. elegans* mutants provide models for many human diseases including neurological disorders, congenital heart disease and kidney disease. By conducting experiments in pure fluid and in a bed of glass beads, it has been shown that the swimming motion of this tiny worm is more efficient in wet granular media than it is in a pure fluid [124, 125]. Although, swimming efficiencies in these environments have been quantified, the forces generated by *C. elegans* have not been measured, which are necessary to determine the precise mechanisms used by the worm to increase its efficiency.

To maintain repeatability, it is found that 12% gelatin spheres are more consistent in maintaining the properties and are immersed in 0.33 M NaCl solution [126] for storage and experimentation to take care of the hydration of gelatin. The use of gelatin spheres is demonstrated for measuring forces generated by an earthworm by using spheres of 3 mm diameter and the practically achieved sensitivity is 1 μN with a precision of 60 μN [126]. The challenge lies in more consistent fabrication of gelatin spheres to really measure the forces generated by *C. elegans*, which the future research must focus on.

5.11 Photoelasticity in developing biomaterials

Biomechanics offers the maximum challenge in terms of materials and their behaviour. They are mostly nonlinear, viscoelastic and many times the material systems are functionally graded. Soft tissues are highly tolerant to defects as they could withstand cracks up to several millimetres without losing much strength. One requires suitable materials for repair as well as surrogate materials to perform model studies to understand complex phenomena in biomechanics. In this section, some of the recent developments on a functionally graded tendon for repair in shoulder joint and a suitable surrogate material to understand the mechanics involved in needle insertion are brought out.

Rotator cuff is a common name for the group of distinct muscles and their tendons that surround the shoulder joint, keeping the head of one's upper arm bone firmly within the shallow socket of the shoulder. Injuries to rotator cuff are common and increase with age. Development of a novel biomaterial that has tissue-like mechanical properties is vital for restoring the biomechanical function of the joint. The cortical bone and the fibrocartilaginous translational regions experience both tensile and compressive stresses. The most common tendon that tears off in rotator cuff injuries has tensile strength ranging from 4 to 22 MPa and the tensile Young's modulus ranging from 0.2 to 0.6 GPa. The cortical bone has a tensile strength ranging from 66 to 170 MPa and compressive strength from 167 to 213 MPa. It has tensile Young's modulus ranging from 11 to 29 GPa and compressive modulus ranging from 14.7 to 34.3 GPa [127].

The functional attachment of bone to tendon requires tissue architectures that mitigate increased stress concentrations due to their mechanical mismatch. In healthy, native attachment sites, this is accomplished, in part, by gradations in mineral content and collagen fibre orientation along the bone–tendon interface to create a resilient and functionally graded interface. Clinically available grafts are not mechanically tuneable. Even though they have comparable tensile strength in the range of 11.9–32.7 MPa, the Young's modulus is quite low of the order of 14–71 MPa. The challenge is in the development of biomaterials that can better approximate the stiffness gradient (variation of Young's modulus) of the natural bone–tendon tissue properties.

Ker *et al* [127] developed a novel phototunable polyurethane polymer that mimics closely the natural bone- and tendon-like mechanical properties. This is achieved by a combination of chemical crosslinking, photo-crosslinking, and heat-curing. The polymer thus developed is labelled as QHM consisting of three monomers, where Q stands for quadrol, H stands for hexamethylene diisocyanate and M for methacrylic anhydride. The phototunable capability arises from vinyl methine groups of M and the compressive strength (58–121 MPa) and the Young's modulus (1.5–3 GPa) increased with longer exposure to UV light. By selecting a desired ratio of Q, H, and M monomers as well as varying UV exposure, the desired properties of the final polymer are obtained. As polyurethane is birefringent, it is easier to demonstrate the reduction in stress concentration by gradation in Young's modulus by using the UV exposure appropriately. Figure 5.37 shows the interface formed by two different Young's moduli of polyurethane by adjusting the UV exposure [127]. The colour change at the interface is smooth when there is slow gradation of Young's moduli and this has also been numerically verified by a 3-D FE modelling of the interface. The material stress fringe value also changes as a function of exposure and the colours must be interpreted based on that. Photoelasticity is directly applicable on the prototype of the tendon as the material is birefringent.

Development of flexible birefringent surrogate tissue is another emerging area in understanding the biomechanics of needle insertion for biopsy extraction or deep penetration injections such as epidural injections [128–130]. Mechanical responses of soft tissues are highly variable, nonlinear, display time and strain rate dependent and may show resistance to tearing. The mechanical properties also differ for types of tissues and the tangent shearing modulus ranges from 8 to 340 kPa for lung, stomach, heart, and liver tissues with the lung being the softest and the liver being the stiffest [131].

Experiments have been conducted on various types of surrogate materials with main ingredients as gelatin, konjac and agar. Konjac gel is derived from konjac plant, which has a fibrous microstructure that improves its fracture toughness and has shown greater resistance to tearing [128]. On the other hand, the gels of gelatin

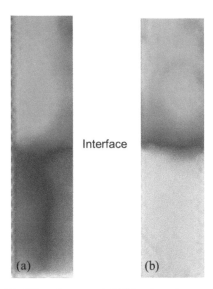

Figure 5.37. Interface formed with differently exposed QHM polyurethane polymer with the top half exposed to 90 s of UV light (a) gradually graded with bottom half exposed to 120 s showing less stress concentration, (b) steeply graded with bottom half exposed to 300 s showing more stress levels. Adapted from [127] with permission from John Wiley and Sons.

and agar do not show any resistance to tearing. The birefringence of konjac is considerably poor in comparison to gelatin gel. Although konjac has shown a strongly nonlinear and hysteretic response like human tissue, for simplistic first level analysis, various combinations of gelatin (weakly nonlinear) with other additives are tried to mimic the tissue properties. Two combinations of gelatin:water:glycerine in the ratios 2:10:10 ($F_\sigma = 0.082$ N/mm/fringe, $E_C = 47.5$ kPa) and 3:10:10 ($F_\sigma = 0.085$ N/mm/fringe, $E_C = 75.5$ kPa) are found to mimic stomach/lung and heart/liver tissues respectively [128]. The properties are sensitive to the source of the ingredients and hence need to be established based on specific tests.

When one normally looks at model studies, usually the models are much smaller than the prototypes for convenience. In needle insertion studies, it is the other way! A scaled up model is used for clarity of fringe evolution and the insertion of a 25-deg bevel tip needle (Westcott) [128] of 5 mm diameter steel rod into a 3:10:10 gelatin mixture (approximately 15% of gelatin by weight) is recorded carefully in a bright field polariscope. When the needle touches the skin, it initially deflects the skin then it punctures and then the needle starts moving in by cutting the tissue during which there is resistance from the tissue as well as frictional resistance due to interaction of the needle shaft with the tissue. Every 0.5 mm insertion of the needle is recorded and at 3 mm, the force drops abruptly indicating the puncture and then the cutting of the tissue takes place that requires lesser force. This force increases for subsequent steps due to higher friction as more of the needle shaft enters the specimen (figure 5.38). With advancements in digital photoelasticity, further studies on using konjac have been initiated [128, 132] as data could be extracted even with very feeble retardation and the future research can be more realistic in modelling tissue mechanics.

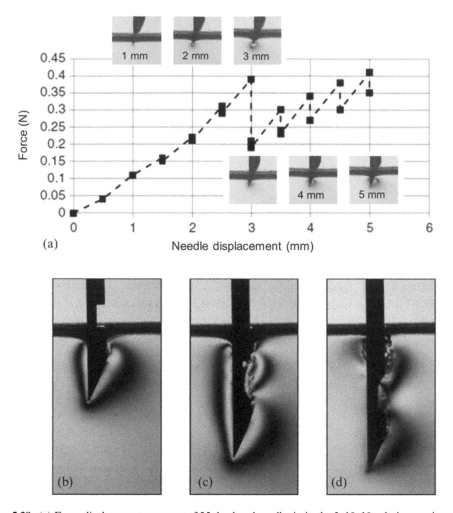

Figure 5.38. (a) Force-displacement response of 25-deg bevel needle tip in the 3:10:10 gelatine specimen and the corresponding bright field isochromatics: before (upper row) and after (lower row) penetration. (b) and (c) Isochromatics observed while the needle is being inserted, (d) relaxation of isochromatics as the needle is held stationary. Adapted from [128] with permission from SPIE. Coloured insets from [129]. Adapted by permission from Springer Nature.

5.12 Applications of Infrared Photoelasticity

Silicon is birefringent [133] and transparent to wavelength greater than 1150 nm and offers scope for *in situ* investigation by photoelasticity in the IR regime [134]. The birefringence developed is too small due to the high material stress fringe value of silicon and use of higher wavelength for visualization. Initial IR polariscopes were capable of only point-by-point analysis. With phase shifting techniques, it is possible to get whole field information using four [135, 136], six [137–140] and ten-step [141] phase shifting techniques (section 3.3). Stress analysis using IR polariscopes is quite involved as the material stress fringe value is anisotropic [142, 143]. Use of IR

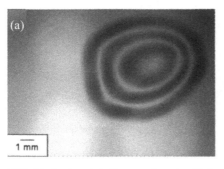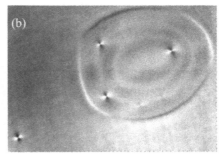

Figure 5.39. Fusion bonded silicon wafer pair with bonding defects: (a) IR image showing a Newton's ring (b) IR-GFP (grey field polariscope) image showing multiple defects in a single Newton's ring. Courtesy: [144]. Reprinted by permission from Springer Nature.

photoelasticity for defect detection and stress field evaluation in multi stack IC chips is discussed in this section.

The need for larger silicon wafers, thinner films and smaller semiconductor packages places growing demands on process control. The major concerns in process control are the residual stresses caused by the temperature gradient in crystal growth, precipitates and other defects, cracks due to mechanical processing, and thermo-mechanical loads during their operation. Defect detection is one of the earliest applications of IR photoelasticity and has several advantages over the direct use of IR cameras for imaging. In IR imaging, Newton's rings (figure 5.39(a)) are used to detect defects. Unlike the visible cameras, where the pixel size can be as small as 4 μm, in thermal cameras the pixel size ranges from 20 to 30 μm and hence the defect detection is limited by this. In IR photoelasticity, a defect or trapped particles can be distinguished by the shear stress field generated around them that appears like a 'butterfly' pattern in transmission (figure 5.39(b)) as well as in reflection modes [144, 145]. A single Newton's ring may be caused by more than one defect (figure 5.39(b)) [144], but a single butterfly pattern represents only one defect [146].

Integrated circuits (ICs) can pack a great deal of functionality into a small footprint by stacking silicon wafers and interconnecting them vertically by means of through silicon vias (TSVs). It also leads to faster operation by drastically shortening the critical electrical paths through the device. The CMOS (complementary metal oxide semiconductor) image sensors are the first applications to adopt TSVs, and they are also extensively used for various types of compact memories such as DRAM (dynamic random-access memory). The TSVs are usually filled with copper by electroplating and serve as signal paths between different IC layers.

Figure 5.40(a) shows a cost effective TSV integration process in which the redistribution layer (RDL) is fabricated before TSV filling that helps to form the front-side micro bumps along with TSV electro-chemical deposition [147]. After TSV etching and filling (figure 5.40), TSVs will undergo thermal cycling at 400 °C with metallization and/or at 200 °C with the bonding process. The coefficient of thermal expansion (CTE) of copper (17.5×10^{-6} °C^{-1}) is a few times higher than

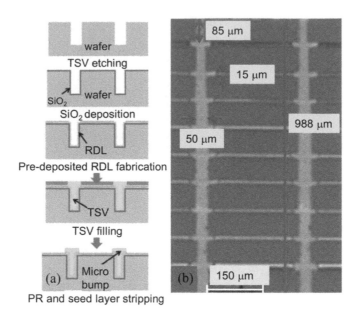

Figure 5.40. (a) Typical steps in the formation of TSV, (b) details of geometric dimensions of a ten-layer stack with TSV. © 2016 IEEE. Adapted, with permission, from [147].

that of silicon (2.5×10^{-6} °C^{-1}) [148]. The Cu area (inside the TSV) has two stress-generating mechanisms: extrinsic thermal stress due to CTE mismatch and intrinsic stress due to rapid grain growth during the thermally loaded processes. The CTE mismatch causes failure modes such as TSV extrusion or pop-up, debonding between the bump-attached chips, and bump-crack or delamination. Rapid grain growth causes void formation and crack propagation due to void growth and coalescence. This presents a significant reliability issue on the functioning of TSVs, which are subjected not only to continuous thermal cycling but also to large electric current densities during service.

A single TSV creates radial mechanical tension and tangential compression in the surrounding Si [149]. Understanding how an array of TSVs will induce the stress pattern is needed to estimate the size of the keep away zone (KAZ) where the integrated circuit cannot be fabricated [140]. To design systems involving TSVs, thermo-mechanical modelling using numerical methods like FE is performed to understand how stress distribution in the structure influences defect formation and growth. Experimental verification of the thermo-mechanical modelling is desirable. For example, micro-Raman spectroscopy [150] has been widely used for stress assessment of chip around TSV, synchrotron micro-beam X-ray diffraction has been applied to measure the stress of the copper filler in a TSV [151], nano-or pico-indentation has been applied to evaluate the stress of both, the copper filler and silicon chip [152]. However, these methods have a common drawback: full field and real time stress evaluation cannot be performed and here, IR photoelasticity becomes a useful tool.

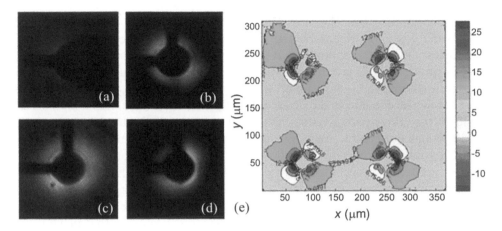

Figure 5.41. Stress evolution on a chip around TSVs during thermal cycling: (a) before experiment; (b) after 80 cycles; (c) after 160 cycles; (d) after 240 cycles; and (e) distribution of quantitative stress in terms of phase difference expressed in degrees. Reprinted from [140] with the permission of AIP Publishing.

The sizes involved in analysing TSVs are quite small (figure 5.40(b)) and only micro polariscopes are useful [140, 153]. The integrated infrared photoelastic microscope (IPEMS) [140] employing six-step phase shifting consists of a bromine tungsten lamp emitting light in the wavelength range 400–2500 nm as light source, a near infrared linear polariser, an achromatic quarter-wave plate between 920 nm and 1240 nm (product series #49–235, Edmund Optics C. Ltd), and a mini thermal chamber to heat the sample up to 500 °C. A CMOS camera (spectral sensitivity of 400–1700 nm, CONTOUR-IR, Belarus) with infrared flat field apochromatic lenses (400–1200 nm, Mitutoyo) with long working distances is used so that the lenses are not affected during high temperature measurements. The range of magnification values is 5× to 50×, focussing an area from 0.2 mm × 0.2 mm to 2 mm × 2 mm. For quantitative data reduction, the minor shift due to vibration while capturing the six phase shifted images is corrected using a texture based matching algorithm. The precision of the system in measuring retardation is estimated to be 2° [140].

Figures 5.41(a–d) show the evolution of fringes observed on the TSV due to thermal cycling between 55 °C and 125 °C. One observes only very faint white annular patches as fringes due to low birefringence of silicon. As six-step phase shifting has been employed, the retardation maps are quantified and are shown in figure 5.41(e). These results are quite valuable as measurement is possible without any external disturbances by utilizing the birefringent nature of silicon profitably. The results are to be considered as integrated retardation patterns as there could be changes in the principal stress direction through the thickness of TSV. Accounting for such effects needs to be the focus of future research. As in precision glass lens moulding [19, 34], inputs from IR photoelasticity experiments need to be used profitably for improving numerical modelling using FE, which again should be the focus of future research.

5.13 Photoelasticity in solid mechanics education

Photoelasticity is an indispensable tool to train students in the concept of stress and allied topics. One of the concepts that is easy to illustrate by photoelasticity is Saint Venant's principle. This requires any photoelastic model preferably a simple rectangular specimen made of polyurethane, which can be pulled, bent, and comes in handy to show how the end effects die down gradually over a distance. Being a whole field technique, the fringe patterns give the overall response of the model to the applied loads. This has been amply used in many of the case studies seen in this chapter.

It is possible to prove mathematically that the stress tensor at a free outward corner is zero and hence the fringe order is zero. A sharp free outward corner represents an intersection of two free surfaces as in point A of figure 5.42(a). Point B is also a free outward corner and three more such points can be identified. If $\{n\}$ be the outward normal of the free surface, then the stress vector on that plane is zero, i.e., $\{\overset{n}{T}\} = 0$, where the stress vector is given by

$$\{\overset{n}{T}\} = [\tau]\{n\} \tag{5.6}$$

On a free surface, the stress tensor need not be zero, but the stress vector is necessarily zero. One knows from mechanics of solids the general result that shear cannot cross a free boundary. This is why, on a bent beam, as the top surface is free, the shear stress at the top and bottom are zero. On the other hand, bending stress exists at the top and bottom, which is a free surface but the bending stress acts on a plane perpendicular to the free surface. This means that for a point lying on the free

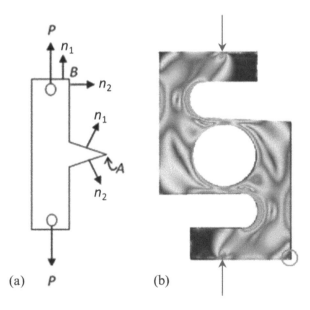

Figure 5.42. (a) Free outward corners with outward normals indicated for the planes involved, (b) isochromatics recorded in white light showing dark fringes at free outward corners indicating fringe order as zero. Courtesy: [154].

surface of a bent beam, the stress vector direction is tangential to the free surface in view of only bending stress existing at that point. A generalisation of shear stress cannot cross a free boundary is that the stress vector direction on a free surface can at best be tangential to the surface, as seen from the forgoing example of a bent beam. On a sharp free corner, this would create an anomaly as two free surfaces meet and either of the tangential vectors will cross the other boundary, which is free, which contradicts the definition of the free surface. Consequently, only if the stress tensor is zero can the stress vectors on these planes remain zero. Consider point B and let,

$$[\tau] = \begin{bmatrix} a & 0 & 0 \\ 0 & b & 0 \\ 0 & 0 & c \end{bmatrix} [n_1] = \begin{bmatrix} 1 \\ 0 \\ 0 \end{bmatrix} [n_2] = \begin{bmatrix} 0 \\ 1 \\ 0 \end{bmatrix} [n_3] = \begin{bmatrix} 0 \\ 0 \\ 1 \end{bmatrix} \qquad (5.7)$$

As each one of these planes is free, the stress vector on these planes must be zero. Considering the plane $\{n_1\}$ one gets,

$$\{\overset{n_1}{T}\} = \begin{bmatrix} a & 0 & 0 \\ 0 & b & 0 \\ 0 & 0 & c \end{bmatrix} \begin{bmatrix} 1 \\ 0 \\ 0 \end{bmatrix} = \begin{bmatrix} a \\ 0 \\ 0 \end{bmatrix} = 0 \Rightarrow a = 0 \qquad (5.8)$$

Proceeding this way, it is possible to show that the stress tensor must be zero at the free outward corner. In a properly conducted photoelastic test with no machining stresses introduced or carefully annealed, the fringe order must be zero at all outward corners as in figure 5.42(b) indicated by a black fringe at these corners in white light recording [154]. The figure is representative of the spring element used in many force transducers.

The concept of stress tensor being zero on a free outward corner is a useful result not only in photoelastic fringe ordering but also to verify the FE modelling of any structure—as a quick verification to test the boundary conditions. Although the importance of re-entrant corners as stress raisers is well taught in courses like fracture mechanics, the importance of stress tensor being zero at the free outward corner has been overlooked in usual solid mechanics training.

Beam under bending is one of the simplest problems taught in any first level course on solid mechanics. Theory of bending is basically developed for a beam under pure bending, but the results are used for any general loading as an engineering approximation, which is justified in slender beams as there is no coupling between shear and bending. Although shear stress is exceedingly small in comparison to bending stress, for example, in a rectangular cantilever beam it is usually 20 times lower than the bending stress, however, the small value of the shear stress at the neutral axis will have its influence on the colour of the fringe in a cantilever beam! [155]. The influence of Poisson's ratio is usually glossed over in beam analysis—all these subtle concepts can be easily illustrated using photo-elasticity as the fringes on the compression side in a beam under pure bending will be slightly more due to Poisson's effect as optical retardation is influenced by a slight increase in the thickness.

Variation of shear stress over the depth of the beam is taught and many know it to be parabolic. The question is, is it parabolic for all sections? What happens when one goes closer to loading points? This is an important aspect to be taught but is generally glossed over and most beams fail in such zones! Although strength of materials (SOM) helps in many ways for analysing and modelling complex systems in a simplified manner, the need for theory of elasticity (TOE) in this instance is particularly important. Figure 5.43 shows the isochromatics in colour for a three-point bent beam simulated by using SOM as well as TOE solution and variation of shear stress at mid-section and closer to load application point. The TOE solution uses 150 terms of the series solution and is plotted using P_Scope® [156]. It is a virtual polariscope with several useful features for training students in photo-elasticity, digital photoelasticity as well as nuances of solid mechanics. The SOM solution in the zone near the load application point does not predict the fringe features compared to the experimental fringes and the shear stress prediction is still parabolic, which is far from reality. The maximum shear stress occurs just below the surface (figure 5.43) and it can be closer in magnitude to the maximum bending stress, thus precipitating a failure. Nevertheless, the area under the curve indicating the shear force remains the same in both the cases. Such concepts are easy to illustrate using a simple photoelastic experiment of three-point bending in the laboratory.

The concept of stress concentration is appreciated instantly by making a model representing the features such as hole, fillet radii or steps and loaded and viewed appropriately in the polariscope. It is well known that if there is a discontinuity, it leads to some form of stress concentration and even a simple uniaxial loading can lead to biaxial stress state. If one takes a finite plate, and pulls it, one gets a uniform stress away from the points of loading. However, if one puts a hole, it introduces

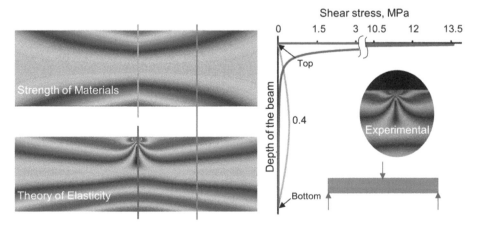

Figure 5.43. Actual variation of shear stress in a beam under three-point bending at various sections. Isochromatics simulated by P_Scope® [156] in colour using solution from SOM are compared with TOE, which matches with the experimental fringes shown for the load application point. Closer to load application points, the shear stress variation is no longer parabolic, and the maximum value can be comparable to the maximum bending stress leading to failure. Courtesy: [156].

stress concentration and produces a nice flowery pattern of fringes (figure 5.44(a)). These fringes are generated by P_Scope® [156] and the fringes correspond to a small hole in an infinite plate. What is interesting to note is, by applying tension one may get compressive stresses developed at certain regions and this becomes important in fracture toughness testing of thin plates and it can lead to local buckling! Although, the concept of stress concentration is well taught, the importance of sign change of tangential stress is glossed over as there are no convenient tools hitherto to illustrate.

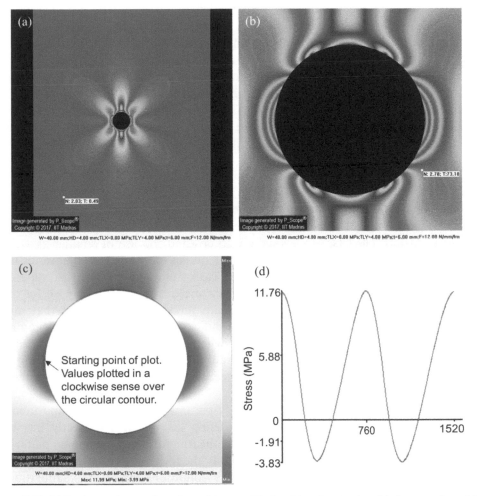

Figure 5.44. (a) Isochromatics seen in a plate with a small hole subjected to tension, (b) close-up view of the isochromatics near the hole boundary, (c) variation of tangential stress near the hole. On close observation one can see a red circle drawn close to the hole boundary with a blue dot at the left. (d) Variation of tangential stress over the circular hole starting from the blue dot in a clockwise sense showing maximum negative value of the stress at two points. P_Scope® [156] has a feature to get such plots at the press of a button to facilitate easy education and illustration of subtle concepts. Courtesy: [156].

The software P_Scope® has excellent tools to magnify and to plot the variation of shear stress on selected contours (figures 5.44(b–d)). This utility is used to draw the variation of tangential stress and it brings out clearly that one gets maximum compressive stress equivalent in magnitude to the applied far-field tensile stress at two points on the circular hole boundary (figure 5.44(d)). This knowledge is important to evaluate stress concentration value in the case of biaxial tension or tension–compression combinations.

Many nuances of solid mechanics such as in 2-D analysis with constant body forces, material constants do not influence stress magnitude or its distribution; in the linear loading regime, that the principal stress directions do not change with increase in load etc, can be easily illustrated. Photoelasticity is an efficient tool to bring out several such subtle concepts and it is left to the ingenuity of the user to reap its full potential. Further, among all the experimental techniques—this is unique in providing the fringes in colour, which adds to enthusiasm in learning concepts of solid mechanics. If establishing a full-fledged photoelastic laboratory is expensive, for teaching the concepts of solid mechanics, the simulation software P_Scope® can be used to supplement the same as it is quite user friendly and cost effective.

5.14 Closure

The case studies discussed have brought out diverse applications of photoelasticity by covering a range of materials that are homogeneous, heterogeneous, granular, and soft matter. The case studies have also covered the use of smaller as well as bigger models than the prototypes, and direct studies on prototypes that range from plastics to glass to silicon. In some of the case studies, work has just started to unravel the mysteries and in a few others, new insight gained can be used for other situations easily and directly. Study on chain links has brought out that minor deviations in interference fits can be quite significant in the field, and it is a common problematic area that one must address whenever interference fits are used in machine assemblies. In the burrowing example, the aspect that burrowing can happen as crack propagation is quite a nice generalisation useful to comprehend nature with a new perspective.

Although FE simulations greatly simplify parametric studies, the importance of specifying the boundary conditions that define the problem correctly has been brought out with the simple example of edge heating of a polycarbonate plate. The intelligent use of birefringence of glass to study the residual stress evolution has lent credibility in evolving the thermal boundary conditions for process simulation in glass moulding. This is an innovative example on the use of photoelasticity to address complex problems. Photoelasticity has been extensively used in glass manufacturing [157] (sections 2.14, 2.15 and 5.4.1) and is also finding applications in manufacturing of 3-D IC chips (section 5.12), and other applications mentioned in the review given in reference [158] establish that photoelasticity is an indispensable tool in the manufacturing sector for process control and quality monitoring [159]. The use of rapid prototyping in the study of a complex model of concrete has generated new experimental results to fine-tune the macro scale theoretical predictions hitherto used by civil engineers. This is also the case in the study of masonry structures where the knowledge of how heterogeneity can lead to complex load transfer needs to be profitably used for conservation of historical monuments.

Innovations of loading and development of suitable experimental rigs have been highlighted in the examples of seals, granular materials, burrowing and plant root stresses. The studies on PET bottles highlight how a common everyday item can benefit from a systematic scientific study. Finally, how photoelasticity can help in communicating subtle ideas of solid mechanics to students has been brought out.

Photoelasticity has always embraced new technologies whenever they are introduced in other fields such as use of digital computers, digital image processing tools for image acquisition and processing, use of rapid prototyping for model fabrication etc. Of late, deep learning has been found to be useful in many fields. Photoelasticity is not lagging behind and already interesting papers have started appearing on this [160, 161]. The current decade will see the use of more and more deep learning tools for various aspects of photoelastic analysis.

In any field, advancements in materials have always acted as a catalyst for new applications. Gelatin that was traditionally used to simulate the effect of self-weight of the structures in civil engineering applications has now been explored in its new form for neurobiology, surrogate material for tissue engineering etc. The promise of VeroClear for stress-freezing studies holds new hope to explore complex three-dimensional problems. Engineers normally use stress relieving holes as seen in the study of the chain offset plate for reducing stresses, but nature has adopted a functionally graded material approach to reduce stress concentration in tendons! This can be explored for practical structures too. It is hoped that the examples provided although not exhaustive, serve as an inspiration for scientists and researchers to benefit from photoelasticity and expand its scope to newer and unexplored areas through their ingenuity and curiosity.

Appendix: Simplified solution for stress field in a circular disc with self-equilibrated forces

In granular media, the challenge is to determine the forces acting on the grains that are arbitrary in number, magnitude, and direction. If the whole system is under equilibrium, the individual grains will also be in equilibrium and hence, the force system will be self-equilibrated. The simplest system that can be analysed is of grains comprising circular discs. Although the focus here is on the inverse problem of finding out the forces from the observed photoelastic fringes, the forward problem of finding out the fringe patterns for a given force set is a necessary step in that direction. This appendix provides an easy-to-use readymade solution for an arbitrarily loaded self-equilibrated disc.

In chapter 1, the stress field equations for a diametrically loaded circular disc are constructed by the method of superposition using the basic radial distribution of the Boussinesq's solution. The diametral loads are the simplest, and it was shown by using the principles of solid mechanics that, to make the outer surface of the circular disc traction-free, a uniform radial tension needs to be applied. A similar simplification is possible, if multiple forces that are self-equilibrated are applied on a disc, the radial tension (R_T) to be applied to make the circular disc traction-free is simply given as [162]

$$R_T = \sum_{i=1}^{n} P_i \cos \theta_i \tag{5.9}$$

where, θ_i is the angle of inclination of the line of action of the force with respect to the radial line joining the point of action of the force and the centre of the disc (figure 5.45).

To determine the stress field, one must make a summation of the simple radial distribution of all the forces as:

$$\sigma_r = \sum_{i=1}^{n} \frac{-2P_i}{\pi h} \frac{\cos \theta_i^B}{r_i} \tag{5.10}$$

It is noteworthy that the utility of Boussinesq's solution is elegant as the radial component of stress value for any force can be determined, at a point of interest, if the angular orientation of the point is measured with respect to the line of action of the force. Figure 5.45 shows for an arbitrary point Q, the definition of the angle θ_i^B

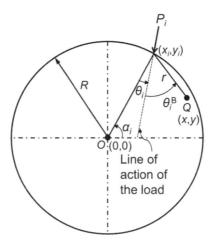

Figure 5.45. Nomenclature of various angles needed to evaluate the stress components contributed by a single arbitrary force P_i acting in a general fashion on the boundary of a circular disc, which is forming part of the set of self-equilibrated loads acting on the disc.

(the superscript 'B' denotes that it is the angle to be used for the stress calculation of Boussinesq's solution) for the force P_i.

The stress components expressed in Cartesian coordinates for a set of 'n' forces is given as [163]:

$$\sigma_x = \sum_{i=1}^{n} \left[\frac{-2P_i}{\pi h} \right.$$

$$\frac{((R - [x \cos \alpha_i + y \sin \alpha_i]) \cos \theta_i + (x \sin \alpha_i - y \cos \alpha_i) \sin \theta_i)(x - R \cos \alpha_i)^2}{((x - R \cos \alpha_i)^2 + (y - R \sin \alpha_i)^2)^2} \qquad (5.11)$$

$$\left. + \frac{P_i}{\pi h D} \cos \theta_i \right]$$

$$\sigma_y = \sum_{i=1}^{n} \left[\frac{-2P_i}{\pi h} \right.$$

$$\frac{((R - [x \cos \alpha_i + y \sin \alpha_i]) \cos \theta_i + (x \sin \alpha_i - y \cos \alpha_i) \sin \theta_i)(R \sin \alpha_i - y)^2}{((x - R \cos \alpha_i)^2 + (y - R \sin \alpha_i)^2)^2} \qquad (5.12)$$

$$\left. + \frac{P_i}{\pi h D} \cos \theta_i \right]$$

$$\tau_{xy} = \sum_{i=1}^{n} \left[\frac{2P_i}{\pi h} \right.$$

$$(5.13)$$

$$\left. \frac{((R - [x \cos \alpha_i + y \sin \alpha_i]) \cos \theta_i + (x \sin \alpha_i - y \cos \alpha_i) \sin \theta_i)(x - R \cos \alpha_i)(R \sin \alpha_i - y)}{((x - R \cos \alpha_i)^2 + (y - R \sin \alpha_i)^2)^2} \right]$$

This is the most general solution for a system of self-equilibrated arbitrary forces acting on the boundary of the disc. The solution for the problem of circular disc under diametral compression (equation (1.19)) is easily obtained by first recognising the fact that θ_i for the forces is zero as these are forces acting along a radial line and the angle α_i acting on the top diametral end is 90° and for the diametrically opposite point is 270°. This is left as an exercise to the reader.

Figure 5.46 depicts the isochromatic fringe patterns plotted theoretically for various numbers of arbitrary forces using equations (5.11)–(5.13) with the angles α_i and θ_i defined as listed in table 5.2 with the disc diameter of 60 mm, the material stress fringe value of 12 N/mm/fringe with the thickness of the disc being 6 mm. The plots demonstrate the usefulness of equations (5.11)–(5.13).

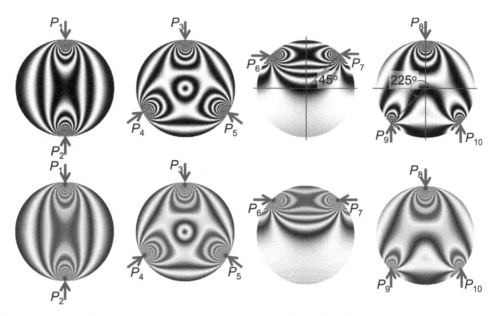

Figure 5.46. Dark field isochromatics plotted using equations (5.11)–(5.13) for various combinations of self-equilibrated loads. Top row: greyscale plotting. Bottom row: fringes plotted in white light.

Table 5.2. Loads with corresponding values of θ and α for plotting figure 5.46.

Load (N)		θ (degrees)	α (degrees)
P_1	600	0	90
P_2	600	0	270
P_3	600	0	90
P_4	600	0	210
P_5	600	0	330
P_6	600	45	135
P_7	600	−45	45
P_8	600	0	90
P_9	300	45	225
P_{10}	300	−45	315

Exercises

1. What are the difficulties in analytically modelling problems dealing with residual stresses, assembly stresses and stresses induced due to manufacturing?
2. What is the simplest way to reveal stresses in PET bottles using photoelasticity?
3. Carry out a literature search on water diffusion into epoxies and list the challenges involved in understanding the phenomenon.

4. What do you know about hygrothermal ageing and hydrothermal ageing?
5. Why is the fracture toughness of epoxy high during absorption as well as at saturation? In which class of epoxies is this phenomenon readily seen?
6. What are the similarities between the residual stresses in glass and stresses due to swelling in epoxies?
7. In precision glass moulding, which key issue has been ably addressed by photoelasticity? Discuss your answer with appropriate sketches of fringe patterns.
8. What was the surprising aspect witnessed in the study of a large rectangular polycarbonate plate coming into contact with the hot plate?
9. Explain in detail how temperature variation recorded from an IR camera in the case of a rectangular plate with edge heating falls short of capturing the physics of the problem and how photoelasticity has helped in obtaining a reliable solution.
10. A photoelastic experiment in two crack configurations is carried out till failure for a model of a mechanical component subjected to planar loading to determine the fracture parameters and make crack growth angle predictions. Isochromatics show that the problem pertains to a mixed mode loading case. After checking several values of the critical distance from crack-tip, a specific value of r_c is selected based on satisfactory match with experimental data. The experimental results were tabulated but some data got missed out. Complete the data entry based on the recorded information.

| K_I | K_{II} | T-stress | Crack growth angle | | |
| | | | Exp. | MTS | GMTS |
$(MPa\sqrt{m})$	$(MPa\sqrt{m})$	(MPa)	(deg)	(deg)	(deg)
3.314	1.355	7.43	−48.49	*	−50.72
4.415	0.991	11.98	−43.50	−23.23	*

*Fill in this missing data

11. In chapter 2, it was shown that the flow induced residual stress in a polycarbonate sheet have the maximum influence on the values of T-stress? Do you think the crack growth theories using only singular terms are sufficient to predict the crack growth directions? Elaborate on your answer.
12. Summarise the general recommendations that could be framed on interference fits from the study of failure of chains? Support your recommendations based on neat sketches of fringe patterns.
13. Elaborate on the challenges involved in designing loading jigs for photoelastic experiments involving the stress freezing technique by taking the case study on seals as an example.
14. Discuss the benefits of the stress-freezing technique in photoelasticity with the help of some case studies. Highlight the inherent suitability of materials used in rapid prototyping for photoelastic analysis.

15. Compare the fringe patterns observed for a slice cut from a 3-D model of heterogeneous concrete model using rapid prototyping and a 2-D fabricated slice. How can you explain the discrepancies?

16. What are the differences between a granular medium and the solids and fluids that we know of from physics?

17. What should be the speed of a missile to penetrate in meteor impact?

18. How do the forces get distributed in a granular medium? Is there any quantification of its distribution available? Explain through neat sketches of the grains with photoelastic fringes.

19. Why does one use either discs/cylinders of different sizes or a mixture of different shapes like circles and ellipses in forming a granular medium to test its response?

20. Summarise the steps involved in automated reconstruction of fringes seen in a granular medium with neat sketches.

21. Observe the schematic diagram of a loaded disc as shown. The vectorial forces in Newtons acting on the disc are listed as: $P_1 = -1.5i - 1.788j$; $P_{2,4} = \mp 1.25j$; $P_3 = 3i$; $P_5 = -1.5i + 1.788j$. The table gives the coordinates of the points of application of the forces acting on the disc with the disc centre as the origin. For each load, compute the values of α and θ with appropriate signs to be substituted in equations (5.11)–(5.13) to determine the stress field inside the disc. Report your answer rounded to the nearest integer degree.

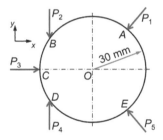

Point	x (mm)	y (mm)
A	19.284	22.981
B	−24.575	17.207
C	−30	0
D	−24.575	−17.207
E	19.284	−22.981

22. Discuss the challenges involved in masonry analysis. Discuss the challenges in using reflection photoelasticity.

23. What is the inspiration behind the study of root stresses? Discuss at least two uses.

24. What is turgor pressure and what is its common value? At what values of stress is the root growth arrested?

25. Sketch the optical arrangement needed for the study of burrowing by worms. What is one of the possible mechanisms for burrowing? Is it prevalent in general?

26. How do snakes slither and what is the key finding about slithering which has been brought out using photoelasticity?

27. Gelatin is used in many of the case studies discussed. Tabulate the different compositions of gelatin used in different applications.

28. What is the least force that can be measured using a gelatin sheet and a gelatin sphere?

29. Human tissue is a soft matter. Summarise its various characteristics. How is stress concentration in tendons managed efficiently in a human body?

30. In what way is IR photoelasticity advantageous to identify defects in silicon. With neat sketches, explain your answer.

31. What is a TSV? How do stresses get developed and how can it be measured?

32. Identify and explain a case of photoelastic analysis where the model is scaled up as compared to the prototype.

33. Provide an account of how photoelasticity can be used in the analysis of residual stresses and assembly stresses by citing some example case studies. Give instances where such analysis has resulted in improvements in the production line.

34. Explain in detail, the difference between the shear stress variation obtained using the strength of materials approach and theory of elasticity approach for a three-point bent specimen. Use the results from photoelastic experiment presented in the chapter to support your findings.

35. Show how a stress tensor is zero at a free outward corner. What is the use of this result in photoelasticity and in numerical studies?

References

[1] Lyu M-Y and Pae Y 2003 Bottom design of carbonated soft drink poly(ethylene terephthalate) bottle to prevent solvent cracking *J. Appl. Polym. Sci.* **88** 1145–52

[2] Marcus P 1971 Injection blow molding method *US Patent* no. **158** 161

[3] Prasath R G R, Newton T and Danyluk S 2018 Stress monitoring of PET beverage bottles by digital photoelasticity *Manuf. Lett.* **15** 9–13

[4] Allahkarami M, Bandla S and Hanan J C 2016 Residual stress in injection stretch blow molded PET bottles *Conf. Proc. of the Society for Experimental Mechanics Series* vol 9 pp 285–90 http://link.springer.com/10.1007/978-3-319-21765-9_34

[5] Prasath R, Danyluk S and Zagarola S 2019 Non-contact stress measurement in PET preforms *Polym. Technol. Mater.* **58** 1802–9

[6] Zhou J and Lucas J P 1999 Hygrothermal effects of epoxy resin. Part I: the nature of water in epoxy *Polymer (Guildf)* **40** 5505–12

[7] Zhou J and Lucas J P 1999 Hygrothermal effects of epoxy resin. Part II: variations of glass transition temperature *Polymer (Guildf)* **40** 5513–22

[8] Scafidi M, Pitarresi G, Toscano A, Petrucci G, Alessi S and Ajovalasit A 2015 Review of photoelastic image analysis applied to structural birefringent materials: glass and polymers *Opt. Eng.* **54** 081206-1–9

[9] Pitarresi G, Scafidi M, Alessi S, Di Filippo M, Billaud C and Spadaro G 2015 Absorption kinetics and swelling stresses in hydrothermally aged epoxies investigated by photoelastic image analysis *Polym. Degrad. Stab.* **111** 55–63

[10] Toscano A, Pitarresi G, Scafidi M, Di Filippo M, Spadaro G and Alessi S 2016 Water diffusion and swelling stresses in highly crosslinked epoxy matrices *Polym. Degrad. Stab.* **133** 255–63

[11] Alessi S, Toscano A, Pitarresi G, Dispenza C and Spadaro G 2017 Water diffusion and swelling stresses in ionizing radiation cured epoxy matrices *Polym. Degrad. Stab.* **144** 137–45

[12] Pitarresi G, Toscano A and Alessi S 2018 Fracture toughness of synthesised high-performance epoxies subject to accelerated water aging *Polym. Test.* **68** 248–60

[13] Stellrecht E, Han B and Pecht M G 2004 Characterization of hygroscopic swelling behavior of mold compounds and plastic packages *IEEE Trans. Components Packag. Technol.* **27** 499–506

[14] Park S, Zhang H, Zhang X, Lung Ng S and Lee H C 2009 Temperature dependency of coefficient of hygroscopic swelling of molding compound *2009 59th Electronic Components and Technology Conf.* (Piscataway, NJ: IEEE) pp 172–9 https://doi.org/10.1109/ECTC.2009.5074012

[15] Jackson M B, Heinz S R and Wiggins J S 2012 Fluid ingress strain analysis of glassy polymer networks using digital image correlation *Polym. Test.* **31** 1131–9

[16] Tarkes Dora P and Ramesh K 2014 Measurement of residual birefringence in thin glass plates using digital photoelasticity *Proc. SPIE* ed S Sirisoonthorn vol 9234 (Bellingham, WA: SPIE) pp 92340J-1–6 https://doi.org/10.1117/12.2054230

[17] Wang F 2014 Simulating the precision glass molding process apprimus. https://apprimus-verlag.de/simulating-the-precision-glass-molding-process.html

[18] Dambon O 2013 Development of a synergistic computational tool for material modeling, process simulation and optimization of optical glass molding (SIMUGLASS final report) https://cordis.europa.eu/project/id/233524/reporting

[19] Dora Pallicity T, Ramesh K, Mahajan P and Vengadesan S 2016 Numerical modeling of cooling stage of glass molding process assisted by CFD and measurement of residual birefringence *J. Am. Ceram. Soc.* **99** 470–83

[20] Joshi D 2014 *Thermo-Mechanical Characterization of Glass and its effect on Predictions of Stress State, Birefringence and Fracture in Precision Glass Molded Lenses* (Clemson, SC: Graduate School of Clemson University) https://tigerprints.clemson.edu/all_dissertations/1463

[21] Ananthasayanam B, Joseph P F, Joshi D, Gaylord S, Petit L, Blouin V Y, Richardson K C, Cler D L, Stairiker M and Tardiff M 2012 Final shape of precision molded optics: part I—computational approach, material definitions and the effect of lens shape *J. Therm. Stress.* **35** 550–78

[22] Mosaddegh P 2010 *Friction Measurement in Precision Glass Molding* (Clemson, SC: Graduate School of Clemson University) https://tigerprints.clemson.edu/all_dissertations/613/

[23] Sarhadi A, Hattel J H and Hansen H N 2014 Evaluation of the viscoelastic behaviour and glass/mould interface friction coefficient in the wafer based precision glass moulding *J. Mater. Process. Technol.* **214** 1427–35

[24] He P and He P 2014 *Design and Fabrication of Nonconventional Optical Components by Precision Glass Molding* (Columbus, OH: Graduate School of Ohio State University) https://etd.ohiolink.edu/apexprod/rws_etd/send_file/send?accession=osu1408548156&disposition

[25] Su L, Wang F, He P, Dambon O, Klocke F and Yi A Y 2014 An integrated solution for mold shape modification in precision glass molding to compensate refractive index change and geometric deviation *Opt. Lasers Eng.* **53** 98–103

[26] Dambon O, Wang F, Klocke F, Pongs G, Bresseler B, Chen Y and Yi A Y 2009 Efficient mold manufacturing for precision glass molding *J. Vac. Sci. Technol. B Microelectron. Nanom. Struct.* **27** 1445–9

[27] Su L, Chen Y, Yi A Y, Klocke F and Pongs G 2008 Refractive index variation in compression molding of precision glass optical components *Appl. Opt.* **47** 1662–7

[28] Jain A and Yi A Y 2005 Numerical modeling of viscoelastic stress relaxation during glass lens forming process *J. Am. Ceram. Soc.* **88** 530–5

[29] Mahajan P, Tarkes Dora P, Sandeep T S and Trigune V M 2015 Optimized design of optical surface of the mold in precision glass molding using the deviation approach *Int. J. Comput. Methods Eng. Sci. Mech.* **16** 53–64

[30] Sellier M, Breitbach C, Loch H and Siedow N 2007 An iterative algorithm for optimal mould design in high-precision compression moulding *Proc. Inst. Mech. Eng. Part B J. Eng. Manuf.* **221** 25–33

[31] Yi A Y and Jain A 2005 Compression molding of aspherical glass lenses—a combined experimental and numerical analysis *J. Am. Ceram. Soc.* **88** 579–86

[32] Ramesh K, Kasimayan T and Neethi Simon B 2011 Digital photoelasticity—a comprehensive review *J. Strain Anal. Eng. Des.* **46** 245–66

[33] Ramesh K 2016 *Digi*Photo digital photoelasticity software for employing phase shifting technique *Photomechanics Lab* (IIT Madras) https://home.iitm.ac.in/kramesh/dphoto.html (accessed Apr. 06, 2021)

[34] Pallicity T D, Vu A T, Ramesh K, Mahajan P, Liu G and Dambon O 2017 Birefringence measurement for validation of simulation of precision glass molding process *J. Am. Ceram. Soc.* **100** 4680–98

[35] Vivekanandan A and Ramesh K 2020 Study of crack interaction effects under thermal loading by digital photoelasticity and finite elements *Exp. Mech.* **60** 295–316

[36] Miskioglu I, Gryzagorides J and Burger C P 1981 Material properties in thermal-stress analysis *Exp. Mech.* **21** 295–301

[37] Neethi Simon B and Ramesh K 2010 A simple method to plot photoelastic fringes and phasemaps from finite element results *J. Aerosp. Sci. Technol.* **62** 174–82 http://www.aerojournalindia.com/2010 contents/Content -Aug-10.html

[38] Ramesh K 2007 Lec-23 Validation of multi-parameter field equations *NPTEL Video Lectures on Engineering Fracture Mechanics* 1st edn (Chennai: NPTEL and MHRD) https://youtube.com/watch?v=e7oBNSSZVWI&list=PLbMVogVj5nJTTGvVQL5ks-woI_vqDTKXF&index=23

[39] Fuller J J and Marotta E E 2001 Thermal contact conductance of metal/polymer joints: an analytical and experimental investigation *J. Thermophys. Heat Transf.* **15** 228–38

[40] Papanikos P and Meguid S A 1994 Theoretical and experimental studies of fretting-initiated fatigue failure of aeroengine compressor discs *Fatigue Fract. Eng. Mater. Struct.* **17** 539–50

[41] Gandhi A 2017 *A Study of Errors in the Stress Intensity Factor, Estimated using Different Experimental Techniques, due to Three Dimensional and Corner Singularity Effects* (Kanpur: : IIT Kanpur)

[42] Shi L, Wei D S, Wang Y R, Tian A M and Li D 2016 An investigation of fretting fatigue in a circular arc dovetail assembly *Int. J. Fatigue* **82** 226–37

[43] Forshaw J R, Taylor H and Chaplin R 1956 Alternating pressures and blade stresses in an axial flow compressor *Aeronaut. Res. Counc. Rep. No.* 2846 https://reports.aerade.cranfield. ac.uk/handle/1826.2/3403

[44] Ramesh K 2000 PSIF software—Photoelastic SIF evaluation *Photomechanics Lab* (IIT Madras) https://home.iitm.ac.in/kramesh/psif.html (accessed Apr. 06, 2021)

[45] Ramesh K and Thomre M 2020 Evaluation of critical crack orientations and crack growth analysis in compressor blade root subjected to fatigue *J. Eng. Gas Turbines Power* **142** 1–9

[46] Brown N E, White S R and Sottos N R 2006 Fatigue crack propagation in microcapsule-toughened epoxy *J. Mater. Sci.* **41** 6266–73

[47] ASTM 2008 Standard test method for strain-controlled fatigue testing, E606/E606M https://www.astm.org/Standards/E606

[48] Smith D J, Ayatollahi M R and Pavire M J 2001 The role of T-stress in brittle fracture for linear elastic materials under mixed-mode loading *Fatigue Fract. Eng. Mater. Struct.* **24** 137–50

[49] Ayatollahi M R and Aliha M R M 2009 Mixed mode fracture in soda lime glass analyzed by using the generalized MTS criterion *Int. J. Solids Struct.* **46** 311–21

[50] Swain D, Philip J and Pillai S A 2014 Assessment of a complex aerospace design through optical techniques *Appl. Mech. Mater.* **592–594** 1006–10

[51] Wright J 2006 *Standard Handbook of Chains: Chains for Power Transmission* (Boca Raton, FL: Taylor & Francis)

[52] Anand V, Dasari N and Ramesh K 2011 Innovative use of transmission and reflection photoelastic techniques to solve complex industrial problems *Exp. Tech.* **35** 71–5

[53] Dasari N and Ramesh K 2011 Analysis of a complex shape chain plate using transmission photoelasticity *Appl. Mech. Mater.* **110–16** 2623–27

[54] Nam J, Hawong J, Han S and Park S 2008 Contact stress of O-ring under uniform squeeze rate by photoelastic experimental hybrid method *J. Mech. Sci. Technol.* **22** 2337–49

[55] Mose B R, Shin D K and Shin D C 2017 Investigating the performance of seals using photoelastic experimental hybrid method and finite element analysis *J. Adv. Mech. Des. Syst. Manuf.* **11** 1–12

[56] Mose B R, Hawong J S, Shin D K, Lim H S and Shin D C 2017 Stress and fracture analysis of D-ring by photoelastic experimental hybrid method *J. Mech. Sci. Technol.* **31** 3657–60

[57] Nam J H, Hawong J S, Kim K H, Liu-Yi, Kwon O S and Park S H 2010 A study on the development of a loading device using a photoelastic stress freezing method for the analysis of o-ring stress *J. Mech. Sci. Technol.* **24** 693–701

[58] Bernard A O, Hawong J S, Shin D C and Dong B 2015 Contact behavior analysis of elastomeric x-ring under uniform squeeze rate and internal pressure before and after forcing-out using the photoelastic experimental hybrid method *J. Mech. Sci. Technol.* **29** 2157–68

[59] Shin D C, Nam J H and Kim D W 2016 Experimental interior stress fields of a constantly squeezed O-ring modeling from hybrid transmission photoelasticity *Exp. Tech.* **40** 59–72

[60] Hawong J, Nam J, Han S, Kwon O and Park S 2009 A study on the analysis of O-ring under uniform squeeze rate and internal pressure by photoelastic experimental hybrid method *J. Mech. Sci. Technol.* **23** 2330–40

[61] Shin D C, Hawong J S, Lee S W, Bernard A O and Lim H S 2014 Contact behavior analysis of X-ring under internal pressure and uniform squeeze rate using photoelastic experimental hybrid method *J. Mech. Sci. Technol.* **28** 4063–73

[62] Ouma A B, Nam J, Seok L H and Hawong J S 2012 A study on the contact stresses of square ring under uniform squeeze rate and internal pressure by photoelastic experimental hybrid method *J. Mech. Sci. Technol.* **26** 2617–26

[63] Ramesh K, Kumar M A and Dhande S G 2000 Fusion of digital photoelasticity, rapid prototyping and rapid tooling technologies BT—digital photoelasticity: advanced techniques and applications *Exp. Tech.* **32** 36–48

[64] Nandini H 2015 *Photoelastic Analysis of Stress Distribution around Implants with Two Different Geometric Designs—An invitro Study* (Chennai: The Tamil Nadu Dr. MGR Medical University)

[65] Rakshagan 2016 *Influence of Platform Switching on Stress Distribution around Implant with Cantilever Prosthesis—A Photoelastic Study* (Chennai: The Tamil Nadu Dr. MGR Medical University)

[66] Ashokan K, Prasath R G R and Ramesh K 2012 Noise-free determination of isochromatic parameter of stereolithography-built models *Exp. Tech.* **36** 70–5

[67] Ramakrishnan V and Ramesh K 2017 Scanning schemes in white light photoelasticity—part II: novel fringe resolution guided scanning scheme *Opt. Lasers Eng.* **92** 141–9

[68] Ramesh K 2008 Photoelasticity *Springer Handbook of Experimental Solid Mechanics* ed J William and N Sharpe (New York: Springer) pp 701–42 https://doi.org/10.1007/978-0-387-30877-7_25

[69] Umezaki E, Okano A and Koyama H 2014 Stress and flow analyses of ultraviolet-curable resin during curing *Int. Conf. on Experimental Mechanics 2013 and Twelfth Asian Conf. on Experimental Mechanics* ed S Sirisoonthorn vol 9234 (Bellingham, WA: SPIE) pp 923408-1–8 https://doi.org/10.1117/12.2053962

[70] Karalekas D E and Agelopoulos A 2006 On the use of stereolithography built photoelastic models for stress analysis investigations *Mater. Des.* **27** 100–6

[71] Wang L, Ju Y, Xie H, Ma G, Mao L and He K 2017 The mechanical and photoelastic properties of 3D printable stress-visualized materials *Sci Rep.* **7** 10918

[72] Ju Y, Xie H, Zheng Z, Lu J, Mao L, Gao F and Peng R 2014 Visualization of the complex structure and stress field inside rock by means of 3D printing technology *Chinese Sci. Bull.* **59** 5354–65

[73] Ju Y, Wang L, Xie H, Ma G, Mao L, Zheng Z and Lu J 2017 Visualization of the three-dimensional structure and stress field of aggregated concrete materials through 3D printing and frozen-stress techniques *Constr. Build. Mater.* **143** 121–37

[74] Ju Y, Wang L, Xie H, Ma G, Zheng Z and Mao L 2017 Visualization and transparentization of the structure and stress field of aggregated geomaterials through 3D printing and photoelastic techniques *Rock Mech. Rock Eng.* **50** 1383–407

[75] Ju Y, Zheng Z, Xie H, Lu J, Wang L and He K 2017 Experimental visualisation methods for three-dimensional stress fields of porous solids *Exp. Tech.* **41** 331–44

[76] Ju Y, Ren Z, Li X, Wang Y, Mao L and Chiang F 2019 Quantification of hidden whole-field stress inside porous geomaterials via three-dimensional printing and photoelastic testing methods *J. Geophys. Res. Solid Earth* **124** 5408–26

[77] Ju Y, Ren Z, Wang L, Mao L and Chiang F-P 2018 Photoelastic method to quantitatively visualise the evolution of whole-field stress in 3D printed models subject to continuous loading processes *Opt. Lasers Eng.* **100** 248–58

[78] Ju Y, Ren Z, Mao L and Chiang F-P 2018 Quantitative visualisation of the continuous whole-field stress evolution in complex pore structures using photoelastic testing and 3D printing methods *Opt. Express* **26** 6182–201

[79] Ju Y, Xie H, Zhao X, Mao L, Ren Z, Zheng J, Chiang F-P, Wang Y and Gao F 2018 Visualization method for stress-field evolution during rapid crack propagation using 3D printing and photoelastic testing techniques *Sci Rep.* **8** 4353-1–10

[80] Ju Y, Wang Y, Ren Z, Mao L, Wang Y and Chiang F 2020 Optical method to quantify the evolution of whole-field stress in fractured coal subjected to uniaxial compressive loads *Opt. Lasers Eng.* **128** 106013

[81] Ju Y, Guo W, Ren Z, Zheng J, Mao L, Hu X and Liu P 2021 Experimental study on mechanical and optical properties of printable photopolymer used for visualising hidden structures and stresses in rocks *Opt. Mater.* **111** 110691

[82] Wakabayashi T 1950 Photo-elastic method for determination of stress in powdered mass *J. Phys. Soc. Japan* **5** 383–5

[83] Dantu P 1957 *Proc. of the 4th Int. Conf. on Soil Mechanics and Foundations Engineering* (Oxford: Butterworth's Scientific Publications) pp 144–8

[84] Drescher A and de Josselin de Jong G 1972 Photoelastic verification of a mechanical model for the flow of a granular material *J. Mech. Phys. Solids* **20** 337–40

[85] Abed Zadeh A *et al* 2019 Enlightening force chains: a review of photoelasticimetry in granular matter *Granul. Matter* **21** 1–12

[86] From solid to liquid and back again https://researchblog.duke.edu/2017/07/07/from-solid-to-liquid-and-back-again/

[87] Meteor impact https://today.duke.edu/2015/04/meteorimpact

[88] Daniels K E, Coppock J E and Behringer R P 2004 Dynamics of meteor impacts *Chaos An Interdiscip. J. Nonlinear Sci.* **14** S4

[89] Daniels K E and Hayman N W 2008 Force chains in seismogenic faults visualized with photoelastic granular shear experiments *J. Geophys. Res.* **113** B11411

[90] Animations 1 and 2 available in the HTML page https://agupubs.onlinelibrary.wiley.com/doi/full/10.1029/2008JB005781%0A

[91] Majmudar T S and Behringer R P 2005 Contact force measurements and stress-induced anisotropy in granular materials *Nature* **435** 1079–82

[92] Daniels K E, Kollmer J E and Puckett J G 2017 Photoelastic force measurements in granular materials *Rev. Sci. Instrum.* **88** 051808-1–13

[93] Zhao Y, Zheng H, Wang D, Wang M and Behringer R P 2019 Particle scale force sensor based on intensity gradient method in granular photoelastic experiments *New J. Phys.* **21** 023009

[94] Abed-Zadeh A *et al* Photoelastic methods wiki https://git-xen.lmgc.univ-montp2.fr/PhotoElasticity/Main/-/wikis/home

[95] Kollmer J E Photoelastic grain solver (pegs) https://github.com/jekollmer/PEGS

[96] Thomas A L and Vriend N M 2019 Photoelastic study of dense granular free-surface flows *Phys. Rev.* E **100** 1–14

[97] Hurley R, Marteau E, Ravichandran G and Andrade J E 2014 Extracting inter-particle forces in opaque granular materials: beyond photoelasticity *J. Mech. Phys. Solids* **63** 154–66

[98] Kumar P, Hariprasad M P, Menon A and Ramesh K 2020 Experimental study of dry stone masonry walls using digital reflection photoelasticity *Strain* **56** 1–16

[99] Naik P and Menon A 2019 *Structural Behaviour of Dry Stack Stone Corbelled Vaults Under Lateral Support Movement RILEM Bookseries* vol 18 ed R Aguilar, D Torrealva, S Moreira, M A Pando and L F Ramos (Cham: Springer International Publishing) pp 540–8 https://link.springer.com/content/pdf/10.1007%2F978-3-319-99441-3_58.pdf

[100] Bigoni D and Noselli G 2010 Localized stress percolation through dry masonry walls. Part I —experiments *Eur. J. Mech. A/Solids* **29** 291–8

[101] Bigoni D and Noselli G 2010 Localized stress percolation through dry masonry walls. Part II—modelling *Eur. J. Mech. A/Solids* **29** 299–307

[102] Baig I, Ramesh K and Hariprasad M P 2015 Analysis of stress distribution in dry masonry walls using three fringe photoelasticity *Int. Conf. on Experimental Mechanics* ed C Quan, K Qian, A Asundi and F S Chau (Bellingham, WA: SPIE) p 93022P https://doi.org/10.1117/12.2081235

[103] Hariprasad M P, Ramesh K and Prabhune B C 2018 Evaluation of conformal and non-conformal contact parameters using digital photoelasticity *Exp. Mech.* **58** 1249–63

[104] Colla C and Gabrielli E 2017 Photoelasticity and DIC as optical techniques for monitoring masonry specimens under mechanical loads *J. Phys. Conf. Ser.* **778** 012003

[105] Ramesh K 2017 *Digi*TFP®-Software for digital twelve fringe photoelasticity *Photomechanics Lab* (IIT Madras) https://home.iitm.ac.in/kramesh/dtfp.html (accessed Apr. 06, 2021)

[106] Dupuy L X, Mimault M, Patko D, Ladmiral V, Ameduri B, MacDonald M P and Ptashnyk M 2018 Micromechanics of root development in soil *Curr. Opin. Genet. Dev.* **51** 18–25

[107] Kolb E, Genet P, Lecoq L E, Hartmann C, Quartier L and Darnige T 2009 Root growth in mechanically stressed environment: *in situ* measurements of radial root forces measured by a photoelastic technique *Sixth Plant Biomechanics Conference* pp 322–7 http://www.iap.tuwien.ac.at/~gebeshuber/Proceedings_PBM_2009.pdf

[108] Sadeghi A, Mondini A, Del Dottore E, Mattoli V, Beccai L, Taccola S, Lucarotti C, Totaro M and Mazzolai B 2016 A plant-inspired robot with soft differential bending capabilities *Bioinspir. Biomim.* **12** 015001

[109] Wendell D M, Luginbuhl K, Guerrero J and Hosoi A E 2012 Experimental investigation of plant root growth through granular substrates *Exp. Mech.* **52** 945–9

[110] Barés J, Mora S, Delenne J Y and Fourcaud T 2017 Experimental observations of root growth in a controlled photoelastic granular material *EPJ Web Conf.* **140** 14008

[111] Harris J K 1978 A photoelastic substrate technique for dynamic measurements of forces exerted by moving organisms *J. Microsc.* **114** 219–28

[112] Full R, Yamauchi A and Jindrich D L 1995 Maximum single leg force production: cockroaches righting on photoelastic gelatin *J. Exp. Biol.* **198** 2441–52

[113] Ceniceros E C 2013 *Development of Photoelastic Gelatin Spheres for Force Measurement and Visualization in Granular Media* (Reno, NV: University of Nevada, Reno Development) http://hdl.handle.net/11714/3250

[114] Hu D L and Shelley M 2012 Slithering locomotion *Natural Locomotion in Fluids and on Surfaces* vol 155 (New York: Springer) pp 117–35 https://doi.org/10.1007/978-1-4614-3997-4_8

[115] Gray J 1946 The mechanism of locomotion in snakes *J. Exp. Biol.* **23** 101–20

[116] Johnson B D, Boudreau B P, Gardiner B S and Maass R 2002 Mechanical response of sediments to bubble growth *Mar. Geol.* **187** 347–63

[117] Boudreau B P, Algar C, Johnson B D, Croudace I, Reed A, Furukawa Y, Dorgan K M, JuMars P A, Grader A S and Gardiner B S 2005 Bubble growth and rise in soft sediments *Geology* **33** 517–20

[118] Dorgan K M, JuMars P A, Johnson B, Boudreau B P and Landis E 2005 Burrow extension by crack propagation *Nature* **433** 475

[119] Horton T *et al* 2021 World Register of Marine Species (WoRMS) https://doi.org/10.14284/170

[120] Dorgan K M, Arwade S R and JuMars P A 2007 Burrowing in marine muds by crack propagation: kinematics and forces *J. Exp. Biol.* **210** 4198–212

[121] Dorgan K M, Arwade S R and JuMars P A 2008 Worms as wedges: effects of sediment mechanics on burrowing behavior *J. Mar. Res.* **66** 219–54

[122] Dorgan K M, Arwade S R and JuMars P A 2007 Burrowing in marine muds by crack propagation: kinematics and forces-movie 1 *J. Exp. Biol.* **210** 4198–212

[123] Murphy E A K and Dorgan K M 2011 Burrow extension with a proboscis: mechanics of burrowing by the glycerid Hemipodus simplex *J. Exp. Biol.* **214** 1017–27

[124] Juarez G, Lu K, Sznitman J and Arratia P E 2010 Motility of small nematodes in wet granular media *Europhys. Lett.* **92** 44002-1–6

[125] Jung S 2010 *Caenorhabditis elegans* swimming in a saturated particulate system *Phys. Fluids* **22** 031903-1–6

[126] Mirbagheri S A, Ceniceros E, Jabbarzadeh M, McCormick Z and Fu H C 2015 Sensitively photoelastic biocompatible gelatin spheres for investigation of locomotion in granular media *Exp. Mech.* **55** 427–38

[127] Ker D F E *et al* 2018 Functionally graded, bone- and tendon-like polyurethane for rotator cuff repair *Adv. Funct. Mater.* **28** 1707107-1–16

[128] Tomlinson R A and Taylor Z A 2015 Photoelastic materials and methods for tissue biomechanics applications *Opt. Eng.* **54** 081208-1–9

[129] Tomlinson R A, Aui Yong W K, Morton G and Taylor Z A 2015 Development of tissue surrogates for photoelastic strain analysis of needle insertion *Conf. Proc. of the Society for Experimental Mechanics Series* vol 68 ed F Barthelat, C Korach, P Zavattieri, B C Prorok and K J Grande-Allen (Cham: Springer International Publishing) pp 37–45 https://doi.org/10.1007/978-3-319-06974-6_6

[130] Vaughan N, Dubey V N, Wee M Y K and Isaacs R 2013 A review of epidural simulators: where are we today? *Med. Eng. Phys.* **35** 1235–50

[131] Saraf H, Ramesh K T, Lennon A M, Merkle A C and Roberts J C 2007 Mechanical properties of soft human tissues under dynamic loading *J. Biomech.* **40** 1960–7

[132] Falconer S E, Taylor Z A and Tomlinson R A 2019 Developing a soft tissue surrogate for use in photoelastic testing *Mater. Today Proc.* **7** 537–44

[133] Dash W C 1955 Birefringence in silicon *Phys. Rev.* **98** 1536

[134] Redner S 1979 Infrared photoelasticity *Strain* **15** 58–60

[135] Horn G, Lesniak J, MacKin T and Boyce B 2005 Infrared grey-field polariscope: a tool for rapid stress analysis in microelectronic materials and devices *Rev. Sci. Instrum.* **76** 045108-1–10

[136] Ng C S and Asundi A K 2009 Defect inspections using infrared phase shift field polariscope *Fourth Int. Conf. Exp. Mech.* **7522** 75220A-1–11

[137] Zheng T and Danyluk S 2002 Study of stresses in thin silicon wafers with near-infrared phase stepping photoelasticity *J. Mater. Res.* **17** 36–42

[138] He S 2005 *Near Infrared Photoelasticity of Polycrystalline Silicon and it's Relation to In-Plane Residual Stresses* (Atlanta, GA: Georgia Institute of Technology)

[139] Li F 2010 *Study of Stress Measurement Using Polariscope* (Atlanta, GA: Georgia Institute of Technology)

[140] Su F and Li T 2019 Development of an infrared polarized microscope for evaluation of high gradient stress with a small distribution area on a silicon chip *Rev. Sci. Instrum.* **90** 063108-1–6

[141] Prasath R G R, Skenes K and Danyluk S 2013 Comparison of phase shifting techniques for measuring in-plane residual stress in thin, flat silicon wafers *J. Electron. Mater.* **42** 2478–85

[142] He S, Zheng T and Danyluk S 2004 Analysis and determination of the stress-optic coefficients of thin single crystal silicon samples *J. Appl. Phys.* **96** 3103–9

[143] Pinky L J R, Islam S, Alam M N K, Hossain M A and Islam M R 2014 Modeling of orientation-dependent photoelastic constants in cubic crystal system *Mater. Sci. Appl.* **05** 223–30

[144] Horn G, Mackin T J and Lesniak J 2005 Trapped particle detection in bonded semi-conductors using gray-field photoelastic imaging *Exp. Mech.* **45** 457–66

[145] Bosseboeuf A, Rizzi J, Coste P and Bessouet C 2017 Wafer bonding defects inspection by IR microphotoelasticity in reflection mode *Proc. of 2017 5th Int. Workshop on Low Temperature Bonding for 3D Integration, LTB-3D* (JSPS 191st Committee on Innovative Interface Bonding Technology) p 19 https://doi.org/10.23919/LTB-3D.2017.7947415

[146] Ng C S and Asundi A K 2011 Rapid defect detections of bonded wafer using near infrared polariscope *Instrumentation, Metrology, and Standards for Nanomanufacturing, Optics, and Semiconductors V* vol 8105 (Bellingham, WA: SPIE) pp 81050P1–11 https://doi.org/10.1117/12.894402

[147] Guan Y, Zhu Y, Ma S, Zeng Q, Chen J and Jin Y 2016 Fabrication, characterization, and simulation of a low-cost TSV integration without front-side CMP process *IEEE Trans. Semicond. Manuf.* **29** 70–8

[148] Selvanayagam C S, Lau J H, Zhang X, Seah S K W, Vaidyanathan K and Chai T C 2009 Nonlinear thermal stress/strain analyses of copper filled TSV (through silicon via) and their flip-chip microbumps *IEEE Trans. Adv. Packag.* **32** 720–8

[149] Mercha A *et al* 2010 Comprehensive analysis of the impact of single and arrays of through silicon vias induced stress on high-k/metal gate CMOS performance *Tech. Dig.—Int. Electron Devices Meet. IEDM* 26–9 https://doi.org/10.1109/IEDM.2010.5703278

[150] Trigg A D, Yu L H, Cheng C K, Kumar R, Kwong D L, Ueda T, Ishigaki T, Kang K and Yoo W S 2010 Three dimensional stress mapping of silicon surrounded by copper filled through silicon vias using polychromator-based multi-wavelength micro Raman spectroscopy *Appl. Phys. Express* **3** 086601-1–3

[151] Okoro C, Levine L E, Xu R, Hummler K and Obeng Y S 2014 Nondestructive measurement of the residual stresses in copper through-silicon vias using synchrotron-based microbeam X-ray diffraction *IEEE Trans. Electron Devices* **61** 2473–9

[152] Lee G, Choi M J, Jeon S W, Byun K Y and Kwon D 2012 Microstructure and stress characterization around TSV using *in situ* PIT-in-SEM *Proc.—Electronic Components and Technology Conf.* (Piscataway, NJ: IEEE) pp 781–6 https://doi.org/10.1109/ECTC.2012.6248921

[153] Su F, Lan T and Pan X 2015 Stress evaluation of through-silicon vias using micro-infrared photoelasticity and finite element analysis *Opt. Lasers Eng.* **74** 87–93

[154] Ramesh K 2009 *e-book on Experimental Stress Analysis* (Chennai, India: IIT Madras) https://home.iitm.ac.in/kramesh/ESA.html

[155] Ramesh K 2011 Lec-3 Stress, strain and displacement fields *Video Lectures on Experimental Stress Analysis* (Chennai: NPTEL and MHRD) https://youtube.com/watch?v=r8KzP7G7Uks&list=PL21BB25670CDC2AEB&index=3

[156] Ramesh K 2017 P_Scope®—a virtual polariscope *Photomechanics Lab* (IIT Madras) https://home.iitm.ac.in/kramesh/p_scope.html (accessed Apr. 06, 2021)

[157] Ramesh K and Vivek R 2016 Digital photoelasticity of glass: a comprehensive review *Opt. Lasers Eng.* **87** 59–74

[158] Ramesh K and Sasikumar S 2020 Digital photoelasticity: recent developments and diverse applications *Opt. Lasers Eng.* **135** 106186

[159] Roadmap to improving quality of optical elements using d'polariscope https://doptron.com/newsletter/polariscopes/improving-quality-of-optical-elements-using-dpolariscope

[160] Sergazinov R and Kramár M 2020 Machine learning approach to force reconstruction in photoelastic materials *arXiv* 2010.01163

[161] Sachin S and Ramesh K 2021 Deep learning approach to evaluate fracture parameters from photoelastic images *Proc. ASME 2021 Int. Mechanical Engineering Congress and Exposition* pp 1–12

[162] Timoshenko S and Goodier J N 1982 *Theory of Elasticity* (New York: McGraw-Hill)

[163] Shins K and Ramesh K 2021 A revisit to the stress field equations for a disk subjected to self-equilibrated loads *Virtual Seminar on Applied Mechanics*, May 28–29 and June 4–5, Organized by the Indian Society for Applied Mechanics, https://sites.google.com/view/virtualseminarappliedmechanics/home?authuser=0

CPSIA information can be obtained
at www.ICGtesting.com
Printed in the USA
BVHW011623011221
622075BV00037B/42